Springer Series in Advanced Manufacturing

The **Springer Series in Advanced Manufacturing** includes advanced textbooks, research monographs, edited works and conference proceedings covering all major subjects in the field of advanced manufacturing.
The following is a non-exclusive list of subjects relevant to the series:

1. Manufacturing processes and operations (material processing; assembly; test and inspection; packaging and shipping).
2. Manufacturing product and process design (product design; product data management; product development; manufacturing system planning).
3. Enterprise management (product life cycle management; production planning and control; quality management).

Emphasis will be placed on novel material of topical interest (for example, books on nanomanufacturing) as well as new treatments of more traditional areas.

As advanced manufacturing usually involves extensive use of information and communication technology (ICT), books dealing with advanced ICT tools for advanced manufacturing are also of interest to the Series.

Springer and Professor Pham welcome book ideas from authors. Potential authors who wish to submit a book proposal should contact Anthony Doyle, Executive Editor, Springer, e-mail: anthony.doyle@springer.com.

More information about this series at https://link.springer.com/bookseries/7113

Vytautas Ostaševičius

Digital Twins in Manufacturing

Virtual and Physical Twins for Advanced Manufacturing

Springer

Vytautas Ostaševičius
Institute of Mechatronics
Kaunas University of Technology
Kaunas, Lithuania

ISSN 1860-5168 ISSN 2196-1735 (electronic)
Springer Series in Advanced Manufacturing
ISBN 978-3-030-98274-4 ISBN 978-3-030-98275-1 (eBook)
https://doi.org/10.1007/978-3-030-98275-1

This Springer imprint is published by the registered company Springer Nature Switzerland AG
The registered company address is: Gewerbestrasse 11, 6330 Cham, Switzerland

I would like to dedicate this book to the centenary of Kaunas University of Technology (1922–2022)

Preface

Digital twins, the Internet of Things (IoT), and the electronic physical system are key concepts in the "Industry 4.0". Digital twins use cloud-connected machine sensors to load real-time operational data, creating state-of-the-art virtual simulations of real-world machines, where IoT is the key to modeling large-scale digital twins. The term "digital twin" still lacks a common understanding, leading to differences in its technological implementation and objectives. This term covers virtual and physical replicas of a product, a machine, or an entire manufacturing process that are used as a specific test-bed for the process or product, where the changes made can be simulated before being implemented in real life. By entitiling virtual and physical copies as virtual and physical twins, respectively, and linking them to modeling and experiments that result in digital output, applications can be attractive in industrial, academic, or scientific segments. The desired result is the quality of the product, often related to the dynamics of the cutting tools, which could be assessed using virtual or physical twins and predicted by artificial intelligence (AI) methods. In the development of production systems, the implementation of such processes is usually associated with modeling to determine parameter settings and machine design. However, the use of modeling in the technological process is inefficient and expensive, as it is time consuming and slows down the manufacturing process itself due to the long simulation run-time. The main idea of this book is to apply modeling in the production preparation stage, using the effects of nonlinear mechanics revealed by the author, and to apply them for product quality improvement. It is needed for existing production systems as a retrofit to ensure a sustainable manufacturing process and a shortened product path to market. This contributes to higher incomes and job security in the manufacturing sector. To this end, an integrated methodology for data acquisition, processing, and analysis using innovative solutions has been developed. The design, development, and process monitoring stages, explored in the book, are among the key elements of the ISO 13399 standard, which defines the principles and methods for creating definitions of the items and their properties that relate to cutting tools. Readers of this book will benefit from examples of artificial neural networks and machine learning supported by process modeling in digital twins, based on technological solutions and data validation. The use cases and detailed methodologies developed in this book facilitate the deployment of various technologies in the manufacturing industry.

Kaunas, Lithuania
December 2021

Vytautas Ostaševičius

Contents

1 The State-of-the-Art in the Theoretical and Practical Applications of the Digital Twins Components 1
1.1 Numerical Modeling and Prediction of Manufacturing Processes 1
1.2 Vibration Cutting for Smart Manufacturing 3
1.3 Vibration Energy Harvesting 4
1.4 Internet of Things Devices for Manufacturing Process Control 5
1.5 The Relationship Between Edge and Cloud-Based Computing 6
References 8

2 Digital Twins for Smart Manufacturing 11
2.1 Digital Twin Emphasis on Cutting Tool Vibration Control Through Design Parameters 11
2.1.1 Optimally Designed Self-Exciting Drill for Vibration Cutting 11
2.1.2 Modified Boring Tool Structures for Effective Cutting 18
2.2 Cutting Tool Physical Entities and Their Virtual Counterparts Synchronization 25
2.2.1 Simulation and Analysis of Vibration Turning Tool 26
2.2.2 Evaluation of Workpiece Surface Roughness and Tool Wear 34
2.2.3 Influence of Boundary Conditions on the Vibration Turning Tool Eigen Modes 39
2.3 Evaluation of Technological Features of Macro- and Micro-drilling 44
2.3.1 Virtual Twin of Vibration Drilling Tool 44
2.3.2 Physical Twin of Vibration Drilling Tool 53
2.3.3 Characterization of Vibration Drilling Process and Workpiece Surface Quality 59
2.3.4 Drilling Process Simulation Using Smoothed Particle Hydrodynamics Method 62

2.3.5 Micro-drill Stiffness Amplification by Buckling Mode Control ... 65
2.3.6 Experimental Study of Micro-drill Physical Twin ... 72
2.4 Quality Improvement of Grinding Operations ... 75
2.4.1 An Excitation Approach to Ultrasonically Assisted Cylindrical Grinding ... 75
2.4.2 Development of Actuator for Back Grinding ... 85
2.5 Artificial Neural Networks Approaches for Quality Prediction in Robotized Incremental Sheet Forming ... 96
2.5.1 Determination of Friction Force Between the Tool and Forming Sheet ... 96
2.5.2 Evaluation Methodology of Metal Sheet Forming Process ... 97
2.5.3 Cupping Test for the Material Model Calibration ... 104
2.6 FE Simulations ... 105
2.6.1 FE Simulations of Cupping Test for Material Model Calibration ... 106
2.6.2 Numerical Simulations of SPIF Process ... 107
2.6.3 Evaluation Methodology of Polymer Sheet Forming Process ... 129
References ... 145

3 Integration of Digital and Physical Data to Process Difficult-to-Cut Materials ... 149
3.1 Digital Twin for Excited Cutting Tool ... 149
3.1.1 Prevention of Chemical Interactions Between Tool and Workpiece Materials by Contact Time Reduction ... 149
3.1.2 Vibration Milling for Surface Finish Improvement ... 166
3.1.3 Vibration Drilling of Brittle Materials ... 174
3.2 Physical Twin of Vibrationally Excited Workpiece Drilling ... 181
3.2.1 Vibration Excitation of a Workpiece for Drilling Force Reduction ... 181
3.2.2 Development of Actuator Enabling a Brittle-Ductile Transition of Warkpiece Material ... 186
References ... 200

4 Wireless Connectivity Options for Tool Condition Monitoring IoT Applications ... 203
4.1 The New Principles of Energy Harvesting in Macro Level ... 203
4.1.1 Virtual Twin of Piezoelectric Energy Harvester ... 203
4.1.2 Enhanced Harvester Configuration ... 234
4.1.3 Investigation of Optimized Cantilever Beam ... 241
4.1.4 Appropriate Way to Extract the Low-Frequency Vibration Energy ... 250

4.2 The New Principles of Energy Harvesting in Micro Level 255
4.2.1 Enhanced Vibration Energy Harvesting Configuration 256
References 264

5 Digital Twin-Driven Technological Process Monitoring for Edge Computing and Cloud Manufacturing Applications 267
5.1 Edge Computing-Enabled Wireless Vibration Sensor Node 267
5.1.1 Use Case of the Non-rotating Tool 267
5.1.2 Use Case of the Rotating Tool 284
5.2 Wireless IoT Vibration Sensor for Cloud Manufacturing Applications 303
5.2.1 Virtual Twin of Piezoelectric Transducer 304
5.2.2 Rotating Shank-Type Tool Condition Monitoring 325
5.3 Tool Wear Status Recognition Based on Machine Learning 348
5.3.1 Support Vector Machines Algorithm Adaptation for Milling Force Prediction 352
References 357

Chapter 1
The State-of-the-Art in the Theoretical and Practical Applications of the Digital Twins Components

1.1 Numerical Modeling and Prediction of Manufacturing Processes

Numerical modeling is defined as a problem-solving approach to better understand real-world physical processes. Numerical modeling plays an important role in the development and optimization of any manufacturing process to avoid the time and cost associated with traditional trial and error. Many literature sources describe different modeling and simulation approaches, but only limited sources cover the complete modeling and simulation cycle, including both the design and development of the model and the verification and validation of the model for use in industrial systems. Modeling of metal cutting has proven to be challenging due to the variety of physical phenomena involved, such as thermo-mechanical, contact/friction, and material failure. Over the last few decades, significant progress has been made in the development of numerical methods for modeling machining operations. The authors in [1] provide an overview of digital modeling techniques and methods used to simulate metal cutting processes. The main objective is to make these techniques more reliable and computationally less expensive. A methodology for micro-grinding process modeling is presented in [2]. The number of works related to composite cutting is increasing, especially in the aerospace, renewable energy, and military industries. An amended friction model was proposed in [3] to calculate the tangential contact between the tool and the composite for a digital rectangular cutting model. Piezoelectric composites play an important role in high-precision material processing technologies. For piezoelectric ceramics, a nonlinear, velocity-dependent constitutive model involving polarization switching behavior is considered in [4], and a linear viscoelastic model is applied to the polymer matrix.

Artificial intelligence (AI) is a cognitive science with a wealth of research in robotics, machine learning areas, etc. Historically, machine learning and AI have been perceived as black-box techniques, and it has been difficult to convince industry that these methods can work repeatedly and consistently, along with the return on investment. At the same time, the performance of machine learning algorithms is

V. Ostaševičius, *Digital Twins in Manufacturing*, Springer Series in Advanced Manufacturing, https://doi.org/10.1007/978-3-030-98275-1_1

highly dependent on the experience and preferences of the developer. The goal of machine learning is to select the right data that provides insight and develops state-of-the-art techniques. As AI becomes the frontier of world-changing technologies, there is an urgent need to systematically develop and deploy AI to see its real impact on the next-generation industrial systems, namely Industry 4.0. The main elements of industrial AI can be described as "A B C D E". These key elements include Analytics technology (A), Big Data Technology (B), Cloud or Cyber Technology (C), Domain know-how (D), and Evidence (E) [5]. Acoustic Emission (AE) is a well-known indirect approach that measures the characteristics of acoustic waves generated during the manufacturing process [6]. AE signals were evaluated using machine learning methods to detect wear anomalies, and a deep learning method based on convolutional neural networks was used for multi-class failure analysis. Quality has become an important success factor for manufacturing companies [7]. This paper compares the use of machine learning algorithms in quality control with statistical process monitoring as a classical method of quality management. Random forests, support vector machines, and Naive Bayes algorithms were used to predict when a manufacturing process is out of control. Management, decision support, and forecasting tools are Key Performance Indicators (KPIs) as they reflect the strategy and vision of the company in terms of objectives and allow it to always meet the expectations of its stakeholders. Authors in [8] apply different machine learning methods, i.e. support vector regression, optimized support vector regression (using a genetic algorithm), random forest, extreme gradient boosting, and deep learning to predict the overall equipment performance. Several configurations of the listed models are used to provide the comparison field. The results showed that the configuration using the cross-validation method and properly split data provides better results of the prediction models. In [9], a real machining experiment was carried out to verify the ability of the proposed model to predict and optimize surface roughness. The forming limit is a major concern in sheet metal forming. The incremental forming process, as an easily modifiable manufacturing technology that does not require dies, clearly enhances the formability of metal sheet.

A machine learning-based approach to predict the occurrence of defects in sheet metal deformation by single-point incremental forming due to the material properties and the sources of variance of the process parameters is analyzed in [10]. An empirical analysis of the performance of the machine learning techniques is presented, taking into account both single-learning and ensemble models. Comparing single learning and ensemble models, the latter can be an effective alternative. The results obtained are promising for applications in industrial environments.

Beyond the state-of-the-art, a modern meshless technique known as smooth particle hydrodynamics method can replace microscopic simulation in solving high deformation problems. The influence of the cutting edge position and tool fixing on the quality of the workpiece, as well as the instability of the higher curvature regime in the micro-drill hole, are addressed. The possibility of using cylindrical vibration grinding is demonstrated. Efforts to grind or cut difficult-to-manufacture materials would be more successful if high-frequency vibrations were used to create the conditions for brittle-ductile transition. The possibility of replacing costly 3D printing with incremental forming of polymer sheets, as well as replacing environmentally unfriendly lubrication in the incremental forming process of metal sheets, is highlighted. The predictive capabilities of different machine learning algorithms are compared.

1.2 Vibration Cutting for Smart Manufacturing

Vibration cutting is a process in which the cutting tool is subjected to high vibrations to increase the efficiency of material removal. The basic principle of this process is to turn the interaction of the tool and the workpiece into a non-monotonic operation in order to facilitate chip separation and reduce machining forces. It can also reduce the deformation zone of the workpiece during machining, thus improving the surface integrity of the machined component. Due to its potential and industrial expectations, two vibration-assisted methods are commonly used: modulated machining and ultrasonic machining. Modulation-assisted machining assigns a controlled low-frequency oscillation to the cutting process, typically up to 1000 Hz, and an amplitude of up to 500 μm. This changes the mechanics of chip formation, and splits the cutting into several events by controlling the modulation parameters. Ultrasonic vibration-assisted machining with high-frequency and low-amplitude cutting performance is appropriate for most advanced materials [11]. Ultrasonic machining is the incorporation of high-frequency excitation, which can reach 100 kHz, and low vibration amplitudes of $3 \div 30$ μm into a tool or workpiece to improve the cutting process. Ultrasonic assisted machining technology offers such advantages as high precision machining of difficult-to-cut alloys or brittle materials. The authors in [12] review the devices used in vibration-assisted manufacturing processes, dividing them into two groups: resonant and non-resonant devices. Resonant vibratory devices usually vibrate at their eigen frequency, and their amplitude is set to certain values. Due to variable vibration amplitudes and frequencies, the non-resonant vibration systems are more flexible. The study [13] observes the cutting process with elliptical vibration using internal data in an ultrasonic elliptical vibration device under test without external sensors. The internal data used here are the change in the resonant frequency of the device, the voltages and electric currents applied to the piezoelectric actuators. In order to efficiently and economically produce a micro-textured surface, a method of radial ultrasonic vibration-assisted turning texturing was proposed in [14]. A theoretical texture generation model has been developed to describe the layout and geometry of micro-textures. The variation of milling force in ultrasonic end milling process is investigated in [15]. The main objective of this study is the effect of externally excited vibration on the milling force, and the influence of milling and vibration parameters on the value of the milling force. Piezoelectric actuators are used to excite oscillations with a frequency of $\geq$15 kHz and an amplitude of (5–25 μm). The natural frequencies of piezoelectric actuators vary depending on the cutting force changes during machining. Consequently, the amplitudes of the transmitted vibrations differ from the desired reference values. In contrast to the use of vibration sensors, the research in [16] introduces a sensorless method for monitoring and controlling the frequency and amplitude of ultrasonic vibration during machining. The actuator model helps to obtain the transfer function between the driving current and the supply voltage. A proportional–integral controller is used as a feedback to maintain the estimated vibration amplitude level during machining.

Beyond the state-of-the-art, the matching of the tool's excitation frequency to the frequency of its higher eigenmodes has a decisive influence on cutting performance. In line with this trend, new design requirements and theoretical concepts for virtual and physical twins are proposed and consistently investigated.

1.3 Vibration Energy Harvesting

The industrial wireless IoT market is growing rapidly, but at the same time, it is constrained by deployment costs and compatibility with existing wireless networks. Typical wireless sensors and transmitters depend on batteries that need periodic maintenance to be replaced. Maintenance costs for batteries used in wireless sensor applications should include not only the cost of the power source and its replacement, but also the cost of equipment downtime during the replacement. One of the three general mechanisms for the electrical conversion of mechanical vibration is piezoelectric effect, electrostatic and electromagnetic transduction. Since piezoelectric material can convert mechanical vibration into electrical energy, the piezoelectric energy harvester is emphasized as a source of energy for wireless sensor network systems. Piezoelectricity indicates pressure electricity and contains certain crystalline materials such as quartz, Rochelle salt, tourmaline, and barium titanate electricity when the pressure is applied. This is called the direct effect. The reverse effect occurs when these crystals are deformed by an electric field and can be used as actuators when direct effect is used as a sensor or energy transducer. Piezoelectric transducers are best suited for mechanical to electrical energy conversion because they have the highest energy density (35.4 mJ cm^{-3}). Therefore, vibrational energy harvesters based on a piezoelectric transmission mechanism are very efficient and generate the maximum power output for a certain transducer size. In [17], a piezoelectric multimodal power harvester with multiple mechanical degrees of freedom is proposed and experimentally validated. The experimental results imply that the proposed energy harvester can obtain four peak values in the range of $10 \div 20$ Hz. In [18], an energy harvesting device with adjustable nonlinearity for ultralow frequency excitation is proposed and analyzed. The energy harvesting device consists of a mass attached to a base with elastic steel slices and a pair of square connecting structures. Theoretical research is carried out by applying a mathematical model. The analysis shows that the device can harvest ultralow-frequency vibration energy because it has an adjustable resonant frequency band. Bistable energy harvesting devices are more efficient than traditional mono-stable devices [19]. This paper introduces a stochastic averaging procedure to evaluate the stationary random response of a bistable energy harvester to additive and multiplicative white noises. The removal of irregular energy from various environmental impacts that capture and transmit high impact pressures to deform piezoelectric components, which can generate the desired electricity to power infrastructure sensors, has received considerable attention in [20]. This method is simple in terms of configuration, but inadequate in terms of lifetime and cannot guarantee stable, long-lasting output at large intervals between target effects. The energy harvesting effect of a composite material is analyzed in [21]. The composite material is simulated by the FE method and fabricated with a pre-impregnated thermosetting epoxy resin matrix reinforced with high-strength R-type glass fibers, and is designed as a beam structure exposed to mechanical vibration. New effects identified in this work are the total bending intensity in the rigid and flexible beam directions for a selected sequence of fiber layers, and the influence of the layer configuration on the energy harvesting efficiency of the composite piezoelectric element. The design and manufacturing process of a packaged micro-piezoelectric vibration energy harvester is presented in [22]. A double *L*-shaped tungsten resistive mass part and a thick copper-based PZT bimorph are proposed to provide a high power output of the energy harvester. An autonomous method for monitoring the condition of rotating machines during operation based on the transmission of RF pulses using energy harvesting from operational vibration is introduced in [23].

Beyond the state-of-the-art, it is possible to obtain significantly higher energy content by exciting the flexible links with higher eigenmodes and by localizing the segments of the macro or micro-piezoelectric elements where the sign of the dynamic deformation field of the structure changes. Information on how to select cutting energy combinations for different cutting tool groups is included.

1.4 Internet of Things Devices for Manufacturing Process Control

The IoT is profitable in areas where faster development and product quality are crucial factors in achieving a higher return on investment. Manufacturing industry is one such area, and the industrial IoT has transformed it with contemporary Big Data, AI, and machine learning technologies. The IoT has many applications in manufacturing companies. It can facilitate the flow of production in a manufacturing plant, as IoT devices automatically monitor development cycles and manage warehouses and inventories. The industrial IoT is evolving rapidly as manufacturers adapt to digitalization and automation. The IoT is one of the main drivers of digital transformation in manufacturing due to its wide range of applications. 5G is designed to meet the requirements of the industrial IoT and to be a catalyst for a new industrial transformation. 5G technology will reinforce the Industry 4.0 activities, allowing further increases in flexibility, cost control, and quality.

Companies committed to digital transformation are starting to draw inspiration from the IoT for their factories of the future. Efficient allocation of computing resources and carrier is crucial for smart manufacturing systems in the industrial IoT.

The current allocation method for smart manufacturing systems cannot guarantee that resources will meet the complex and volatile user requirements in a timely manner. The authors in [24] discuss the allocation of resources in an auction format, taking into account the real-time demand and supply of resources, security, privacy, and trust evaluation issues. [25] presents an IoT-based industrial data management framework that can manage huge amounts of industrial data, support online monitoring, and control smart manufacturing. The framework consists of five main layers, i.e. physical, database, middleware, network, and application layers, in order to provide a service-oriented architecture to the end users. There are many studies on industrial IoT and Big Data processing, but only a few studies have examined the convergence of these two paradigms [26]. This paper investigates the state-of-the-art in Big Data analytics technologies, algorithms and methods that can support the development of intelligent industrial IoT systems. A taxonomy is proposed to classify subject-related literature sources according to relevant parameters (e.g. data sources, analysis tools and techniques, requirements, industrial analysis programs, and types). It also lists the opportunities that Big Data analytics has brought to the industrial IoT. IoT cybersecurity is a complex task and is at the heart of IoT development [27]. This article provides a systematic overview of IoT cybersecurity. Key aspects are the protection and integration of heterogeneous smart devices and information communication technologies. This paper provides useful information and insights for researchers and practitioners interested in IoT cybersecurity, including current research on IoT cybersecurity architecture and taxonomy, key enabling responses and strategies, key industry applications, research trends and challenges.

Beyond the state-of-the-art a methodology has been proposed to develop the most suitable self-powering IoT devices that wirelessly send information about specific manufacturing processes for the purpose of edge computing and cloud manufacturing.

1.5 The Relationship Between Edge and Cloud-Based Computing

Edge and cloud computing are two different technologies, but they are often used together because of their compensating advantages. Edge computing processes data in time-sensitive actions, while cloud-based computing stores data in a centralized location that does not require timely response. Across the IoT infrastructure, manufacturers can take advantage of the edge computing for real-time data collection, predictive analytics and autonomous decision-making, while the cloud is beneficial for aggregate data analysis, benchmarking, and trend analysis. The main objective of industrial IoT is to apply advanced analytics to large volumes of machine data, to reduce unplanned downtime and overall machine maintenance costs, as well as aspiration to take advantage of machine learning opportunities. The cloud has been useful in such massive data acquisition, transmission, and analysis. Edge computing and analytics simply takes this concept to a new level by reducing the physical distance between the machines and the data processing itself. For companies that distribute and collect data remotely from multiple sources, this proximity helps them cope with the cost of massive ongoing data transmission, the ability to perform quick analysis, enabling real-time decision making, as well as ensuring data security and network integrity. All these realities, when addressed within edge computing, allow manufacturers to reduce costs and increase efficiency.

The authors in paper [28] introduce an intelligent computing architecture with cooperative edge and cloud computing for industrial IoT. The main objective is to improve service accuracy. Based on the computing architecture, an AI improved offloading system is proposed, where service accuracy is considered as a new metric without delay. It intelligently distributes traffic to edge servers or the corresponding path to the remote cloud. A case study of transfer learning is conducted to demonstrate the effectiveness of the proposed system. Deep learning for IoTs in the edge computing environment is introduced in [29]. As the processing capabilities of existing edge nodes are limited, a new unloading strategy to optimize the performance of IoT deep learning programs with edge computing is designed. The performance of several deep learning tasks in an edge computing environment is assessed. The assessment results show that the proposed approach outperforms other deep-learning IoT optimization solutions. The smart manufacturing system described in [30] takes into account the edge computing, blockchain, and machine learning methods. Edge computing allows to balance the computational workload and also provides timely responses for the devices. Machine learning method provides an advanced data analysis of a large production dataset, while blockchain technology uses data transfer and transactions between manufacturing systems. For smart manufacturing systems computing environment, the model solves problems using a swarm intelligence-based approach. The results of the experiment present the edge computing mechanism and similarly improve the processing time of many tasks in the manufacturing system. The fog computing paradigm is discussed in [31]. Cyber-physical systems as the coupling of the digital world and physical processes are considered, taking the fog computing paradigm into consideration. The main scope of the article covers industrial IoT domains and Smart Factory environments. The architecture of fog nodes, as well as a more detailed discussion of the instrumentation system and the programmable properties of the fog node, are presented. Based on the simulation model, the advantages of having a programmable fog node supported by an orchestration system is shown. The cyber and physical integration of manufacturing is becoming increasingly important for smart manufacturing [32]. The most appropriate means to achieve interoperability and integration in the physical and cyber worlds are cyber-physical systems and digital twins. From a hierarchical perspective, these means can be divided into unit level, system level and system of system level. The three complementary technologies, such as edge computing, fog computing, and cloud computing are critical to accelerating the development of various cyber-physical systems and digital twins.

Beyond the state - of - the - art, an universal sensor node has been proposed that enables the self-powering of sufficient energy for process monitoring and control devices, regardless of the dynamic characteristics of the machining process. Machine learning algorithm was adapted for cutting force prediction.

References

1. Rodríguez JM, Carbonell JM, Jonsén P (2020) Numerical methods for the modelling of chip formation. Arch of Comp Meth in Eng 27:387–412
2. Rypina L, Lipinski D, Bałasz B, Kacalak W, Szatkiewicz T (2020) Analysis and modeling of the micro-cutting process of Ti-6Al-4V titanium alloy with single abrasive grain. Mater 13:5835
3. Xub J, Deng Y, Wang C, Liang G (2021) Numerical model of unidirectional CFRP in machining: Development of an amended friction model. Compos Struct 256:113075
4. Lin C-H, Muliana A (2019) Micromechanics modeling of hysteretic responses of piezoelectric composites. Cre and Fat in Poly Matr Comp: 121–155
5. Lee J, Davari H, Singh J, Pandhare V (2018) Industrial artificial ntelligence for industry 4.0-based manufacturing systems. Manuf Lett 18:20–23
6. Konig F, Sous C, Ouald Chaib A, Jacobs G (2021) Machine learning based anomaly detection and classification of acoustic emission events for wear monitoring in sliding bearing systems. Trib Int 155: 106811
7. Khoza SC, Grobl J (2019) Comparing machine learning and statistical process control for predicting manufacturing performance. EPIA: Prog in Art Intel 108–119
8. El Mazgualdi C, Masrour T, El Hassani I, Khdoudi A (2021) Machine learning for KPIs prediction: a case study of the overall equipment effectiveness within the automotive industry. Soft Comp 25:2891–2909
9. Hartmann C, Opritescu D, Volk W (2019) An artificial neural network approach for tool path generation in incremental sheet metal free-forming. J Intell Manuf 30:757–770
10. Dib MA, Oliveira NJ, Marques AE, Oliveira MC, Fernandes JV, Ribeiro BM (2020) Single and ensemble classifiers for defect prediction in sheet metal forming under variability. Neur Comp Appl 12335–12349
11. Yang Z, Zhu L, Zhang G, Ni C, Lin B (2020) Review of ultrasonic vibration-assisted machining in advanced materials, Int J Mach Tools and Man 156: 103594
12. Zheng L, Chen W, Huo D (2020) Review of vibration devices for vibration-assisted machining. Int J Adv Man Techn 108:1631–1651
13. Jung H, Hayasaka T, Shamoto E (2018) Study on process monitoring of elliptical vibration cutting by utilizing internal data in ultrasonic elliptical vibration device. Int J of Prec Eng and Manu-Green Techn 5(5):571–581
14. Liu X, Wu D, Zhang J, Hu X, Cui P (2019) Analysis of surface texturing in radial ultrasonic vibration-assisted turning. J Mat Proc Techn 267:186–195
15. Shen X-H, Xu G-F (2018) Study of milling force variation in ultrasonic vibration-assisted end milling. Mat Manu Proc 33(6):644–650
16. Gao J, Caliskan H, Altintas Y (2019) Sensorless control of a three-degree-of-freedom ultrasonic vibration tool holder. Prec Eng 58:47–56
17. Toyabur RM, Salauddin M, Park JY (2017) Design and experiment of piezoelectric multimodal energy harvester for low frequency vibration. Cer Int 43:S675–S681
18. Wang F, Sun X, Xu J (2018) A novel energy harvesting device for ultralow frequency excitation. Ener 151:250–260
19. Xu M (2018) Li X Stochastic averaging for bistable vibration energy harvesting system. Int J Mech Sci 141:206–212
20. Wang H, Mao M, Liu Y, Qin H, Zhang M, Zhao W (2019) Impact energy harvesting system using mechanical vibration frequency stabilizer. Smart Mat Stru 28: 075006

21. Borowiec M, Gawryluk J, Bochenski M (2021) Influence of mechanical couplings on the dynamical behavior and energy harvesting of a composite structure. Polym 13(1):66
22. Wang L, Ding J, Jiang Z, Luo G, Zhao L, Lu D, Yang X, Ryutaro M (2019) A packaged piezoelectric vibration energy harvester with high powerand broadband characteristics. Sen Act A 295:629–636
23. Khazaee M, Rezaniakolaie A, Moosavian A, Rosendahl L (2019) A novel method for autonomous remote condition monitoring of rotating machines using piezoelectric energy harvesting approach. Sen Act A 295:37–50
24. Xu X, Han M, Nagarajan SM, Anandhan P (2020) Industrial internet of things for smart manufacturing applications using hierarchical trustful resource assignment. Comp Comm 160:423–430
25. Saqlain M, Piao M, Shim Y, Lee JY (2019) Framework of an IoT-based industrial data management for smart manufacturing. J. Sensor Actuat Netw 8(2):1–21
26. Yang H, Kumara S, Satish T.S. Bukkapatnam STS, Tsung F (2019) The internet of things for smart manufacturing: A review. 26. Rehman MH, Yaqoob I, Salah K, Imran M, Jayaraman PP, Perera C (2019) The role of big data analytics in industrial Internet of Things, Fut Gen Comp Syst 99, pp. 247-259
27. Lu Y, Xu LD (2018) Internet of Things (IoT) cybersecurity research: a review of current research topics. IEEE Intern Thin J 6(2):2103–2115
28. Sun W, Liu J, Yue Y (2019) AI-enhanced offloading in edge computing: when machine learning meets industrial IoT. IEEE Netw 33(5):68–74
29. Li H, Ota K, Dong M (2018) Learning IoT in edge: deep learning for the Internet of Things with edge computing. IEEE Netw 32(1):96–101
30. Shahbazi Z, Byun Y-C (2021) Improving transactional data system based on an edge computing–blockchain–machine learning integrated framework. Proc 9(92):1–20
31. De Brito MS, Hoque S, Steinke R, Willner A, Magedanz T (2017) Application of the fog computing paradigm to smart factories and cyber-physical systems. Trans Emerg Tel Tech 29(4) e3184: 1–14
32. Qi Q, Zhao D, Liao T-W, Tao F (2018) Modeling of cyber-physical systems and digital twin based on edge computing, fog computing and cloud computing towards smart manufacturing. ASME Int Man Sci Eng Conf Proc 7

Chapter 2
Digital Twins for Smart Manufacturing

2.1 Digital Twin Emphasis on Cutting Tool Vibration Control Through Design Parameters

Since the quality of machining is determined by the dynamic interaction of the cutting tool and the workpiece, it is necessary to anticipate the possibilities of improving this interaction at the tool design stage. The presented research results, aimed at a deeper understanding of the vibrational process during high-frequency vibration cutting, are accomplished by treating the cutting tool as an elastic structure that is characterized by several eigenmodes. An approach for surface quality improvement is proposed by taking into account that the quality of the machined surface is related to the intensity of tool-tip (cutting edge) vibrations. It is based on the excitation of a particular higher eigenmode of a tool, which leads to the reduction of deleterious vibrations in the machine tool–workpiece system through intensification of internal energy dissipation in the tool material. The combined application of virtual analysis with an accurate FE model as well as different physical analysis methods during the investigation of the vibration cutting process allowed us to determine the most favorable vibration mode of the tool. Such a mode can be self-excited due to the selection of the design parameters of the tool or by using a piezoelectric transducer vibrating at the corresponding eigenfrequency, thereby enabling minimization of surface roughness and tool wear.

2.1.1 Optimally Designed Self-Exciting Drill for Vibration Cutting

Higher surface quality of the machined workpiece is obtained when the cutting tool is excited at high-frequency vibrations that are superimposed on its continuous movement. Effective could be the passive way with respect to structural peculiarities of the tool [1]. The aim of optimization is to select such geometrical parameters that

V. Ostaševičius, *Digital Twins in Manufacturing*, Springer Series in Advanced Manufacturing, https://doi.org/10.1007/978-3-030-98275-1_2

would correspond to the technical characteristics of the structure and give a minimum value to a certain quality functional or target function [2]. In optimal design, first of all, the constraints, geometrical, and structure performance should be distinguished. The target functions, most frequently used in optimal design, express the structure mass minimized with respect to the constraints of a prescribed vibration frequency. It is desirable to design a structure, with the natural frequencies which would not fall in a certain interval $(\omega_{\min}, \omega_{\max})$. The twist drill structure could be considered as a cantilever, but accomplishing torsion vibrations. The most applicable method for the optimization of the cantilever structure mass at a prescribed frequency of torsional vibrations could be the gradient projection in the state variables space method of nonlinear programming. For this purpose, constraints are expressed in the form of inequalities. While applying this method, the information only about the first derivatives or gradients is used. Let's denote design variables as $A_1, A_2, \ldots A_n$ or in vector form $A = [A_1, A_2, \ldots, A_n]^T$ where A_i—are cross-sections of the FE of cantilever depending on its FE diameters d_i. During optimization, the cross-sections A_i are constrained by the value $\underline{A_i}$ from below, where $i = 1, 2, \ldots, n$.

The target function of structure optimization could be expressed as follows:

$$\Phi(A) = \min \rho\alpha(A_1 + A_2 + \ldots + A_n). \tag{2.1}$$

If the mesh of FE is chosen of constant length, the target function could be rewritten in the following form:

$$\Phi(A) = \min(A_1 + A_2 + \ldots A_n).$$

The cross-section of the structure is constrained from below, so it is necessary to constrain the cross-sections of all FE:

$$\begin{aligned} \Psi_1(A) &= 1 - \tfrac{A_1}{\underline{A_1}} \le 0; \\ \Psi_2(A) &= 1 - \tfrac{A_2}{\underline{A_2}} \le 0; \\ &\vdots \\ \Psi_n(A) &= 1 - \tfrac{A_n}{\underline{A_n}} \le 0. \end{aligned} \tag{2.2}$$

As the fixed eigenfrequency of the structure is not less than ω^* so the constraints are also applied to it:

$$\Psi_{n+1}(\omega) = 1 - \frac{\omega}{\omega^*} \le 0, \tag{2.3}$$

where ω^*—given frequency of the structure and ω—eigenfrequency of the mode of vibrations under investigation.

The eigenproblem is described as follows:

$$K(A)\theta = \xi M(A)\theta, \tag{2.4}$$

where $K(A)$, $M(A)$—the stiffness and mass matrices of the structure, θ—the matrix of eigenvectors in torsion, and we obtain eigenvalue vector ξ. As Eq. (2.4) is homogenous to the eigenvector, it is possible to express $\theta^T M\theta = 1$. Equation (2.4) describes structure by its eigenvalues. Eigenfrequency can be found by solving expression $\omega = \sqrt{\xi}$. During calculations, the constraints activity should be verified at each iteration. For this purpose, (2.2) and (2.3) expressions are constrained by the constants ε^A and ε^ω. Then

$$\begin{aligned} \Delta\Psi_i(A) &= 1 - \frac{A_i}{\underline{A_i}} \le -\varepsilon^A, \quad i = 1, \ldots, n; \\ \Delta\Psi_{n+1}(\omega) &= 1 - \frac{\omega}{\omega^*} \le -\varepsilon^\omega. \end{aligned} \tag{2.5}$$

Constraints like $\varepsilon^A < \Psi_i(A)$ are called active and should be considered, because they are almost violated. If they are not regarded, the oscillations could occur during other iterations. During each iteration, the matrix of active constraints gradients is constructed. The matrix columns of design variables' active constraints are identified from the following expression:

$$e^i = \frac{\partial \Psi_i^T}{\partial A}, \tag{2.6}$$

where *i*—the index of violated constraints $\Delta\Psi_j$, $j = 1, 2, \ldots, n$. The components of vector e^i are the sensitivity coefficients of corresponding design variables. Based on the design variables, these vectors define the derivatives of the target function and the constraints. They are useful for the designer because they help to identify the impact of design variables on the target function and the constraints. If component e^i is positive, then Ψ_i grows with the increase of A_j. If e^i is negative, A_j reduces Ψ_i. The magnitude of sensitivity coefficients e^i informs the designer which design variables have a significant impact on Ψ_i and vice versa. When the number of violated constraints is *1*, …, *n*, the columns of the matrix acquire the following form:

$$e^1 = \begin{bmatrix} -1/\underline{A_1} \\ 0 \\ 0 \\ \cdots \\ 0 \end{bmatrix}, \quad e^2 = \begin{bmatrix} 0 \\ -1/\underline{A_2} \\ 0 \\ \cdots \\ 0 \end{bmatrix}, \quad \ldots, \quad e^n = \begin{bmatrix} 0 \\ 0 \\ 0 \\ \cdots \\ -1/\underline{A_n} \end{bmatrix}. \tag{2.7}$$

The gradient of state variable $\Delta\Psi_{n+1}$ is defined by dependency as follows:

$$e^{iT} = \left[\frac{\partial}{\partial A}\{\theta^{iT} K(A)\theta^i\} - \omega\frac{\partial}{\partial A}\{\theta^{iT} M(A)\theta^i\}\right]\frac{\partial \Psi_i}{\partial \omega}. \tag{2.8}$$

The vector of the target function can be defined as follows:

$$e^0 = \frac{\partial \Phi}{\partial A}, \tag{2.9}$$

or, in this case, is obtained as follows:

$$e^0 = \begin{bmatrix} 1 \\ 1 \\ \vdots \\ 1 \end{bmatrix}.$$

The Lagrange multipliers, denoted by vectors λ_I and λ_2, are defined by the matrix of active constrained gradients (2.5) and the vector of target function gradients (2.7):

$$\begin{aligned} e^T W^{-1} e\lambda_1 &= -e^T W^{-1} e^0; \\ e^T W^{-1} e\lambda_2 &= -\Delta\overline{\Psi}_n, \end{aligned} \tag{2.10}$$

where W—the matrix of design variables; $\Delta\overline{\Psi}$—the vector of active and violated constraints.

The vectors of variations of design variables are identified based on the Lagrange multipliers:

$$\begin{aligned} \delta A^1 &= W^{-1}\left(e^0 + e\lambda_1\right); \\ \delta A^2 &= -W^{-1} e\lambda_2, \end{aligned} \tag{2.11}$$

where δA^1 corresponds to the decrease direction of the target function under constraints and δA^2 corresponds to the required correction of the constraints.

It is not difficult to find parameter γ, used for estimating the size of the step:

$$\gamma = -e^{0T} \delta A^1 / 2\Delta\Phi, \tag{2.12}$$

where $\Delta\Phi$ is the variation of the target function, estimated by the following expression:

$$\Delta\Phi = -\alpha \sum_{i=1}^{n} A_i,$$

where α is the coefficient of the target function decrease in percent.

The step of target function decrease corresponds to the percentage reduction of target function in every iteration. Further, the vector multiplier η is estimated as follows:

$$\eta = \lambda_1 + 2\gamma\lambda_2.$$

If all components of vector η, corresponding to active constraints, are not negative, then the solution satisfies the Kuhn–Tucker conditions [2]. Otherwise, if some components of ηi are negative, it means that the target function $\Phi(A)$ acquires the value which is bigger than the minimum and the results can be improved by discarding corresponding constraints. This leads to the reduction of a number of active constraints; thus, a new matrix e is composed and the multiplier η, corresponding to the remaining active constraints, is defined. The process is repeated until all η components become positive. Variation δA is estimated from the following expression:

$$\delta A = -\frac{1}{2\gamma}\delta A^1 + \delta A^2$$

and the new vector of the design parameters acquires the following appearance:

$$A^1 = A^0 + \delta A.$$

It is worth noticing the fact that when A^0 satisfies all constraints, no other variation exists at point A^0 which would violate constraints and reduce $\Phi(A)$. It means that A^0 is the relative minimum point in nonlinear programming task. Using this algorithm, the optimal configuration of the cantilever beam-shaped structure is obtained with respect to the second torsional vibration frequency (Fig. 2.1). It can be considered as a periodical structure, where the minimum and maximum cross-sections vary every 1/3 length of the cantilever.

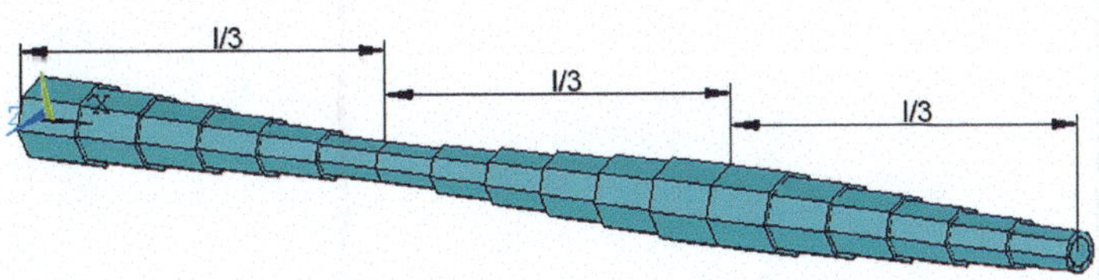

Fig. 2.1 Optimal configuration of the cantilever beam-shaped structure obtained for the given second torsional vibration frequency could be considered as periodical structure in which minima and maxima cross-sections vary every 1/3 length of the cantilever

2.1.1.1 Identification of Dynamical Properties of the Optimally Designed Tool

Suppose that the torsional vibrations of the cantilever are excited. As can be seen from Fig. 2.2, the frequency range of torsional vibrations of constant cross-section structure of the diameter $d = 10$ mm and length of $l = 100$ mm could be described by the sequence of eigenfrequencies of torsional vibrations at 9, 20, 30 kHz, etc.

The intensity of vibrations changes with the cantilever of optimal configuration (Fig. 2.3). The excitation of such structure by the wide frequency range, as occurs during cutting, gives a leap of torsional vibrations on the second eigenmode at 28 kHz frequency, when the amplitudes of the first eigenmode of optimal structure at 6 kHz frequency are minimal. Another phenomenon is observed during simulation, which is related to the reduction of lateral vibrations of the rotating structure. It could be

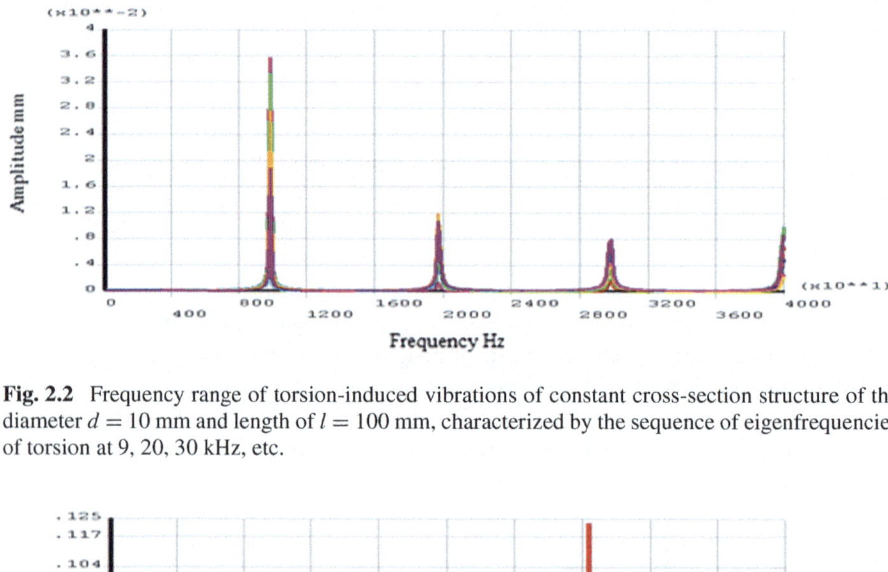

Fig. 2.2 Frequency range of torsion-induced vibrations of constant cross-section structure of the diameter $d = 10$ mm and length of $l = 100$ mm, characterized by the sequence of eigenfrequencies of torsion at 9, 20, 30 kHz, etc.

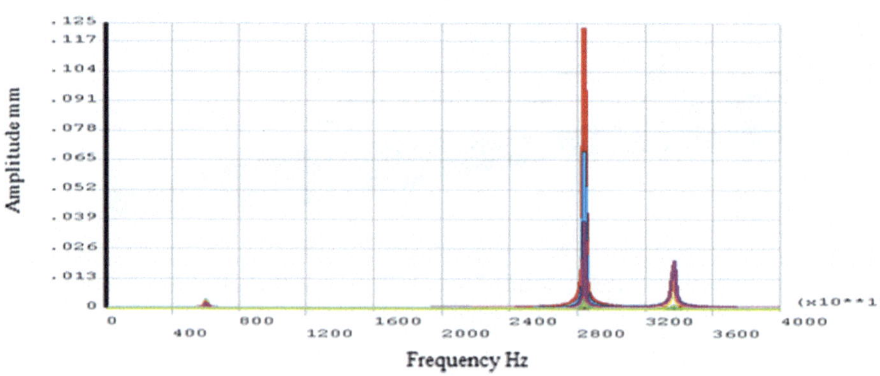

Fig. 2.3 Frequency range of torsion-induced vibrations of optimal configuration cantilever beam-shaped structure obtained for the given second eigenfrequency of torsional vibrations with the pronounced second torsional mode at a frequency of 28 kHz, when the first eigenmode of this structure at 6 kHz is minimal

explained by the increase of the dynamic stiffness of the structure due to the more intensive rotational deformations of FE cross-sections. This phenomenon could be useful for long tool structures with strict constraints for lateral deviations during cutting.

2.1.1.2 Practical Realization of Self-Exciting Twist Drill Structure

Structure optimized with reference to the established criteria and constraints may turn out to be irrational for practical use, especially for drilling tools. For instance, if the technological requirements are neglected and not included in the constraints, there is a possibility that the developed structure will be optimal in terms of mass, but irrational in the technological aspect. For this purpose, the rational structure from the technological point of view of the drilling tool has been recognized (Fig. 2.4) as an invention [3]. This structure, as the optimal one in Fig. 2.4, has periodically changing cross-sections, which coincide with the structural elements of twist drill such as shank 1, neck 2, and body 3, when the outer surfaces of these parts are neither convex nor concave, but cylindrical. As the drill vibrates in torsion, it lengthens and shortens periodically in the axial direction, thus, inducing a vibration cutting regime. Simulation of torsional–axial vibration is conducted by means of Bayly's model, which is based on the fact that when the twist drill "untwists", it extends in length.

The tool, affected during drilling by a wide range of forces acting in the cutting area, self excites in the second mode of rotational vibration with an amplitudes θ_2 at the cutting edge of the drill much smaller than the first mode vibration amplitude θ_1 as shown by dotted lines. The frequency corresponding to the second mode is much higher than to the first mode. Since the drill is like a twisted spiral, in the case of rotational vibrations, the spiral turns and at the same time lengthens. In this case, the

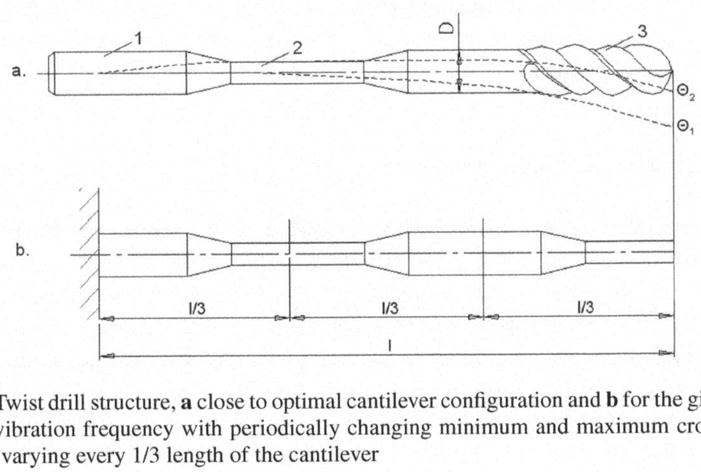

Fig. 2.4 Twist drill structure, **a** close to optimal cantilever configuration and **b** for the given second torsional vibration frequency with periodically changing minimum and maximum cross-sections diameters varying every 1/3 length of the cantilever

cutting edges of the drill vibrate in the longitudinal direction, simulating the case of vibration drilling the advantages of which will be revealed later.

The structural changes of tools and possibilities to excite higher modes are related to the modification of tool structures so as to approach the optimal one with respect to the given higher mode of natural vibrations. This, in turn, has important practical implications because the proposed approach of tool mode control is relatively simple to implement in the industrial environment as it does not require sophisticated control devices.

2.1.2 Modified Boring Tool Structures for Effective Cutting

Traditionally excitation of high-frequency vibrations at the tool cutting edge needs special equipment, which is not relatively simple to implement in the industrial environment. More effective could be the possibility to excite higher vibration modes due to the cutting tool's structural changes [4]. Such a structural approach does not need complicated hardware and the end-user does not need to introduce new handling routines. The main idea is based on treating the cutting tool as a flexible structure which is characterized by several modes of natural vibrations. In such machining processes as boring, the structural configuration of the tool resembles the cantilever beam. The first eigenmode of the cantilever is characterized by maximum amplitudes of free end vibrations. The establishment of structural modifications of the cutting tool as the flexible structure is related to the intensification of the higher eigenmodes, resulting in the reduction of the magnitude of unwanted deleterious vibrations generated during machining. This suggests that excitation of higher eigenmodes could be advantageous for this purpose since it is known that as the amplitude of higher modes becomes more intensive, energy dissipation inside tool material increases significantly and thereby makes the tool a more effective damper, which positively influences the amplitudes of the workpiece or machine tool itself, providing the possibility to reduce chatter.

2.1.2.1 Analysis of Boring Process

The tool vibration during cutting could be described as a vibro-impact process. As any kind of tool has distributed mass, stiffness, and other parameters, it is necessary to consider the dynamics of such elastic structure that is characterized by several eigenmodes. The impact interaction between elastic links is characterized by a rich spectral content of excitation impulses capable to excite a wide range of eigenmodes. In boring, the structural configuration of the tool resembles the cantilever beam. Figure 2.5 presents a computational scheme of the developed virtual FE model of impacting boring tool on the workpiece internal surface expressed by the rheological properties—stiffness and viscous friction. The model consists of $i = 1, 2\ldots, m$ finite elements.

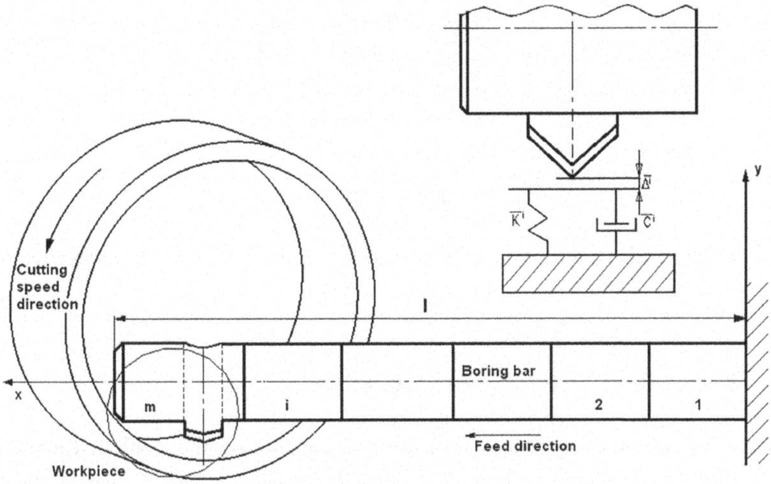

Fig. 2.5 Boring tool and workpiece interaction

Impact modeling is based on the contact element approach and makes use of the Kelvin–Voigt (viscoelastic) rheological model, in which linear spring is connected in parallel with a damper—the former represents the impact force and the latter accounts for energy dissipation during impact. After proper selection of generalized displacements in the inertial system of coordinates, model dynamics is described by the following equation of motion given in a general matrix form:

$$[M]\{\ddot{y}(t)\} + [C]\{\dot{y}(t)\} + [K]\{y(t)\} = \{F(y, \dot{y}, t)\} \tag{2.13}$$

where *[M], [C], [K]* are mass, damping, and stiffness matrices, respectively; $\{y(t)\}$ $\{\dot{y}(t)\}$ $\{\ddot{y}(t)\}$—displacement, velocity, and acceleration vectors, respectively; and $\{F(y, \dot{y}, t)\}$ vector of interaction forces between tool cutting edge and the workpiece, the components of which express the reaction of the tool cutting edge impacting workpiece and acquire the following form:

$$f_i(y_i, \dot{y}_i, t) = \overline{K}^i(\overline{\Delta}^i - |y_i(t)| + \overline{C}^i y_i(t), \tag{2.14}$$

where $\overline{K}^i, \overline{C}^i$—stiffness and viscous friction coefficients of the workpiece material and $\overline{\Delta}^i$—distance from the cutting edge located at the *i*-th nodal point of the tool structure to the surface of the workpiece. In the case of the considered model, the assumption of proportional damping is adequate; therefore, internal damping is modeled by means of the Rayleigh damping approach [5]:

$$[C] = \alpha_{dM}[M] + \beta_{dK}[K], \tag{2.15}$$

where α_{dM}, β_{dK} are mass and stiffness damping parameters, respectively, that are determined from the following equations using two damping ratios ξ_1 and ξ_2 that correspond to two unequal natural frequencies of vibration ω_1 and ω_2:

$$\begin{aligned} \alpha + \beta\omega_1^2 &= 2\omega_1\xi_1 \ ; \\ \alpha + \beta\omega_2^2 &= 2\omega_2\xi_2 \ . \end{aligned} \tag{2.16}$$

The presented FE model of the vibro-impact system was implemented in virtual tween.

2.1.2.2 Boring Tool Virtual Twin Simulation

The vibro-impact process consists of free vibrations of the tool in the intervals between the impacts and its vibration during the impact interaction with the workpiece. Therefore, a profound investigation of the free and impact vibration of the elastic tool is essential. The modes of natural transverse vibrations of elastic cantilever beam-shaped tool presented in Fig. 2.6 consist of transverse displacement Y. Of the whole range of natural vibrations, the first five eigenmodes are distinguished (*I, II, III, IV*, and *V*) which in the intersection with the axis line form nodal points marked by numbers that express the ratio between the distance x from the fixing site of the cantilever beam-shaped tool and its whole length l. The letters Y_{ij} denote the values of the maximum amplitudes (deflections) of the flexural eigenmodes. It is known that the first eigenmode of a cantilever is characterized by maximum amplitudes of free end vibrations as the amplitudes of each higher eigenmode gradually decrease.

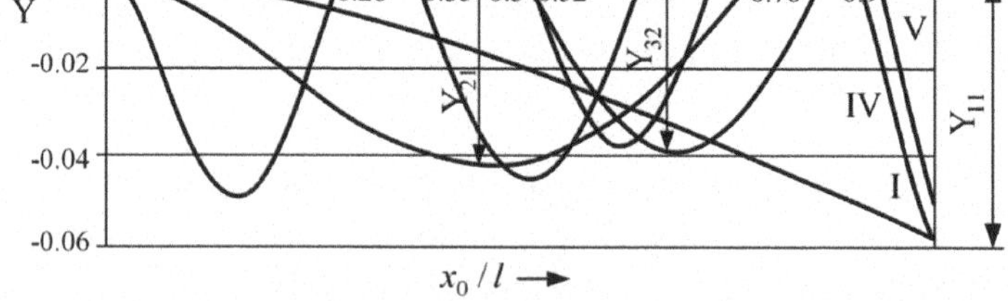

Fig. 2.6 The eigenmodes of transverse flexural vibrations of the tool. x_0/l denotes the ratio between the distance x_0 from the anchor of the tool and its whole length l, Y_{ij} —maximum amplitudes of the flexural vibrations: index i—number of vibration mode, j—sequence number of the maximum amplitude point with respect to the anchor point

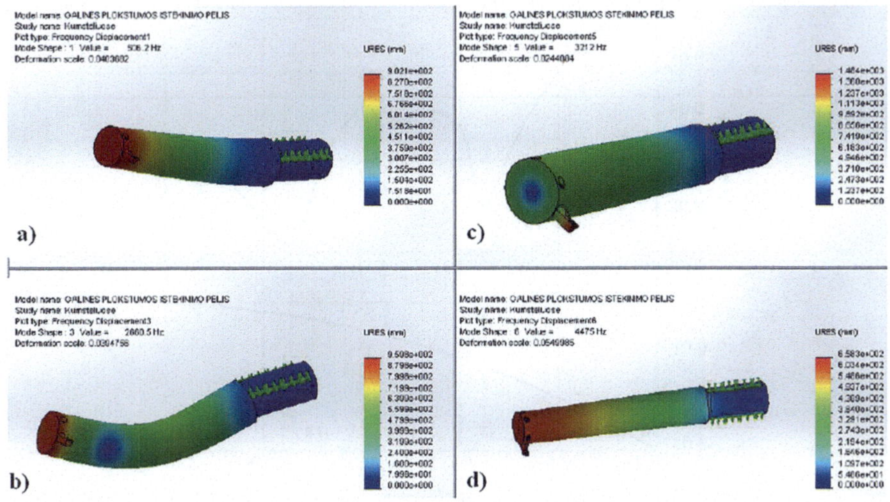

Fig. 2.7 The eigenmodes of boring tool vibration: **a** the first transverse eigenmode, **b** the second transverse eigenmode, **c** the first rotational eigenmode, and **d** the first longitudinal eigenmode

Figure 2.7 represents different eigenmodes of the boring tool. Eigenfrequencies depend on structural parameters by changing which it is possible to approach transverse and rotational frequencies of the boring tool.

Identically to the process of free impact vibrations of cantilever released from statically deformed position bouncing the support, the process of free impact vibrations of boring tool is simulated. The main purpose of such simulation is to imitate the cutting process which is characterized by the wide frequency range of cutting forces during chip formation when the tool contacts with the elastically recovered surface of the workpiece. Figure 2.8 presents the dimensionless dependence of the maximum amplitude of the post-impact rebound $z_{\max} = y_{\max}/l$ on the position of the tool cutting edge, where the smallest rebound amplitudes are obtained when the cutting edge is located at points coinciding with $x_0/l = 0.87$ or $x_0/l = 0.67$. A slight decrease in the rebound amplitude is also observed at $x_0/l = 0.78$. The lower curve in Fig. 2.8, which asymptotically approaches the axis line, corresponds to the deflection of the free end of the tool during the impact with the workpiece. According to Fig. 2.6, points $x_0/l = 0.87$ and $x_0/l = 0.67$ coincide with particular points of the third eigenmode of transverse vibrations of cantilever, when the point $x_0/l = 0.78$ with the nodal point of the second eigenmode.

2.1.2.3 Boring Tool Physical Twin Experimentation

The practical issue of given simulation results could be the invented modified boring tool structure (Fig. 2.9a) in which the cutting edge is located at the distance 0.87 l from

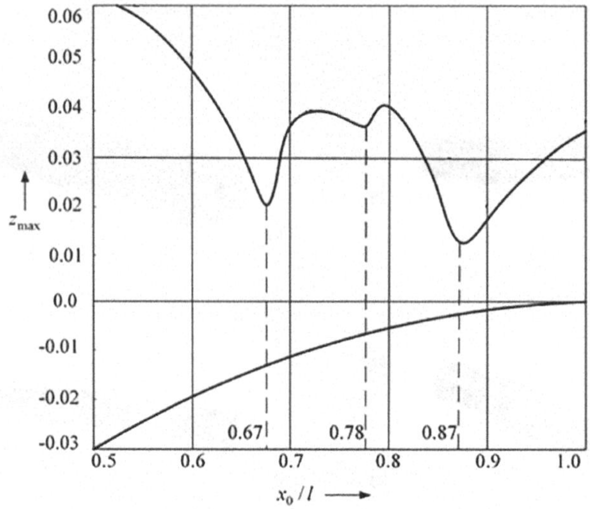

Fig. 2.8 Dependence of dimensionless rebound amplitude of the tool $z_{max} = y_{max}/l$ on the position of the cutting edge expressed as a ratio between the distance x_0 from the anchor of the tool and its whole length l

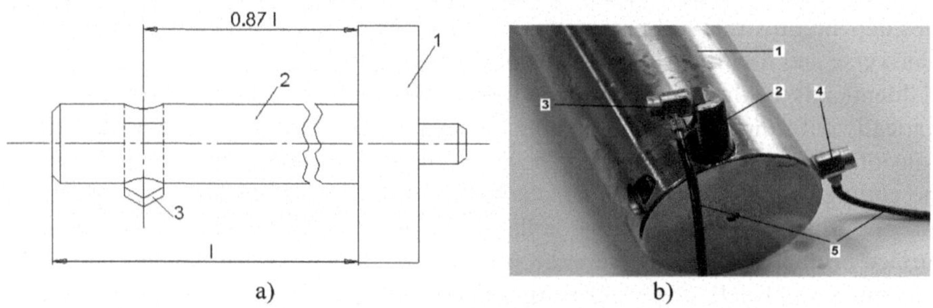

Fig. 2.9 Modified boring tool structure with cutting edge located in the nodal point of the third natural vibration mode at the distance 0.87 l from the anchor point **a** and practical realization of modified tool structure (**b**): 1-boring tool holder, 2-cutting insert, 3-sensor for radial and 4-for tangential vibrations measurements, and 5-cables

the anchor point of the tool [6]. The main purpose of this invention is an improvement of cutting conditions by decreasing tool vibration amplitude and consequently increasing frequency. For example, a desire of being able to perform a cutting operation into pre-drilled holes in a workpiece limits the diameter or cross-sectional size of the boring bar during boring when the vibrations are a cumbersome part of the manufacturing process. Usually, a boring bar is comparatively long and slender and is thereby more sensitive to excitation forces. Vibrations usually dominate by the first resonant frequency in either of the two directions of the boring bar. This process usually is not stationary. The vibrations of the boring bar affect the result of machining and surface finish in particular. The tool life is also likely to be influenced by vibrations. The tool structure consists of the end part 1 for the fixation in the machine tool spindle and the cantilever part 2 of length l constant cross-section tool holder. At the distance 0,87 l, the cutting insert 3 is fixed. When the tool is cutting the variable force excites vibro-impact motion. As the cutting insert is located in the nodal point of the third eigenmode of flexural vibrations of the cantilever-shaped structure, the third eigenmode of construction vibrations is predominated. This mode is distinguished by lower amplitudes and higher frequencies resulting in a vibration cutting regime. Intensification of the vibration energy dissipation in the tool holder material decreases not only the amplitudes of tool, but also the amplitudes of the workpiece and machine tool vibrations.

Practical realization of the modified boring tool is presented in Fig. 2.9b. On the boring tool holder 1, the cutting insert 2 could be fixed at the distance, which coincides with the nodal points of the higher eigenmodes of flexural vibrations in the radial direction. As vibrations usually dominate in either of the two directions of the boring bar, two accelerometer sensors—KD-91 (RFT, Germany) 3 and 4—are placed in the two perpendicular planes. Sensor 3 is attached in the same plane as the cutting part and is capable to measure vibrations in the radial direction as well as sensor 4 in the perpendicular direction for tangential direction to cutting surface measurements.

The experimental study was carried out with the intention to demonstrate tool vibrations. Workpiece from steel 37 was machined using identical cutting parameters with conventional and modified tools: feed $f = 0.14$ mm/rev, spindle rotation $n = 310$ rpm, and cutting depth $a = 0.25$ mm (Fig. 2.10). The vibrations of the conventional tool, when cutting insert is located at the free end of the boring bar (in red-dashed line), are characterized by higher amplitudes and lower frequency than the vibrations of the passive tool, modified by fixing cutting insert in the place of third eigenmode nodal point at the distance 0,87 l (in blue-continues line).

As is indicated in Fig. 2.10 during boring operations with modified tools, the higher eigenmodes of transverse vibrations of cutting tools are expressed by higher

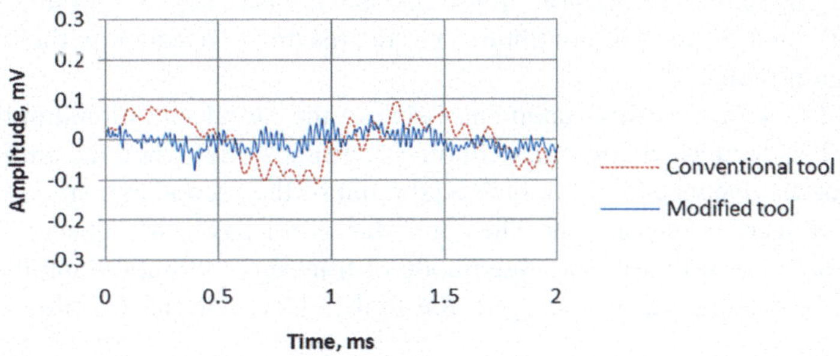

Fig. 2.10 Boring bar vibrations in direction of radial cutting force in the case of conventional (in red) and modified (in blue) tool with cutting insert fixed in nodal point of the third eigenmode of flexural vibrations at 0.87 l

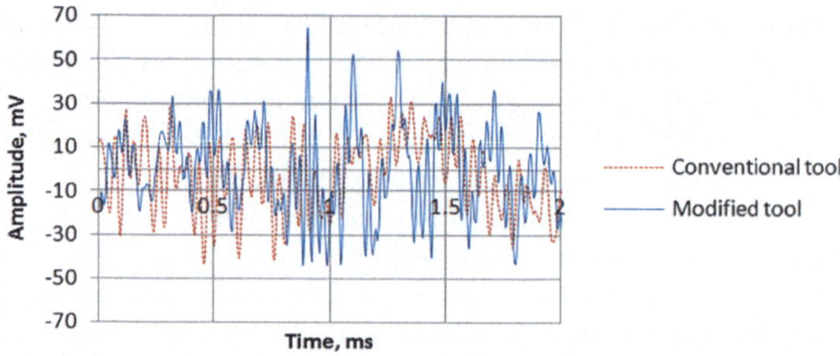

Fig. 2.11 Boring bar vibrations in direction of tangential cutting force case of the conventional and modified tool with cutting insert fixed in nodal point of the second eigenmode of flexural vibrations at 0.78 *l*

eigenfrequencies and by the few times reduced vibration amplitudes that generate vibration cutting effect. As the cutting insert is located in the nodal point of the third eigenmode of flexural vibrations of the cantilever-shaped structure, the third eigenmode of construction is predominated, which is distinguished by lower amplitudes and higher frequencies resulting in a vibration cutting regime.

Another possibility for modifying the boring tool is to modify the dimensions of the structural parameters of the tool holder, which allows the first eigenfrequency of the torsional vibrations to be aligned with the second eigenfrequency of the flexural vibrations. For intensification of the second eigenmode of flexural vibrations of the boring tool, the cutting insert should be placed at the distance 0.78 *l*, which coincides with the nodal point of the second eigenmode of flexural vibrations in the radial direction. Figure 2.11 illustrates a distinguished increase of tangential vibrations amplitudes, when the cutting insert is fixed at the point 0.78 *l* (in blue-continues line) accordingly to vibration amplitudes of conventional tool (in red-dashed line).

It means that this is the rotational resonance case, which could be useful for the reduction of transverse vibrations in the radial direction. The coincidence of two eigenfrequencies initiates the intensification of the first eigenmode of rotational and the second eigenmode of flexural vibrations as well as dissipation of energy in the tool holder material (in blue-continues line) decreasing consequently the transverse vibration amplitudes.

Figure 2.12 confirms this statement, because the radial vibration amplitudes of the boring bar decrease a few times (in blue-continues line) when the cutting insert is located at the distance 0,78 *l* in the nodal point of the second eigenmode of transverse vibrations of the boring bar. The coincidence of two eigenfrequencies initiates the intensification of the second eigenmode of transverse vibration amplitudes and consequently dissipation of energy in the tool holder material (in blue-continues line).

The cutting tool is a flexible structure that is characterized by several eigenmodes of natural vibrations and intensification of some of them. Intensification of the higher vibration modes increases tool vibration frequency, which becomes

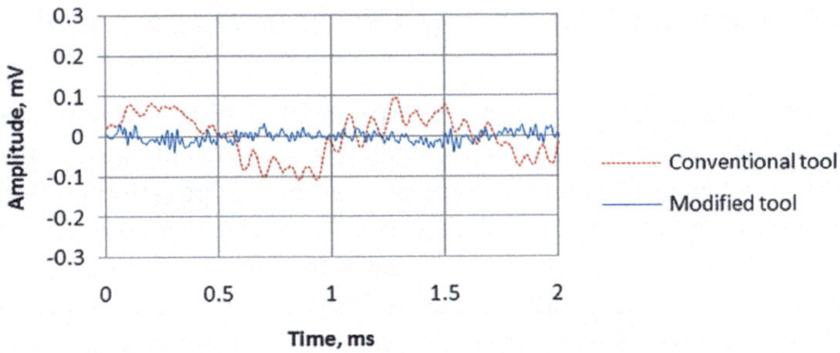

Fig. 2.12 Boring bar vibrations in direction of radial cutting force in the case of a conventional and modified tool with cutting insert in nodal point of the second eigenmode of flexural vibrations at 0.78 l

similar as during vibration cutting, and decreases tools cutting part vibration amplitudes assuring improvement of surface finish. The structural changes of tools and possibilities to excite higher eigenmodes are related to the modification of tool structures by fixing cutting insert in the nodal point of laterally vibrating tool holder or approaching the first eigenfrequency of torsion of the tool holder to the second one of transverse vibrations. This, in turn, has important practical implications since the presented approach of modified tool mode control is relatively simple to implement in the industrial application as it does not require sophisticated control devices.

2.2 Cutting Tool Physical Entities and Their Virtual Counterparts Synchronization

The digital twin is a virtual mirror of the physical world throughout its life cycle. In the case of a manufacturing system, it is a digital model designed to simulate or reproduce the functions of a real production system by providing system modeling information or directly controlled by a real system with proper connections between the real-world system and model. The main research question is how to effectively develop a digital twin model in the design phase of a complex manufacturing system and make it usable throughout the life cycle of the system, such as the production phase. This chapter focuses on modeling techniques for how to quickly create a virtual model and the mechanism for deploying the connection between the physical world production system and the virtual model it reflects.

2.2.1 Simulation and Analysis of Vibration Turning Tool

Control of vibration modes by means of experimental methods is a complex and labor-intensive task. Therefore, it is expedient to apply virtual models for the analysis of ultrasonic transducers. FE method is the predominant approach for the simulation of associated dynamic processes. Piezoelectric transducers, employed in mechanical machining, usually possess higher efficiency in comparison to magnetostrictive transducers owing to lower energy losses and higher qualitative characteristics of the former. Common piezoelectric materials are used in form of stacks, functioning as actuators in vibration cutting. Turning is the most extensively applied machining process worldwide; therefore, it is necessary to thoroughly examine the possibility to control vibration turning operation.

2.2.1.1 Virtual Twin of Vibration Turning Tool

The vibration turning tool consists of a standard tool selected from the ISO-13399 standard with an integrated piezo actuator that excites tool vibrations [7]. Piezoceramic rings were used for the excitation of high-frequency vibrations in the designed and fabricated experimental prototype of the vibration turning tool [8]. FE software Ansys was applied for the development of a detailed 3D virtual model of the turning tool incorporating a piezoelectric transducer (Fig. 2.13). The goal of the FE modeling was to reproduce the actual prototype of the vibration turning tool as accurately as possible. The model is composed of the following main parts: 1—tool insert, 2—concentrator of mechanical vibrations (horn), 3—areas representing places where the tool is fixed in a lathe holder, 4—piezoelectric transducer (piezoceramic rings), and 5—backing (back cylinder). Values of material properties that were used for FE modeling are listed in Table 2.1.

When high-frequency electric impulses from the generator through the power amplifier are supplied to the input of the piezoelectric transducer 4, it starts

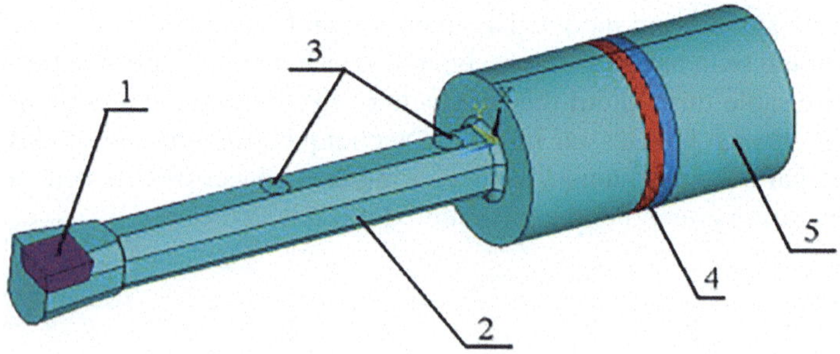

Fig. 2.13 FE model of vibration turning tool: 1—tool insert, 2—concentrator of mechanical vibrations (horn), 3—areas of tool fixing in a lathe, 4—piezoelectric transducer (piezoceramic rings), and 5—backing

Table 2.1 Parameters used for finite element modeling of the vibration turning tool

Mechanical properties of the material used in FE analysis component	Material	Density ρ (kg/m^3)	Poisson's ratio ν	Young's modulus E (GPa)
Turning tool, tool holder, fixing contact areas, horn, backing	Steel C45	7850	0.33	210
Tool insert	Carbide (Tungaloy grade KS05F)	15,000	0.24	640
Piezoceramic transducer (ring-shaped)	PZT5	7800	0.371	66

vibrating due to the inverse piezoeffect. These vibrations induce longitudinal waves in concentrator 2, which intensifies tool-tip vibration amplitudes.

10-node tetrahedral FE Solid98 with up to six degrees of freedom at each node was used for meshing of the model, which consisted of about 80.000 FE and 650.000 degrees of freedom (dof) in total. The element has a quadratic displacement behavior and is well suited to model irregular meshes as well as to simulate mechanical structure under piezoelectric control since it is characterized by large deflection and stress stiffening capabilities. Spring elements Combin14 were applied for modeling of contact interaction in vibration turning tool with the purpose to achieve adequate representation of the actual character of tool fixation in the holder (Fig. 2.14). These elastic elements with stiffness k_i were placed at each node of fixing area 3 (Fig. 2.14) and connected to immovable support. In general, Combin14 elements have either longitudinal or torsional capability in 1D, 2D, or 3D applications. In addition, the longitudinal spring–damper option is also available, which provides a massless uniaxial tension–compression element with up to three degrees of freedom

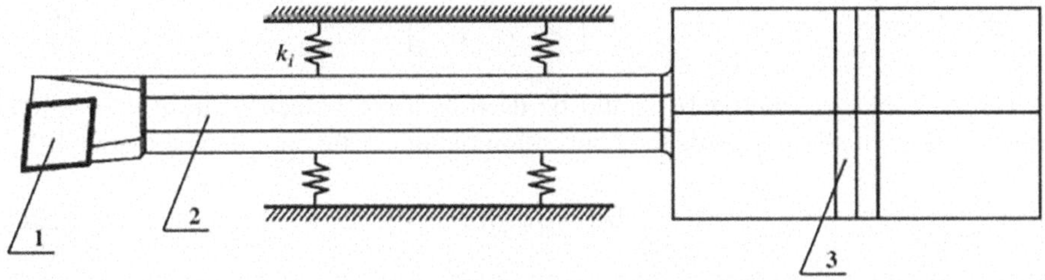

Fig. 2.14 Schematic representation of computation model of vibration turning tool: tool fixing areas are connected to fixed support through springs (with stiffness k_i) located at each node of the fixing area. 1—tool insert, 2—concentrator of mechanical vibrations, and 3—piezoelectric transducer

at each node: translations in the nodal *x*-, *y*-, and *z*-directions. No bending or torsion is considered in the latter case.

Simulation of piezoelectric behavior is based on the following linearized constitutive relations, which are the most widely recognized description. In the case of linear piezoelectricity, the dependencies of elasticity are connected into electrostatic charge dependencies when the piezoelectric constants are averaged:

$$\begin{bmatrix} \{T\} \\ \{D\} \end{bmatrix} = \begin{bmatrix} [c^E] & [e] \\ [e]^T & -[\varepsilon^S] \end{bmatrix} \begin{Bmatrix} \{S\} \\ -\{E\} \end{Bmatrix}, \tag{2.17}$$

where $\{T\}$—stress vector, $\{D\}$—electric flux density vector, $\{S\}$—strain vector, $\{E\}$—electric field intensity vector, $[c^E]$—elastic compliance matrix when subjected to a constant electrical field, $[e]$—matrix of piezoelectric stresses, and $[\varepsilon^S]$—permittivity matrix measured at a constant strain.

$[c^E]$ may be represented in the following form:

$$[c^E] = \begin{bmatrix} c_{11} & c_{12} & c_{13} & c_{14} & c_{15} & c_{16} \\ & c_{22} & c_{23} & c_{24} & c_{25} & c_{26} \\ & & c_{33} & c_{34} & c_{35} & c_{36} \\ & \text{Symmetry} & & c_{44} & c_{45} & c_{46} \\ & & & & c_{55} & c_{56} \\ & & & & & c_{66} \end{bmatrix}. \tag{2.18}$$

Matrix of piezoelectric stresses $[e]$ associates electric field intensity vector $\{E\}$ with stress vector $\{T\}$ and is expressed as follows:

$$[e] = \begin{bmatrix} e_{11} & e_{12} & e_{13} \\ e_{21} & e_{22} & e_{23} \\ e_{31} & e_{32} & e_{33} \\ e_{41} & e_{42} & e_{43} \\ e_{51} & e_{52} & e_{53} \\ e_{62} & e_{63} & e_{64} \end{bmatrix}. \tag{2.19}$$

The piezoelectric matrix $[e]$ could be used as a piezoelectric strain matrix $[d]$. Ansys automatically performs the conversion using elastic compliance matrix $[c^E]$:

$$[e] = [c^E][d]. \tag{2.20}$$

Permittivity matrix $[\varepsilon^S]$ has the following form:

$$[\varepsilon^S] = \begin{bmatrix} \varepsilon_{11} & 0 & 0 \\ 0 & \varepsilon_{22} & 0 \\ 0 & 0 & \varepsilon_{33} \end{bmatrix}. \tag{2.21}$$

Permittivity matrix measured at a constant strain acquires the following form:

$$\left[\varepsilon^S\right] = \begin{bmatrix} \varepsilon_{11} & \varepsilon_{12} & \varepsilon_{13} \\ & \varepsilon_{22} & \varepsilon_{23} \\ \text{Symmetry} & & \varepsilon_{33} \end{bmatrix}. \tag{2.22}$$

Ansys automatically converts permittivity matrix measured at a constant stress $\left[\varepsilon^T\right]$ into permittivity matrix measured at a constant strain:

$$\left[\varepsilon^S\right] = \left[\varepsilon^T\right] - [e]^T[d]. \tag{2.23}$$

Equation of motion of the piezoelectric transducer in a matrix form is written as follows:

$$\begin{bmatrix} M_{uu} & 0 \\ 0 & 0 \end{bmatrix}\begin{Bmatrix} \ddot{u} \\ \ddot{\phi} \end{Bmatrix} + \begin{bmatrix} K_{uu} & K_{u\phi} \\ K_{u\phi}^T & K_{\phi\phi} \end{bmatrix}\begin{Bmatrix} u \\ \phi \end{Bmatrix} = \begin{Bmatrix} F_S \\ Q_S \end{Bmatrix} \tag{2.24}$$

with uncoupled boundary conditions:

$$\{u\} = 0 \tag{2.25}$$

and

$$\{\phi\} = \{\phi_0\}, \tag{2.26}$$

where $\{u\}$—mechanical displacement degrees of freedom, $\{\phi\}$—electric potential degrees of freedom, $[M_{uu}]$—mass matrix, $[K_{uu}]$—mechanical stiffness matrix, $\left[K_{u\phi}\right]$—piezoelectric coupling matrix, $\left[K_{\phi\phi}\right]$—dielectric stiffness matrix, $[F_S]$—mechanical surface forces equal to zero, $[Q_S]$—electric surface forces, and $\{\phi_0\}$—the specified electric potential.

2.2.1.2 Evaluation of Vibration Turning Virtual Twin Accuracy

Development of physical twin adequate to FE model of the vibration turning tool was essential in order to be able to provide reliable clarification of experimental findings (particularly the results indicating marked improvement of surface roughness at particular excitation frequency 17.1 kHz of vibration turning tool (see Fig. 2.22). The degree of conformity between experimental and simulated resonant frequencies of the tool was adopted as a criterion defining the accuracy of the numerical model. The critical step in FE modeling was to accurately reproduce the actual contact interaction in the tool fixing areas by selecting appropriate stiffness k_i values of Combin14 elastic elements, which were defined in the nodal x-, y-, and z-directions. To this aim, the model was subjected to a series of frequency response analyses conducted by varying values of stiffness k_i. During simulations, the piezoelectric transducer was harmonically excited with a voltage of 100 V in a frequency range of 10–30 kHz. Figure 2.15 illustrates several representative amplitude–frequency characteristics of the tool-tip obtained at different stiffness values, which were expressed in terms of the magnitude of Young's modulus E of $C45$ grade steel that was used for fabrication

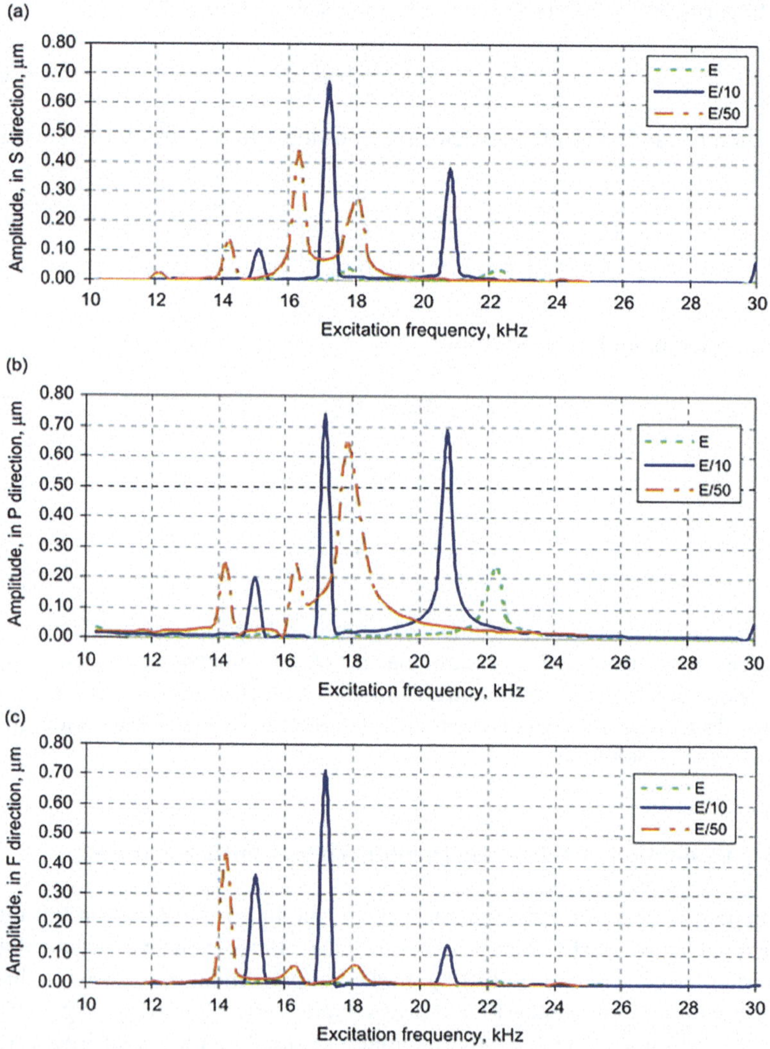

Fig. 2.15 Simulated frequency response characteristics of tool cutting edge obtained at different contact stiffness values, which are expressed in terms of the magnitude of Young's modulus E of steel $C45$: **a** direction of vertical cutting force component S, **b** direction of radial cutting force component P, and **c** direction of axial cutting force component (feed direction) F

of the concentrator and backing of the tool (Table 2.1). It is obvious from Fig. 2.15 that larger stiffness values lead to higher resonant frequencies. The accuracy of the FE model was judged by comparing simulated frequency responses in Fig. 2.15 with the experimental ones (see Fig. 2.18). The objective of adjusting stiffness k_i was to attain a reasonably close agreement between simulated and measured values of resonant frequencies, particularly for the case of the most significant frequency of 17.1 kHz. Thus, stiffness values were modified until simulated resonance peaks in all three directions of the cutting force were located as close as possible to the experimental value of 17.1 kHz. Frequency response curves marked as "E/10" (solid blue line) in Fig. 2.15 demonstrate that the best match between simulation and experimental results is achieved when stiffness k_{i} is equal in magnitude to the 1/10th of Young's modulus of *C*45 grade steel. At this particular stiffness value, simulated resonance peaks at 17.19 kHz are present in all cutting force directions. The accuracy of the final FE model is evaluated quantitatively by calculating relative error using numerical and experimental values of the considered resonant frequency: $\delta = (|17.1 - 17.19|/17.1) \times 100 = 0.53\%$. This value thereby confirms the ability of the developed FE model of the vibration turning tool to provide sufficiently reliable numerical results in the frequency domain.

Subsequently, the FE model was subjected to numerical modal analysis in order to determine the mode shapes that are excited at particular frequencies of interest. Figure 2.16 presents the simulations results. Figure 2.16a reveals that excitation frequency of 15.11 kHz induces tool-tip vibration mode in the direction of axial cutting force component (feed direction) *F* (directions are illustrated in Fig. 2.17). This mode corresponds to the second flexural eigenmode of a cantilever; meanwhile, in Fig. 2.16b, we observe that the second flexural eigenmode in the vertical cutting force direction *S* manifests at the frequency of 17.19 kHz. The next dominant frequency is 20.82 kHz, which corresponds to the tool vibration eigenmode in the direction of radial cutting force component *P*, which is equivalent to the first longitudinal vibration mode of a cantilever (Fig. 2.16c). The results of the numerical modal analysis demonstrate that longitudinal waves generated by the piezoelectric transducer excite both longitudinal and transverse vibrations of the tool-tip, particularly at the resonant frequencies of the concentrator in feed or vertical cutting force directions. A closer look at the vibration mode in Fig. 2.16b reveals that the horn oscillates around immovable point corresponding to the nodal point *O* of the second flexural eigenmode in vertical cutting force direction *S*. This indicates that at the resonant frequency of 17.19 kHz, the tool-tip vibration amplitude is significantly reduced in comparison to the amplitude of the first flexural eigenmode, which frequency is about six times lower with respect to the second eigenmode.

2.2.1.3 Physical Twin of Vibration Turning Tool

In general, vibration turning experiments may be performed for two cases: rough turning, when the goal is to maximize the volume of removed material and the cutting process is more influenced by high-frequency tool-tip vibrations in the feed direction,

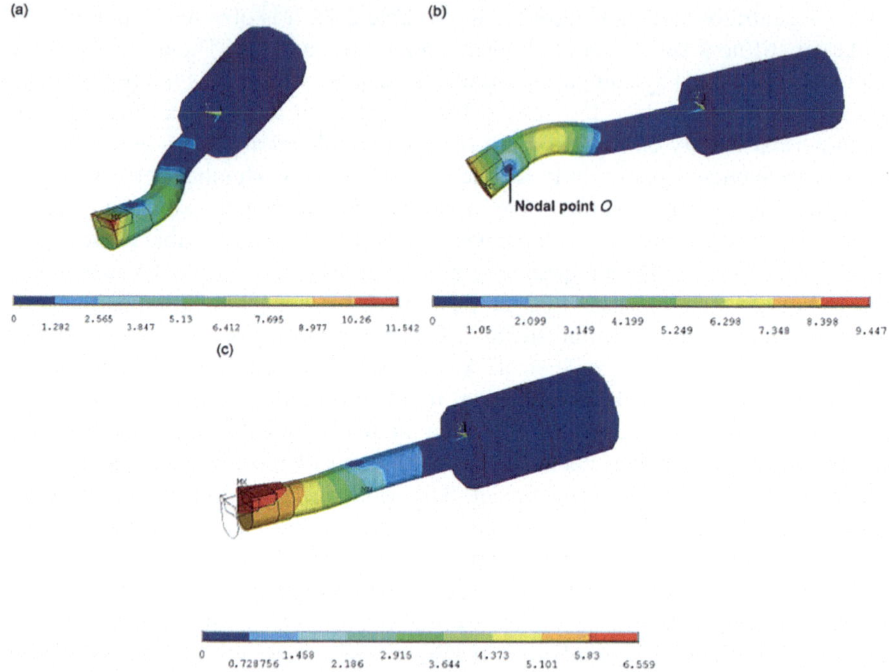

Fig. 2.16 Simulated natural vibration modes of the turning tool: **a** the second flexural eigenmode in the direction of axial cutting force component (feed direction) (15.11 kHz), **b** the second flexural eigenmode in the direction of vertical cutting force component (17.19 kHz), and **c** longitudinal eigenmode in the direction of radial cutting force component (20.82 kHz)

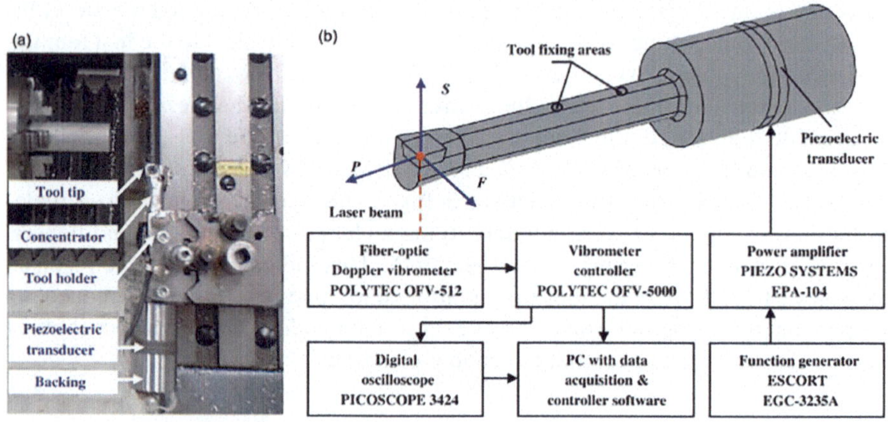

Fig. 2.17 **a** Vibration turning tool mounted in CNC lathe Denford EO4621 and **b** schematics of experimental set-up for measurement of tool-tip frequency response characteristics

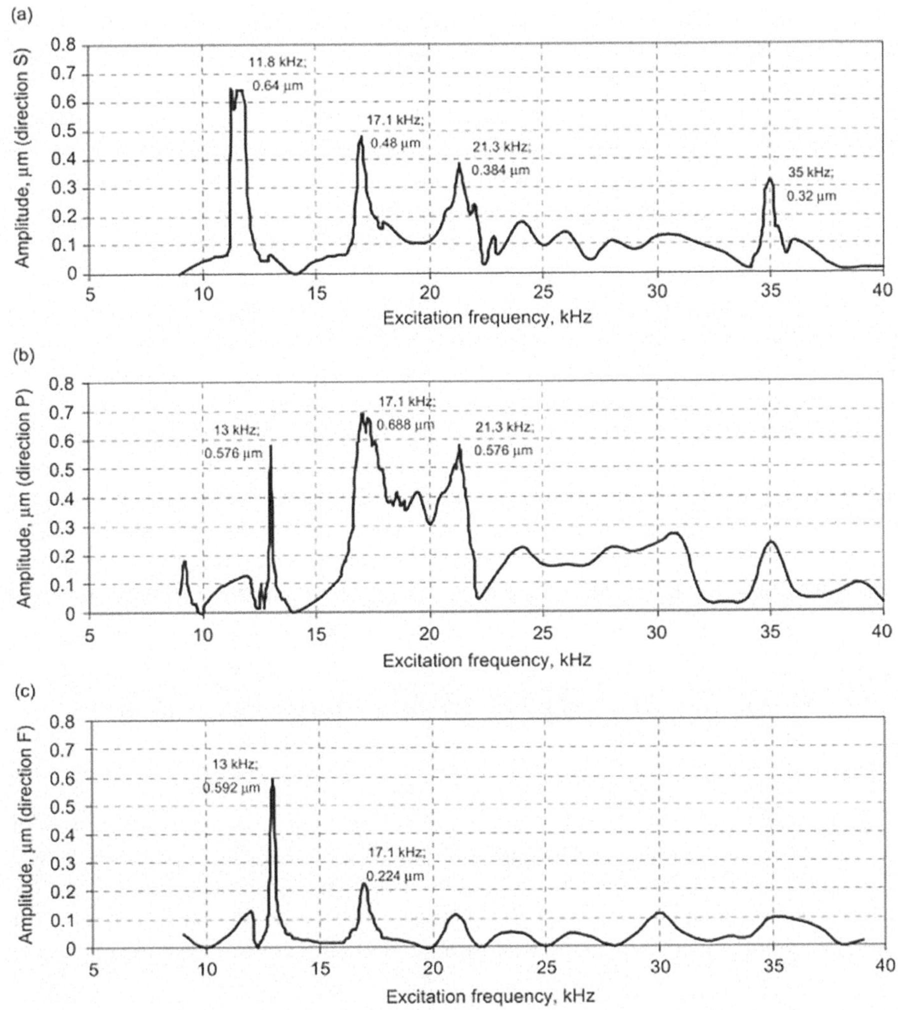

Fig. 2.18 Measured frequency response characteristics of tool cutting edge: **a** direction of vertical cutting force component S, **b** direction of radial cutting force component P, and **c** direction of axial cutting force component (feed direction) F

and finish turning, when surface quality is of primary importance and performance of the process is more determined by tool-tip vibrations generated in the direction of vertical and radial cutting force components. This research is devoted to the latter case and the aim is to characterize the vibration turning process in terms of surface quality improvement. Therefore, different types of experiments were carried out, providing both qualitative and quantitative results on the subject matter.

Laser doppler vibrometry was applied for frequency response measurements of the vibration turning tool mounted in CNC lathe Denford EO4621 (Fig. 2.18a).

The registered amplitude–frequency characteristics of the tool-tip have been used to verify the accuracy of the developed FE virtual model of the tool [9, 10]. Schematic representation of the patented [11] experimental set-up is given in Fig. 2.17b. The tool was fixed in the holder in two places. The conditions of tool fixation strongly affect the results of dynamic measurements, which are particularly sensitive to variations in bolt fastening force as well as location and number of fixation places [12]. The piezoelectric transducer was driven harmonically by using function generator Escort EGC-3235A. A constant excitation voltage of 100 V was maintained by power amplifier Piezo systems EPA-104. Polytec fiber-optic Doppler interferometer OFV-512 in conjunction with controller OFV-5000 were used to measure vibration amplitudes of the tool-tip (insert cutting edge) in three cutting force directions (Fig. 2.18b): *S*—direction of vertical cutting force component, *P*—direction of radial cutting force component, and *F*—direction of axial cutting force component (feed direction). Measured signals were captured by digital oscilloscope PicoScope-3424 and transferred to a personal computer for subsequent processing and visualization of the results. Measurements were conducted in the excitation frequency range of 8–40 kHz. It should be noted that at resonant frequencies of 13, 17.1, and 21.3 kHz, piezoelectric transducer excites tool-tip vibrations in several cutting force directions, which indicates that elliptical tool-tip motion is generated at these frequencies.

2.2.2 Evaluation of Workpiece Surface Roughness and Tool Wear

A suitable approach for the evaluation of contact interaction between a turning tool and workpiece is to employ surface roughness. Measured surface roughness values of machined workpieces confirm that vibration cutting improves surface quality for different types of materials. With the purpose of assuring the surface quality of components machined using vibration turning, a quality control system (Fig. 2.19)

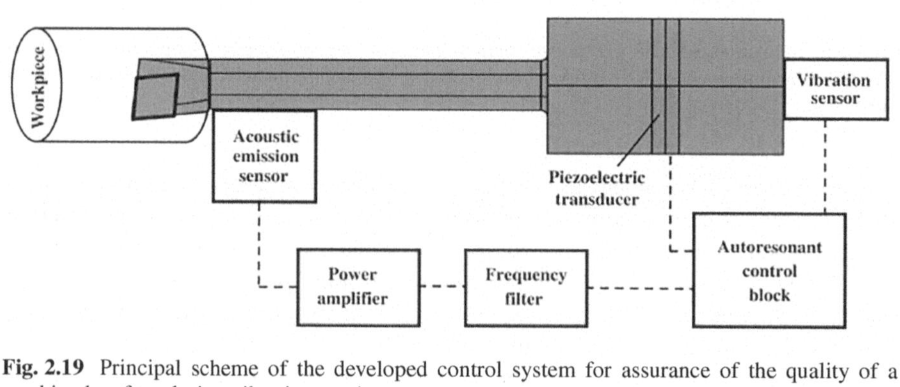

Fig. 2.19 Principal scheme of the developed control system for assurance of the quality of a machined surface during vibration turning

was developed and approved as an invention by patent [12]. The system consists of a vibration turning tool with a vibration sensor, which is rigidly attached onto the backing and connected in parallel to the autoresonant control block, acoustic emission (AE) sensor, which is rigidly attached to the nonworking surface of the tool and is connected to the control block through the power amplifier and frequency filter that are linked in series. Resonant excitation of the system is maintained by the autoresonant control block, which uses signals from the vibration sensor in order to provide the required excitation. During vibration turning operation, the amplified and filtered signal from the AE sensor is fed into the control block, where it is analyzed in order to separate from the frequency spectrum a component that is characteristic of the turning process, which is terminated when the critical value of the separated spectral component is reached. In this way, the system controls tool wear-out, simultaneously allowing to assure the surface quality of the workpiece. Experimental results indicate that the most informative AE signals are located within the range of 5–80 kHz, which provides information about dry friction. Comparison of measured AE signals (Fig. 2.20) reveals that in the case of vibration turning of aluminum workpieces, the signals are lower with respect to the case of the conventional process. This implies that friction is reduced during vibration cutting. Analogous results were also obtained for workpieces made of steel and brass.

Multiple experiments were carried out with the intention to demonstrate surface roughness improvement and tool wear reduction during vibration cutting. Two workpieces were machined using identical cutting parameters: feed $f = 0.1$ mm/rev, spindle rotation $n = 1800$ rev/min, and cutting depth $a = 0.25$ mm. For both workpieces, 12 cutting passes of $\approx$15 mm length were performed. The surface of the workpieces was qualitatively analyzed by means of scanning electron microscope JEOL JSM-IC25S. Obtained images (Fig. 2.21) provide a visual indication that the surface of workpieces machined with vibration turning is smoother in comparison to the conventional process. Furthermore, it was observed that microcracks are

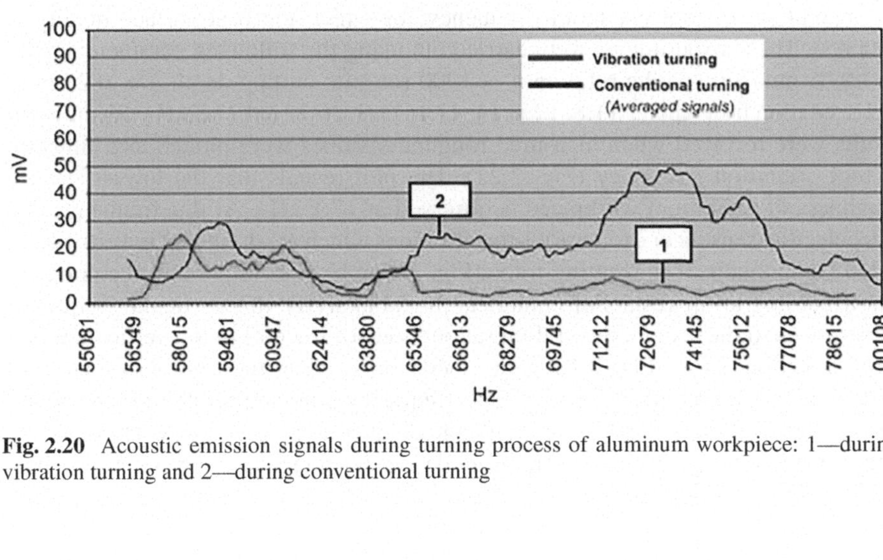

Fig. 2.20 Acoustic emission signals during turning process of aluminum workpiece: 1—during vibration turning and 2—during conventional turning

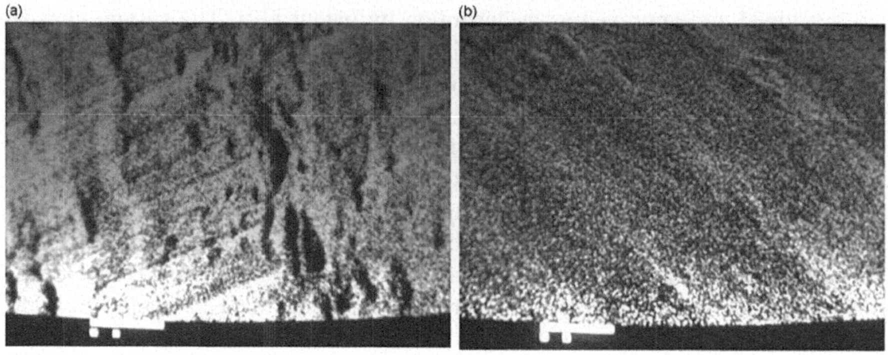

Fig. 2.21 SEM images of the surface of the machined workpiece (2000×, length of white bar = 10 μm): **a** after conventional turning and **b** after vibration turning

absent in the former case. In addition, it is evident from closer inspection of scanning electron microscope images that the workpiece surface after traditional turning (Fig. 2.21a) resembles a "roughly ploughed field", which refers to instability of the cutting process. Meanwhile, vibration turning produces a surface (Fig. 2.21b) that is similar to a "cultivated field", which is characterized by a strictly regular fine-meshed structure.

Roughness meter Mitutoyo Surftest SJ-201 was used to obtain quantitative results about surface roughness of aluminum workpieces turned with and without superimposed vibrations (feed $f = 0.05$ mm/rev, spindle rotation $n = 1900$ rev/min, and cutting depth $a = 0.25$ mm). The value of $R_a = 191 \pm 003$ μm, measured after conventional turning, reveals that surface roughness in the case of vibration turning at 17.1 kHz (Fig. 2.22) is $R_a = 081 \pm 002$ μm and approximately one roughness grade s (according to DIN EN ISO 1302).

A series of turning experiments were performed with the purpose of determining the magnitude of tool excitation frequency for which the best surface quality is obtained. These experiments were carried out using the following parameters: feed $f = 0.05$ mm/rot; spindle rotation $n = 1900$ rot/min; cutting depth $a = 0.25$ mm; and excitation frequencies 11.8, 12.9, 13, 17.1, 21.3, 26.2, and 27.3 kHz. Meaningful results were received when measured roughness values were plotted as a function of tool excitation frequency (Fig. 2.22). The plot reveals that the lowest surface roughness of aluminum workpiece is observed at 17.1 kHz. At this frequency, the piezoelectric transducer excites tool-tip vibrations which are dominant in vertical and radial components of the cutting force (Fig. 2.18a, b). It is important to emphasize that according to the results of numerical modal analysis, this particular frequency corresponds to the second flexural vibration mode of the tool in the vertical cutting force direction S (see Fig. 2.16b). In other words, when the tool is excited with 17.1 kHz, its cutting edge executes high-frequency–low-amplitude motion around nodal point O of the second flexural eigenmode in vertical direction S. These results imply that excitation of this vibration mode is advantageous in achieving the best

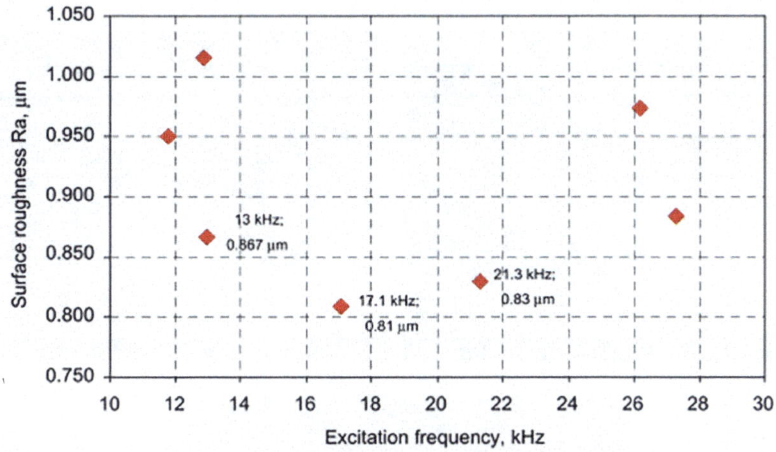

Fig. 2.22 Measured values of surface roughness R_a of the aluminum workpiece after vibration turning by using different excitation frequencies of the cutting tool

surface quality of the workpiece. It could be reasoned that the physical mechanism that is responsible for surface roughness reduction at the 17.1 kHz is related to more intensive energy dissipation inside tool material when the tool vibrates in the second flexural eigenmode since it is considered that energy dissipated by the structure vibrating in the higher eigenmode exceeds energy dissipated by the structure vibrating in its fundamental eigenmode as many times as is the ratio of eigenfrequencies of the modes. In the considered case, the frequency of the second flexural eigenmode is about six times larger than the first flexural eigenmode. Consequently, excitation of the second flexural eigenmode of the tool at 17.1 kHz leads to the corresponding sixfold increase in energy dissipated inside the tool material. This, in turn, suggests that at this particular excitation frequency the cutting tool becomes a more effective damper, which considerably contributes to the suppression of deleterious vibrations generated in the system during the cutting process thereby favoring chatter mitigation. The significant reduction in surface roughness when turning tool is excited in the second eigenmode makes it possible to refuse further finishing operations, such as grinding, thus, reducing the cost of the production process and speeding up the product's time to market. The influence of vibration cutting on tool wear was analyzed by examining cutting inserts with the scanning electron microscope. Obtained images (Fig. 2.23) provide a visual evidence that the cutting edge of the new insert is neither deformed nor damaged. Tool insert that was used during vibration cutting essentially does not differ from the new one (Fig. 2.23a) and is nearly undamaged (Fig. 2.23b), whereas the image of the insert after conventional turning reveals clearly visible deterioration of the cutting edge, which is characterized by irregular and chipped contour (Fig. 2.23c). These findings confirm that vibration cutting leads to reduced tool wear in comparison to the conventional process.

Holographic interferometry set-up (Fig. 2.24a) based on Hytech Prism defor-

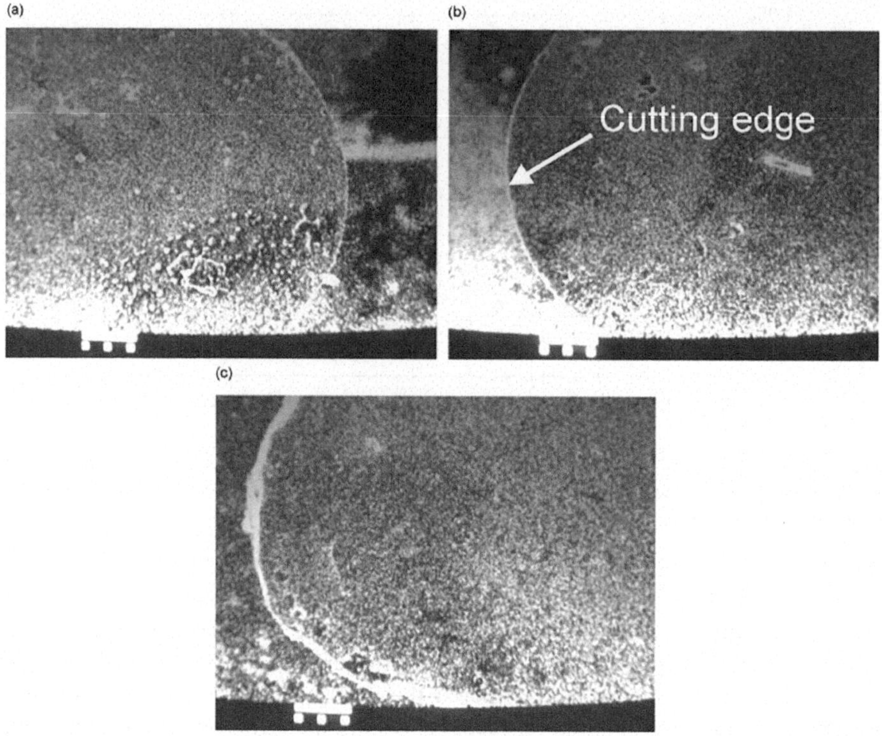

Fig. 2.23 SEM images of cutting edge of turning tool insert (100×, length of white bar = 100 μm): **a** new tool insert, **b** tool insert used in vibration turning, and **c** tool insert used in conventional turning

mation and vibration measurement system was used to determine higher vibration modes of the turning tool. Deformed shapes of the tool were recorded at different harmonic excitation frequencies (Fig. 2.24b–d). Provided results confirm that simulated vibration mode shapes obtained by numerical modal analysis are in agreement with the experimental shapes. Figure 2.24b indicates that at 13 kHz the tool vibrates in the second flexural eigenmode in the direction of axial cutting force component (feed direction) F, which corresponds to the simulated vibration shape at 15.11 kHz in Fig. 2.16a. The holographic image in Fig. 2.24c obtained at 17.1 kHz reveals the second flexural eigenmode in the direction of the vertical cutting force component S, which is in good agreement with vibration shape prediction at 17.19 kHz illustrated in Fig. 2.16b. Finally, vibration mode shape registered at 21.3 kHz and presented in Fig. 2.24d is a manifestation of longitudinal vibration mode in the direction of radial cutting force component P, which well corresponds to the simulated vibration shape at 20.82 kHz in Fig. 2.16c.

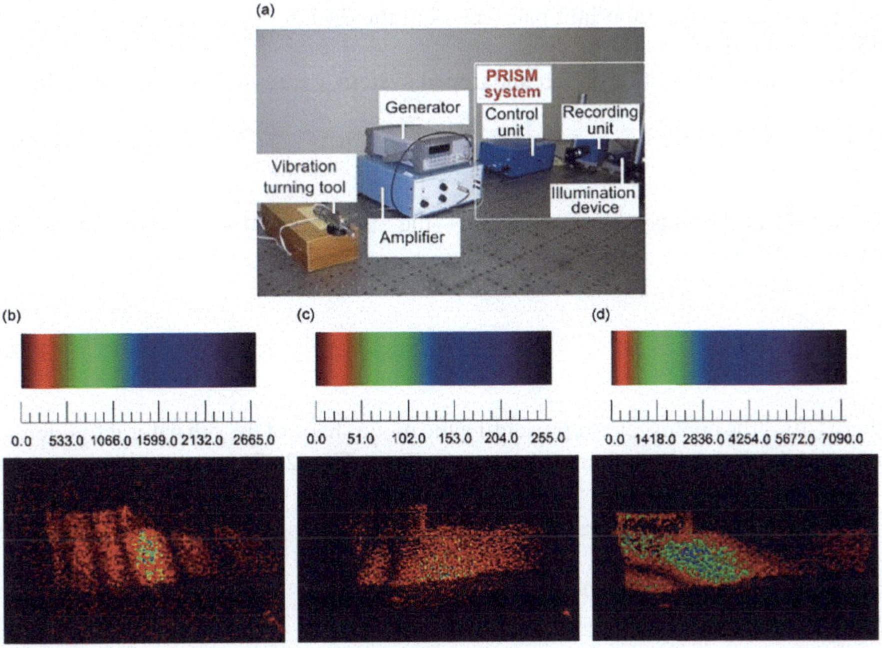

Fig. 2.24 **a** Experimental set-up based on holographic interferometry system Hytech Prism for measurement of vibration modes of the vibration turning tool and **b–d** holographic images of mode shapes of the vibration turning tool measured in axial cutting force component (feed direction) F: **b** mode shape at 13 kHz, **c** mode shape at 17.1 kHz, and **d** mode shape at 21.3 kHz

2.2.3 Influence of Boundary Conditions on the Vibration Turning Tool Eigen Modes

Modal and harmonic analysis of the vibration-assisted turning tool is presented in order to find the useful frequencies for lowering the cutting force and decreasing the surface roughness [13]. Longitudinal and transverse turning tool vibration eigenmodes are investigated while changing the turning tool fixation areas. The goal of modal analysis is to determine the eigenmode shapes and frequencies of a structure. The FE method is commonly used to perform this analysis. The eigenmodes are inherent properties of a structure, if either the material properties or the boundary conditions of a structure change, its modes will change. At or near the eigenfrequency of a mode, the overall vibration shape (operating deflection shape) of a machine or structure will tend to be dominated by the mode shape of the resonance. Resonant vibrations typically amplify the structure response far beyond the level of deflection, stress, and strain caused by static loading. A modal analysis determines the vibration characteristics (eigenfrequencies and mode shapes) of a structure or a machine component. It can also serve as a starting point for another, more detailed, transient, and harmonic response analysis or a spectrum analysis. The eigenfrequencies and mode shapes are important parameters in the design of a structure for dynamic loading conditions.

The equation of motion for an undamped system, expressed in matrix notation is

$$[M]\{\ddot{u}\} + [K]\{u\} = \{0\}, \tag{2.27}$$

where $[M]$ is the mass matrix and $[K]$—the structure stiffness matrix, including prestress effects.

The solution of Eq. (2.27) has the general form:

$$\{u_i\} = \{\Phi\}_i \omega_i t, \tag{2.28}$$

where $\{\Phi\}$—the eigenvector representing the mode shape of the i-th natural frequency and ω_i is the i-th natural circular frequency. Thus, Eq. (2.27) becomes

$$\left(-\omega_i^2[M] + \{K\}\right)\{\Phi\}_i = \{0\}. \tag{2.29}$$

Rather than outputting the circular frequencies ω_i, the eigenfrequencies were outputted:

$$f_i = \frac{\omega_i}{2\pi}. \tag{2.30}$$

The normalization of each eigenvector Φ_i was in respect to the mass matrix:

$$\{\Phi\}_i^T [M]\{\Phi\}_i = 1. \tag{2.31}$$

The eigenvalues and eigenvectors are the solutions of the equation:

$$[K]\left\{\Phi_j\right\} = \lambda_j [M]\left\{\Phi_j\right\}, \tag{2.32}$$

where $[K]$—the structure stiffness matrix, $\left\{\Phi_j\right\}$—the eigenvector, λ_j—the eigenvalue, and $[M]$—the structure mass matrix.

Harmonic response analyses are used to determine the steady-state response of a linear structure to loads that vary sinusoidally (harmonically) with time, thus, enabling to verify whether or not designs will successfully overcome resonance, fatigue, and other harmful effects of forced vibrations. Harmonic response analysis gives the ability to predict the dynamic behavior of cutting tool structures. This technique is used to determine the steady-state response of a linear structure to loads that vary harmonically with time. The general equation of motion for a structural system is

$$[M]\{\ddot{u}\} + [C]\{\dot{u}\} + [K]\{u\} = \left\{F^a\right\}, \tag{2.33}$$

where $[M]$—structural mass matrix; $[C]$—structural damping matrix; $[K]$—structural stiffness matrix; $\{\ddot{u}\}$—nodal acceleration vector; $\{\dot{u}\}$—nodal velocity vector; $\{u\}$—nodal displacement vector; $\{F^a\}$—applied harmonic load vector.

As stated above, all points in the structure are moving at the same known frequency, however, not necessarily in phase. Also, it is known that the presence of damping causes phase shifts. Therefore, the displacements may be defined as follows:

$$\{u\} = \left\{u_{\max} e^{i\,\Phi}\right\} e^{i\,\Omega t}, \tag{2.34}$$

where $\{u_{\max}\}$—maximum displacement; i—square root of -1; $\Omega = 2\pi f$—imposed circular frequency (radians/time); f—imposed frequency (cycles/time); t—time; Φ—displacement phase shift.

The experimental model of the vibration-assisted turning tool is composed of the following main parts (Fig. 2.25a): carbide insert 1, turning tool 2, tool holder 3, and two fixation bolts 4. The Langevin-type piezoelectric transducer contains horn 5, piezoelectric rings 6, and backing 7. Properties of the parts are listed in Table 2.1. The model was transferred to the CAD system Solidworks after the dimensions were taken. The main feature of the model is the four contact areas. These areas act as the representation of bolts used in the real model.

It is known that boundary conditions are very important for the analysis results and for the dynamics of the turning tool. In the model, the boundary condition is the location of the contact areas. These areas are changed during the analysis of the influence of tool dynamics. The optimization problem for six different contact area sets of the tool (Fig. 2.25b) is solved. The 3D model of the turning tool was transferred to FE software Ansys. The Solid98 FE was chosen for the modal and harmonic response analysis of the turning tool which has a quadratic displacement behavior and is well suited to model irregular meshes (such as produced from various

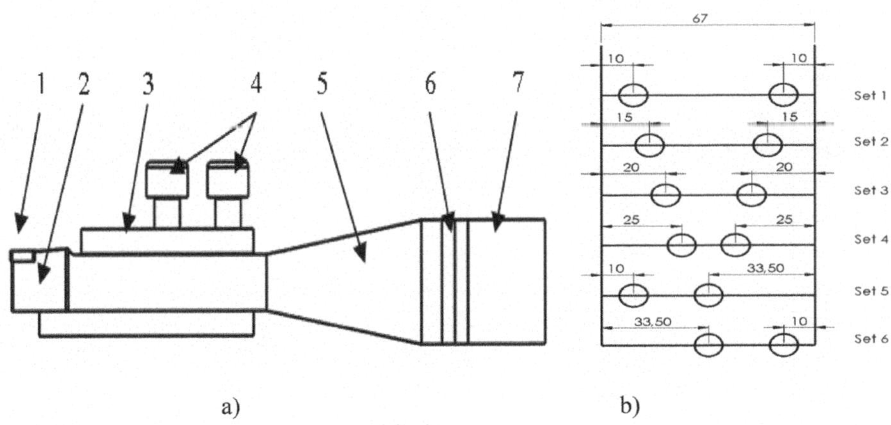

Fig. 2.25 **a** Scheme of turning tool with holder: 1—insert, 2—turning tool, 3—tool holder, 4—bolts, 5—horn, 6—piezoceramic rings, and 7—backing and **b** location of contact area positions

CAD/CAM systems). When used in structural and piezoelectric analyses, Solid98 has large deflection and stress stiffening capabilities. The element is defined by ten nodes with up to six degrees of freedom at each node.

In Fig. 2.26a, the meshed turning tool with the 20,042 nodes and 10,925 elements is presented. Green areas are the highlighted contact areas. Modal analysis of the turning tool was carried out between 0 and 40 kHz frequency. The longitudinal vibrations are very useful for decreasing surface roughness. The first vibration turning tool longitudinal eigenmode at 5.3 kHz and second at 27.6 kHz were determined (Fig. 2.26b). The first transverse eigenmode of the turning tool was determined at 9.9 kHz and the second at 35.1 kHz. Analyzed turning tool transverse and longitudinal frequency dependence from contact area position are showed in Fig. 2.27. It can be seen that eigenmodes most efficiently differ in frequency when the fifth contact area set is used. This is very useful when it's necessary to achieve the needed vibration mode and the modes' frequencies do not overlap each other.

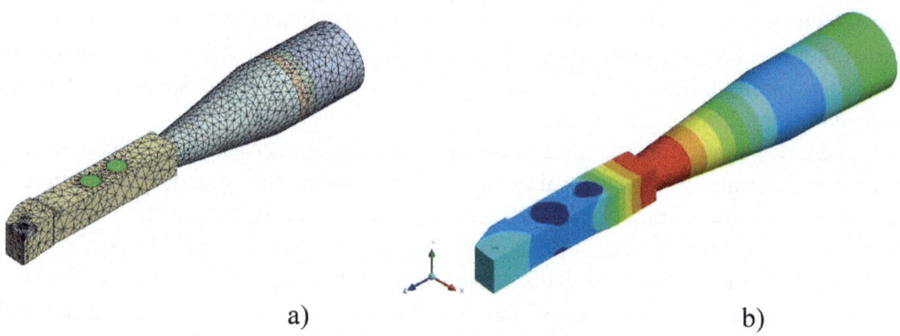

Fig. 2.26 **a** Finite element model and **b** second longitudinal mode of the vibration turning tool

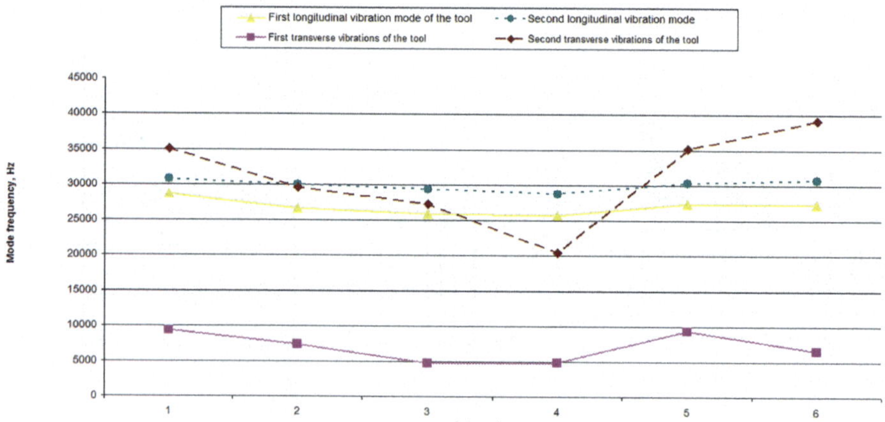

Fig. 2.27 Vibration mode dependence from the contact area set number

Such overlapping is seen in the second and third contact area fixing set. The modal analysis of the turning tool with different contact areas has showed that the position of the contact areas has a large influence on tool dynamics. For example, the eigenfrequency difference between the first transverse eigenmode in Set 1 ($f = 10$ kHz) and the same transverse mode but in Set 4 is two times lower ($f = 4.9$ kHz). The eigenfrequency of the second transverse eigenmode of the turning tool is approximately more than three times greater. The lowest eigenfrequency of the second transverse eigenmode ($f = 27.6$ kHz) at the contact area Set 3 when the highest eigenfrequency of the second transverse mode ($f = 39.6$ kHz) is reached for Set 6 contact area. Longitudinal vibrations are one of the main vibrations which are used to lower surface roughness and cutting forces and to extend tool life. In the model, longitudinal vibrations are achieved in high frequencies. The range in which longitudinal vibrations occur is from $f = 25$–32 kHz. There was no significant difference between the frequencies of the first and second longitudinal eigenmodes. The obtained data gives the possibility to design the tool for efficient machining operation.

The harmonic analysis of the vibration-assisted turning tool was carried out in order to get quantitative results of the analysis. From the modal analysis results (Fig. 2.27) for the contact area Set 5, the turning tool eigenmodes are separated quite evenly; therefore, the contact area Set 5 has been chosen for harmonic analysis. Furthermore, in Set 5, good localization of the vibration modes is evident. This feature is very important because such good localization of the modes will enable to excite interested modes of the turning tool to get useful vibrations. Harmonic analysis was carried out in the range $f = 15$–40 kHz. The material properties and the mesh used were the same as in the modal analysis. The results of the harmonic analysis are presented in Fig. 2.28. Figure 2.28a shows turning tool-tip amplitude–frequency characteristic in the axis Z along the tool holder direction at contact are Set 5. The peak amplitude reaches in when the tool is excited with the frequency equal to 26.5 kHz and the displacement 2.29×10^{-3} mm. This peak amplitude

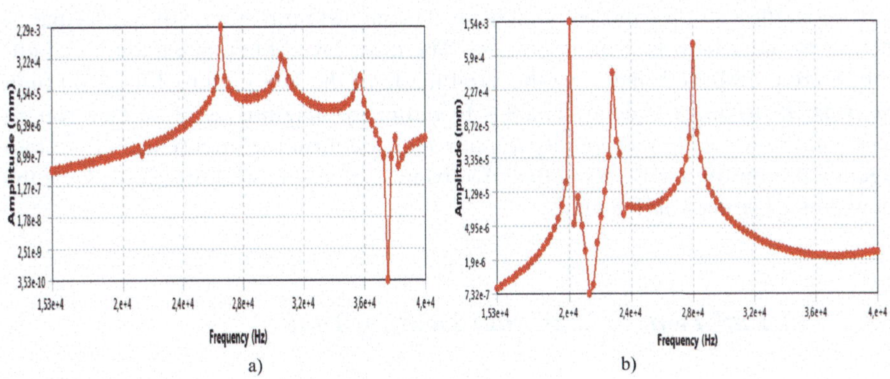

Fig. 2.28 Amplitude–frequency characteristic of the tool-tip displacement in axis Z-direction: **a** Set 5 and **b** Set 4

corresponds to the first longitudinal eigenmode of the tool. The second peak occurs at the frequency equal to 30.5 kHz and the displacement 3.63×10^{-3} mm. At the same frequency as the second peak (Fig. 2.28a) in Fig. 2.27, the second longitudinal vibration eigenmode of the tool is seen. It is clear from this comparison that the amplitudes of the cutting edge can be usefully varied by selecting the contact areas. Three peaks of tool amplitude were observed (Fig. 2.28b) when the tool was excited harmonically using contact area set 4. First peak $f = 20$ kHz and the displacement 1.54×10^{-3} mm corresponds to second transverse eigenmode. The last two peaks occurring, respectively, at 22.8 and 28 kHz represent the first and second longitudinal vibration modes of the turning tool. Harmonic analysis of the vibration-assisted tool has shown that using certain excitation frequencies the useful displacements of the turning tool-tip could be achieved.

2.3 Evaluation of Technological Features of Macro- and Micro-drilling

This study is concerned with the application of virtual–physical approach for characterizing the dynamic behavior of the piezoelectrically excited vibration drilling tool with the aim to identify the most effective conditions of tool vibration mode control for improved cutting efficiency. 3D FE models of the tool were created on the basis of an elastically fixed pre-twisted cantilever (standard twist drill). The models were experimentally verified and used together with tool vibration measurements in order to reveal rich dynamic behavior of the pre-twisted structure, representing a case of parametric vibrations with axial, torsional, and transverse eigenvibrations accompanied by the additional dynamic effects arising due to the coupling of axial and torsional deflections ((un)twisting). Virtual simulation results combined with extensive data from interferometric, accelerometric, dynamometric, and surface roughness measurements allowed us to determine critical excitation frequencies and the corresponding vibration modes, which have the largest influence on the performance metrics of the vibration drilling process. The most favorable tool excitation conditions were established: inducing the axial mode of the vibration tool itself through tailoring of driving frequency enables to minimize magnitudes of surface roughness, cutting force, and torque. Research results confirm the importance of the tool eigenmode control in enhancing the effectiveness of vibration cutting tools from the viewpoint of structural dynamics.

2.3.1 Virtual Twin of Vibration Drilling Tool

Control of vibrational behavior of different technological tools is a widely applied approach enabling improvement of process efficiency. This is particularly true for vibrations of cutting tools generated during machining since the magnitude of induced vibrations has a direct effect on the surface quality of a workpiece. Continuous efforts to increase machining efficiency led to the observation that the cutting process can be enhanced if the tool is assisted with ultrasonic frequency vibrations, which results in reduced cutting forces and surface roughness. It is imperative to have a physically substantiated virtual model enabling accurate predictions of the dynamic behavior of the drill bit driven by the developed vibration drilling tool in order to be able to correctly interpret measurement data acquired during vibration drilling experiments, i.e. explain results that refer to the appreciable decrease of cutting force and surface roughness at specific tool excitation frequencies [14–17]. FE analysis in Ansys software was used for this aim. Design of the developed vibration drilling tool Fig. 2.29a illustrates the main structural elements of a prototype of the vibration drilling tool. A piezoelectric transducer (a stack of two piezoceramic rings 8) is integrated into the vibration tool assembly (consisting of components 2–9) for generating high-frequency vibrations of the cutting edge of the drill bit 10. The transducer converts electrical power from the supply into the mechanical motion of the drill bit. The power is supplied to the device through the collector rings 4. The ultrasonic power supply generates up to 200 W with a sinusoidal waveform. A concentrator 9 with chuck is fitted onto the end of the transducer assembly, which leads to the intensification of the drill-tip vibration amplitude that may reach up to 20 μm. For increase of the vibration amplitude, the drilling tool is designed to operate in the resonance mode, i.e. its length is equal to an integral number of half-wavelengths at a given frequency (Fig. 2.29b). The vibratory system consisting of the

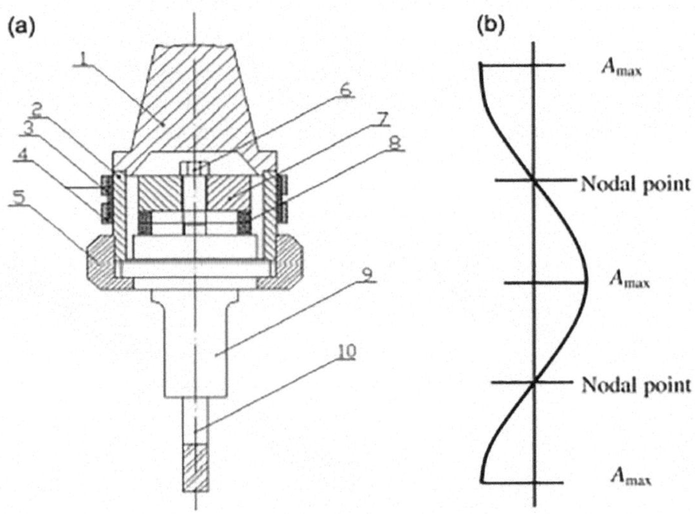

Fig. 2.29 **a** Structural diagram of the developed vibration drilling tool: 1—standard holder (Weldon) DIN 6359, 2—cylinder, 3—textolite cylinder, 4—collector rings, 5—nut, 6—bolt, 7—collet, 8—piezoceramic rings, 9—concentrator, and 10—drill bit and **b** graphical representation of the vibratory operational principle of the tool

transducer assembly, concentrator, and the drill bit is rigidly clamped to the standard Weldon tool holder 1 at a nodal (zero amplitude) point. This solution prevents losses of vibrational energy through dissipation into the body of the machine tool.

Measurement data collected during dynamic testing of the vibration drilling tool revealed that drill bit dynamics has negligible influence on the axial vibrations generated by the vibration drilling. Therefore, it is safe to assume that the actual excitation generated by the piezoelectrically driven vibration tool may be accurately represented as an equivalent kinematic (base) excitation of the drill bit, which is imposed onto zones where the drill bit is clamped in the chuck with the bolts. Furthermore, the considered vibratory system is linear; therefore, the reliable interpretation of the dynamic behavior of the vibration drilling tool by using a virtual model of a single drill bit, which is imposed with the boundary and excitation conditions that accurately reproduce the actual case when the drill bit is excited by the vibration tool, is possible. A meaningful outcome of this structural model reduction is that the final FE model is manageable from a computational point of view, i.e. it does not incorporate bulky and structurally complex transducer assembly, which leads to a marked decrease in the total number of degrees of freedom, thereby enabling to perform computationally intensive dynamic simulations within a reasonable time frame.

The modeling procedure commenced from the construction of a solid model of a drill bit (pre-twisted cantilever) in CAE software SolidWorks (Fig. 2.30a). Subsequently, the model was exported to FE software Ansys, where it was imposed with the appropriate boundary conditions. The following geometrical and material properties were used for the model: drill length l = 132 mm, diameter d = 10 mm, density $\rho = 8000$ kg/m^3, Young's modulus $E = 207$ GPa, and Poisson's ratio $\nu = 0.3$. 10-node tetrahedral finite element Solid92 with up to three degrees of freedom at each node was used for meshing. Solid92 has a quadratic displacement behavior and is well suited for modeling irregular meshes such as those produced from various CAD/CAM systems. The Cartesian coordinate system was used for modal analysis. However, for harmonic and transient simulations, it was more convenient to define the

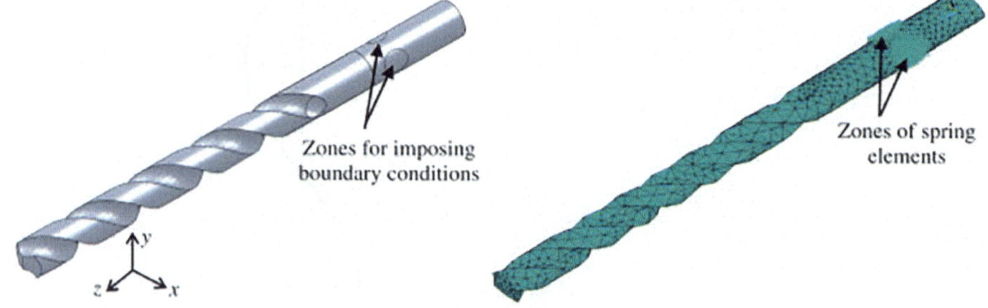

Fig. 2.30 **a** 3D CAD model of a drill bit with the delineated "fixing" zones where the structure is imposed with appropriate boundary conditions and **b** ANSYS FE model with the designated zones that are meshed with spring elements Combin14

coordinates in a cylindrical system with the following degrees of freedom: displacements z along the rotational axis of the drill bit (axial direction), displacements r orthogonal to the axis (transverse direction), and rotations φ about the rotational axis (torsional direction).

A distinctive feature of the developed FE model is that it takes into account the elastic nature of drill bit fixation in the vibration drilling tool. It is crucial to accurately represent this fixation elasticity in the model since it may have an appreciable influence on drill bit dynamics due to varying chuck clamping force, temperature effect, or other factors. Therefore, a sub-model composed of elastic links was implemented into the FE model of the drill, thereby allowing to determine the stiffness of the links. Specifically, spring elements Combin14 were applied for modeling of contact interaction in the places where the drill bit is mounted in a chuck of the vibration tool (drill bit position is secured with three bolts positioned around the shank circumference at 120° angles). In the FE model, the drill bit is fixed elastically by employing longitudinal spring elements k_r, k_z and torsional spring elements k_φ (k_r—stiffness in transverse r-direction, k_z—stiffness in axial z-direction, and k_φ—torsional stiffness). These 1D link elements are positioned in all directions of the cylindrical coordinate system and are placed at each node of the three "fixing" zones located on the shank (these zones are marked by the circles in Fig. 2.30a). One node of a spring element is superposed with the corresponding node located on the zone (Fig. 2.30b), while the other link node ("free node") is connected to the immovable support when running modal analysis and left unconstrained but is imposed with the kinematic excitation in the axial direction when running harmonic and transient simulations. In summary, the developed FE model consists of about 6163 elements Solid92 and 267 elements Combin14 with a total of about 30.000 degrees of freedom. In the considered FE formulation, the dynamics of the drill bit is described by the following equation of motion in a block form by taking into account that base motion law is known and is defined by the nodal displacement vector u_K:

$$\begin{bmatrix} M_{NN} & M_{NK} \\ M_{KN} & M_{KK} \end{bmatrix} \begin{Bmatrix} \ddot{u}_N \\ \ddot{u}_K \end{Bmatrix} + \begin{bmatrix} C_{NN} & C_{NK} \\ C_{KN} & C_{KK} \end{bmatrix} \begin{Bmatrix} \dot{u}_N \\ \dot{u}_K \end{Bmatrix} + \begin{bmatrix} K_{NN} & K_{NK} \\ K_{KN} & K_{KK} \end{bmatrix} \begin{Bmatrix} u_N \\ u_K \end{Bmatrix} = \begin{Bmatrix} 0 \\ \overline{r} \end{Bmatrix}, \tag{2.35}$$

where nodal displacement vectors u_N, u_K correspond to displacements of free nodes and kinematically excited nodes, respectively; M, K, C are mass, stiffness, and damping matrices, respectively; and r is a vector representing reaction forces of the kinematically excited nodes.

The displacement vector of unconstrained nodes is expressed as $u_N = u_{N\mathrm{rel}} + u_{Nk}$, where $u_{N\mathrm{rel}}$ denotes a component of relative displacement with respect to moving base displacement u_{Nk}. Vectors u_K and u_{Nk} correspond to rigid-body displacements which do not induce internal elastic forces in the structure. The proportional damping approach is adopted in the form of $C = \alpha M + \beta K$ with α and β as the Rayleigh damping constants. Hence, the following matrix equation is obtained after the algebraic rearrangements of Eq. 2.35:

$$M_{NN}\ddot{u}_{Nrel} + C_{NN}\dot{u}_{Nrel} + K_{NN}u_{Nrel} = \hat{M}, \tag{2.36}$$

where $\hat{M} = M_{NN}K_{NN}^{-1}K_{NK} - M_{NK}$, the left-hand side of the equation contains matrices of the structure constrained in the nodes of imposed kinematic excitation, while the right-hand side represents a vector of inertial forces that act on each node of the structure as a result of applied kinematic excitation.

Model verification was performed in order to validate the adopted tool modeling approach and thereby ascertain that the constructed FE model is able to accurately predict the dynamic behavior of the drill bit that is driven by the vibration drilling tool. The degree of conformity between the measured and simulated frequency responses was selected as a quantitative criterion characterizing the accuracy of the model. One of the key factors that define the vibrational response of the drill bit is related to its boundary conditions. A major modeling challenge was to achieve an adequate representation of the friction-based fixation of the drill bit in the chuck. Furthermore, many efforts have been directed to ensuring that during dynamic analysis the FE model is subjected to kinematic excitation that is equivalent to the actual excitation generated by the vibration tool. During the model fine-tuning stage, an extensive search for the appropriate stiffness k_z, k_r, and k_φ values was performed: the values of these coefficients were adjusted until a reasonably close agreement between the simulated and measured resonant frequencies was achieved. This procedure was executed by conducting a sequence of frequency response analyses in a range of 0–22 kHz with different values of stiffness coefficients. Figure 2.31 provides a comparison between the simulated frequency response and the experimental response of the drill-tip in the axial direction, which was calculated from the measured frequency responses considering that the drill-tip response (Fig. 2.35b) was registered under conditions when the drill bit was excited by the vibration tool with the variable-amplitude signal. In other words, the actual drill-tip response in Fig. 2.31 was obtained by taking into account that the analyzed system is linear and drill-tip displacement is proportional to the

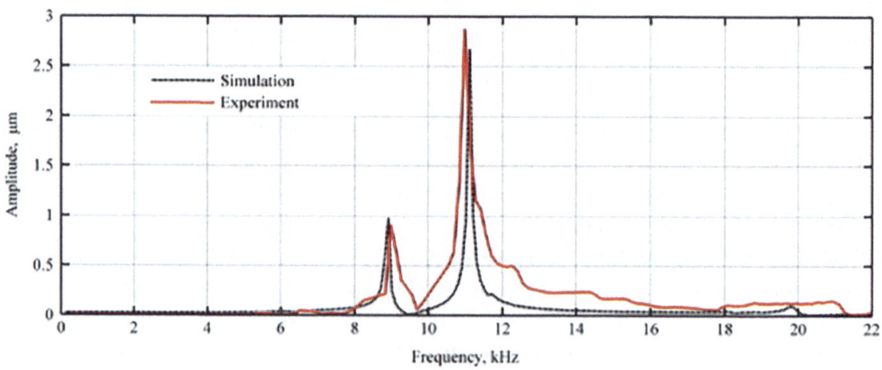

Fig. 2.31 Simulated frequency response of the drill bit in the axial direction and the corresponding experimental response obtained under harmonic excitation with a constant relative amplitude

amplitude of applied excitation. Thereby, the final experimental curve in Fig. 2.31 represents drill-tip response to a harmonic excitation with a constant relative amplitude (therefore, it is appropriate for comparison with the corresponding simulated response computed under analogous excitation conditions).

Comparison of the responses in Fig. 2.31 reveals that the measured and simulated resonance peaks coincide very well. The accuracy of the FE model was evaluated by calculating the relative errors using virtually and physically obtained values of the frequencies of the two resonant peaks in Fig. 2.31. The average relative error was found to be less than 2%. This confirms that the developed FE virtual model of the vibration drilling tool is physically adequate. Moreover, the accuracy of the FE model was additionally demonstrated in the case of transient responses.

The purpose of the numerical modal analysis was to determine vibration modes of the drill bit that are excited at particular resonant frequencies. For these simulations, the free nodes of the spring elements were constrained in all directions. Table 2.2 summarizes the results of the modal analysis, which were carried out in Ansys using the Block Lanczos mode extraction method in a frequency range of 0–22 kHz. In total, five transverse eigenmodes in two different planes, three torsional eigenmodes, and one axial eigenmode were detected in the considered frequency range. A harmonic analysis was subsequently performed to reveal additional peculiarities of the dynamic behavior of the modeled pre-twisted cantilever.

Harmonic analysis was carried out by applying the following boundary conditions to the free nodes of the spring elements: displacements in transverse and torsional directions were fully constrained, while an external load in the form of kinematic excitation was imposed on the nodes in the axial direction. Figure 2.32 illustrates the computed amplitude–frequency characteristics of the cantilever tip in axial z-, transverse r-, and torsional φ-directions.

Inspection of the frequency responses obtained in axial and torsional directions reveal two predominant peaks located at frequencies of 8.9 and 11.1 kHz. A resonant peak at 19.8 kHz is notably smaller in comparison to the latter. Simulation results indicate a larger number of pronounced resonant peaks in the frequency response in the transverse direction. It is evident that the highest peaks in all the presented frequency responses are observed in the frequency range of 11–12 kHz. The preceding modal analysis established the mode shapes of the pre-twisted cantilever, which are associated with the particular resonant frequencies. They will be used here to interpret the results of the harmonic analysis. Figure 2.32a indicates that the highest resonant peak observed in the frequency response in the axial direction corresponds to the first axial mode at 11.1 kHz. However, amplitude peak at this particular frequency is also observed in the frequency response in the torsional direction (Fig. 2.32c). It should be pointed out that modal analysis yielded no torsional mode at this frequency. Computer visualization of the mode shape at 11.1 kHz revealed "twisting" and "untwisting" motion of the cantilever, which implies that torsional oscillations are induced alongside the prevalent axial vibration mode. This coupling of axial and torsional deflections manifests due to the helical geometry of the cantilever. The same effect is also observed in the "opposite" direction: the second torsional mode at 8.87 kHz in Fig. 2.32c is accompanied by a corresponding peak in the frequency

Table 2.2 Summary of simulated natural vibration modes of the drill bit

Vibration mode No	Resonance frequency, kHz	Type of vibration mode and its consecutive number				Mode visualization (displacement vector sum)
		Axial	Torsional	Tranversale		
				Plane xz	Plane yz	
1	0.494			1		
2	0.522				1	
3	2.561			2		
4	2.610				2	
5	3.025		1			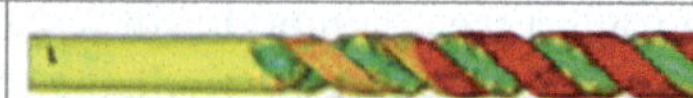
6	6.419			3		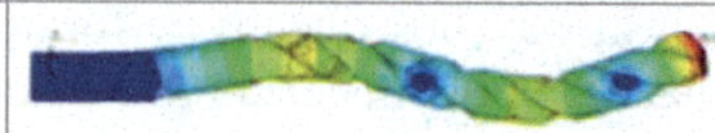
7	6.461				3	
8	8.870		2			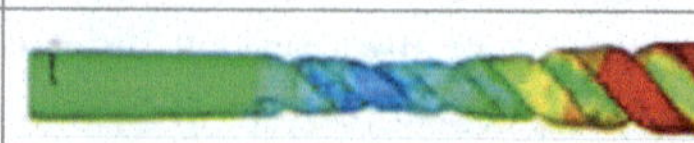
9	11.100	1				
10	11.688			4		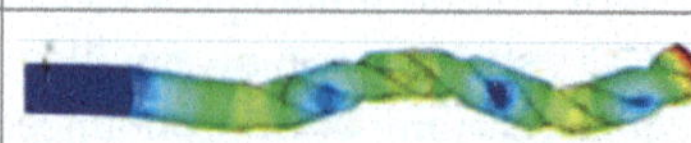
11	12.229				4	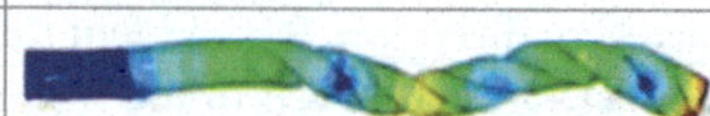
12	18.151			5		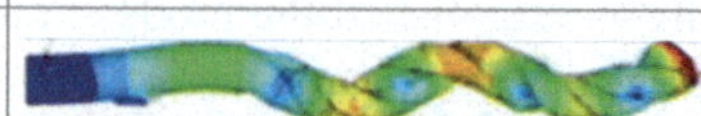
13	18.894				5	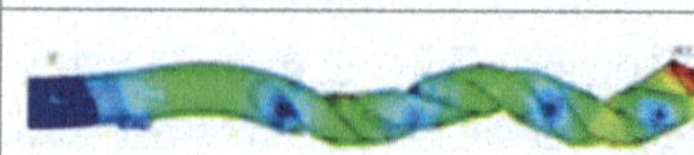
14	19.843		3			

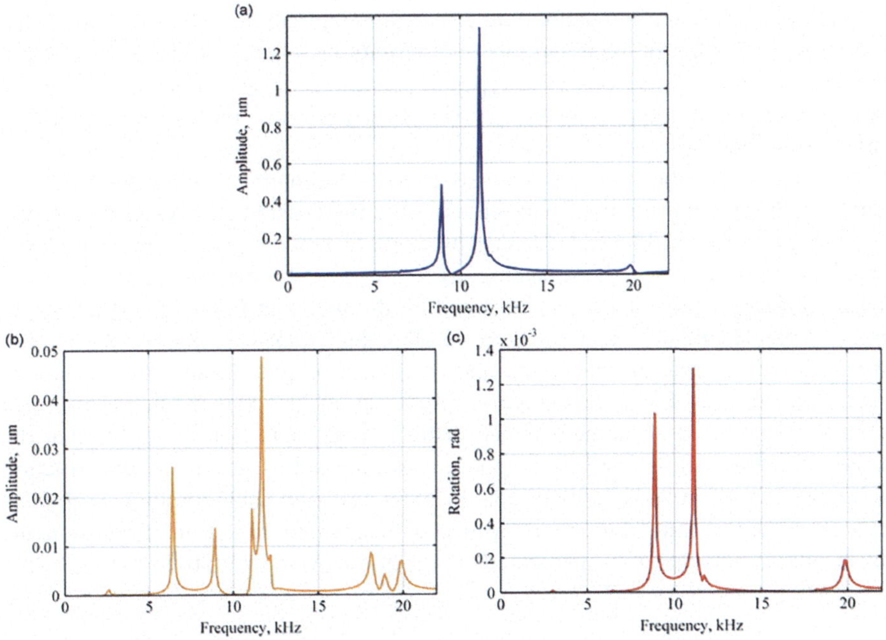

Fig. 2.32 Simulated frequency responses of the drill bit in axial, **a** transverse, **b** and torsional directions. **c** Numerical values for the latter response are provided in radians

response in the axial direction (a smaller peak in Fig. 2.32a), while computed eigenmodes in Table 2.2 indicate no axial mode at 8.87 kHz. Thus, in the latter case, torsional vibrations of the pre-twisted cantilever lead to a simultaneous axial deflection, i.e. elongation and contraction of the structure, which is consistent with Bayly's model [18]. Performed harmonic analysis shows that the coupling effect is indeed pronounced since the frequency response in torsional direction (Fig. 2.32c) reveals that the peak at 11.1 kHz, corresponding to the induced twisting motion, is even larger in amplitude when compared to the peak associated with the torsional eigenmode of the structure at 8.87 kHz. Results of harmonic and modal analysis also resolved the resonant peaks observed in the frequency response in the transverse direction (Fig. 2.32b): the first two peaks represent the second and third eigenmodes at 2.61 kHz and 6.461 kHz, respectively, while the highest peak corresponds to the fourth transverse eigenmode at 11.688 kHz. The preceding smaller peak in a range of 8–9 kHz is associated with the second torsional eigenmode at 8.87 kHz. A group of three smaller peaks in the 18–20 kHz range corresponds to the fifth transverse eigenmodes at 18.151 and 18.894 kHz, while the peak at 19.843 kHz refers to the third torsional eigenmode. Results of these simulations reveal comprehensive dynamic behavior of the drill bit indicating diverse types of vibrations that are generated in the frequency range of interest during kinematic excitation of the structure in axial direction: transverse, torsional, and axial eigenmodes in conjunction with the additional twisting/untwisting and elongation/contraction, which are attributed to the coupled nature of axial and torsional vibrations inherent to the twisted structure. Furthermore, all the observed types of vibrations are generated as a result of excitation of the structure only in the axial direction, thus, representing a case of parametric vibrations.

The purpose of the transient analysis is to additionally examine the effect of coupling of axial and torsional vibrations in the pre-twisted cantilever, which was previously detected at 11.1 kHz. Transient simulations were performed by applying boundary conditions, which are analogous to the case of harmonic analysis, except that the kinematic excitation was imposed on the nodes in the axial z-direction in terms of displacement vector u_K with its elements equal to $A\sin(\omega t)$, where A—amplitude, ω—excitation frequency, and t—time. Computed transient response in Fig. 2.33a confirms that at the excitation frequency of 11.1 kHz the first eigenmode of axial vibrations is generated in the drill bit since it is clearly visible that the excitation curve and the drill-tip response curve are out of phase, i.e. the structure undergoes extension and contraction in the axial direction, and vibration amplitudes at the tip are significantly larger with respect to the applied excitation. In addition, the out-of-phase character of the presented curves of transverse vibrations in Fig. 2.33b,

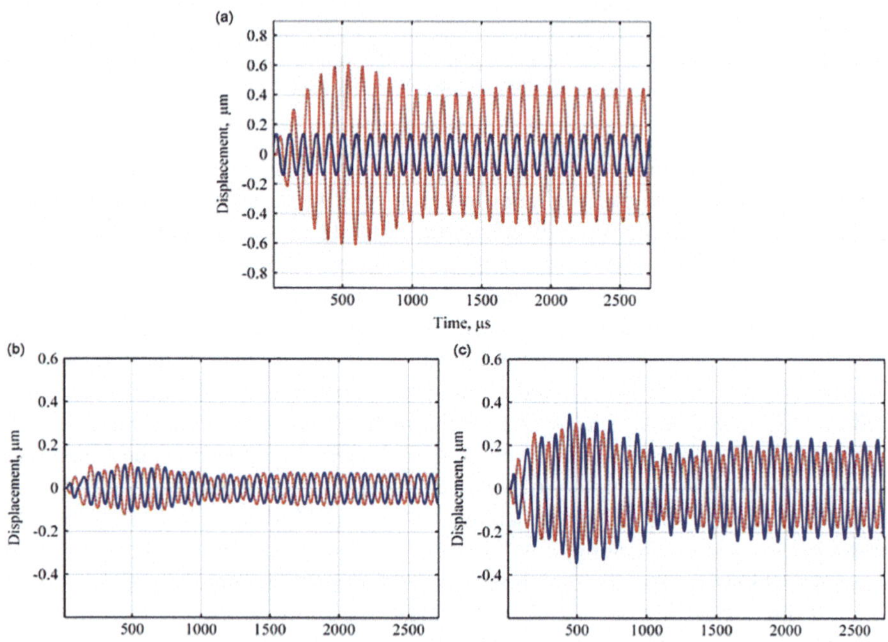

Fig. 2.33 **a** Simulated transient axial vibrations of the drill-tip (red-dashed line) as a response to harmonic kinematic excitation (blue solid line) at 11.1 kHz and **b, c** Transient transverse vibrations of the two points located at the opposite sides of the drill-tip cross-section (**b**—y-direction, **c**—x-direction)

c, which were registered at two points located at the opposite sides of the drill-tip cross-section, refer to the aforementioned (un)twisting motion, which is induced simultaneously with the prevalent axial vibration mode.

2.3.2 Physical Twin of Vibration Drilling Tool

Measurements of dynamic characteristics of the developed vibration drilling tool were performed with the goal to clarify the key aspects of tool vibrational behavior, which are responsible for the observed variations in cutting force, torque, and surface quality at particular tool excitation frequencies during vibration drilling operations. The highest effectiveness of the vibration drilling is expected when the drill-tip is excited with the largest amplitude, which is achieved during resonant operation of the device. Therefore, a series of frequency response measurements were carried out in order to determine dynamic characteristics of the vibration drilling tool. Figure 2.34a provides a schematic representation of the experimental set-up, which is based on application of the laser Doppler vibrometer for registration of the frequency responses. The piezoelectric transducer of the vibration drilling tool was driven harmonically by using function generator Escort EGC-3235A. A constant excitation voltage of approximately 100 V was maintained with the power amplifier Krohn-hite 7500. Politec fiber-optic interferometer OFV-512 and controller OFV-5000 were used to measure the vibration amplitudes of the drill-tip in the axial direction. Measurements were conducted in the excitation frequency range of 2.5–22 kHz. The obtained signal was converted and transmitted to the computer via an analog–digital converter (digital oscilloscope Pico 3424). PicoScope software was used for the processing of measurement results.

Frequency responses in the axial direction were recorded for three different cases (Fig. 2.35): measurement at the drill-tip and at concentrator-tip (with and without the drill bit mounted in the vibration drilling tool). Comparison of the responses at the concentrator-tip in Fig. 2.35a reveals nearly coincident characteristics for two cases when the drill bit is inserted into the tool and when it is removed. Thus, it may be safely assumed that drill bit dynamics has negligible influence on the axial vibrations generated by the vibration drilling tool. This observation was used as a major justification of the tool modeling strategy.

Frequency response registered at the drill-tip (Fig. 2.35b) reveals two pronounced resonances at the excitation frequencies of 11.2 and 16.6 kHz when using a twist drill bit of 10 mm and length of 132 mm. Comparison of this frequency response with those registered at the concentrator-tip (Fig. 2.35a) indicates that the resonant peak at 16.6 kHz corresponds to the axial eigenmode of the vibration drilling tool itself but not the drill bit. This statement is supported by the simulation results in Fig. 2.31, which indicate that the drill bit has no axial vibration mode at 16.6 kHz. Meanwhile, the peak at 11.2 kHz, on the contrary, refers to the excitation of the first axial mode of the drill bit since this peak is not present in the frequency responses of the vibration tool measured at concentrator-tip (Fig. 2.35a), simulation results in Fig. 2.31 confirm

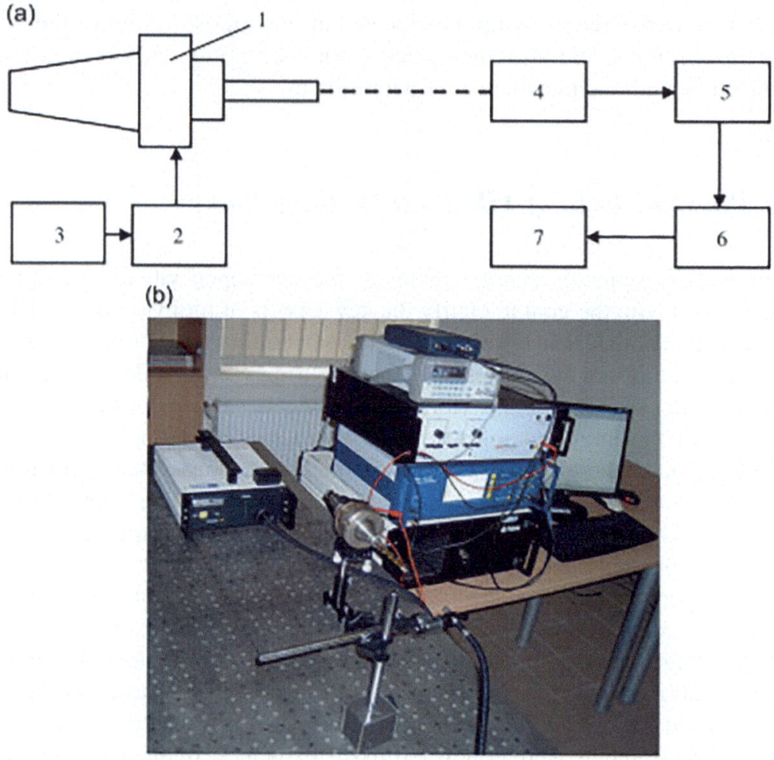

Fig. 2.34 **a** Block diagram of the experimental set-up for measurement of frequency responses with the laser Doppler vibrometer: 1—vibration drilling tool, 2—power amplifier Krohn-hite 7500, 3—signal generator Escort EGC-3235 A, 4—fiber-optic laser interferometer Polytec OFV-512, 5—vibrometer controller Polytec OFV-5000, 6—analog–digital converter Pico 3424, and 7—a computer with data acquisition and management system. **b** Photo of the experimental set-up

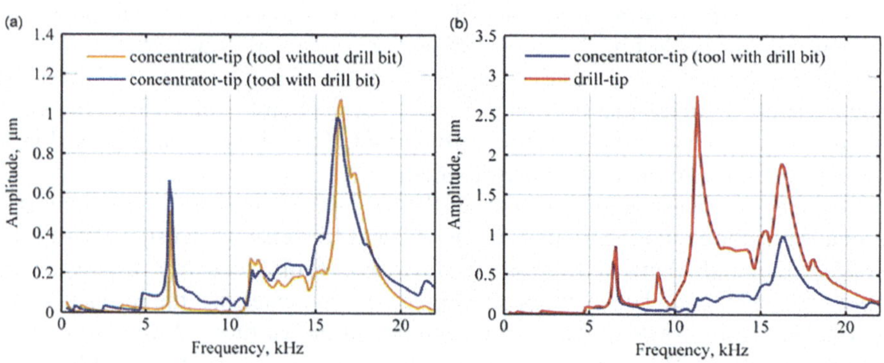

Fig. 2.35 Measured frequency responses: **a** at the concentrator-tip with and without the drill bit mounted in the vibration drilling tool and **b** at the concentrator-tip with the drill bit mounted in the tool and at the drill-tip

that this measured peak corresponds to the numerically determined first axial mode of the drill bit at 11.1 kHz (Table 2.2). The two dominant resonant frequencies of 11.2 and 16.6 kHz were subsequently applied for tool excitation during vibration drilling experiments, which revealed that the lowest values of cutting force, torque, and surface roughness are obtained at 16.6 kHz. In addition, a series of time response measurements were carried out in order to examine in more detail the peculiarities of the vibratory behavior of the tool at 16.6 kHz. Moreover, the results of these experiments served as additional proof confirming the accuracy of the developed FE model.

2.3.2.1 Time Responses to Harmonic Excitation—Study of Axial Vibrations

Several series of time response measurements in axial direction were performed by exciting the tool harmonically. The main goal of these experiments was to evaluate the dynamic interaction between the vibration tool and the drill bit, i.e. to determine how the tool generates and transfers vibrations to the drill bit at two excitation frequencies: 16.6 and 12 kHz (the latter was selected arbitrarily for comparison purposes as a frequency which is close in value to the fourth transverse eigenmode of the drill bit determined from the simulations (Table 2.2)). Figure 2.36a provides a schematic representation of the experimental set-up, which is based on the application of accelerometers for the registration of time responses. These dynamic measurements were conducted with the tool placed on a vibration-isolation table. The piezoelectric transducer of the vibration drilling tool was driven harmonically by using function generator Escort EGC-3235A. A constant excitation voltage of approximately 100 V was maintained by power amplifier Piezo systems EPA-104. Vibrations were registered by using two single-axis piezoelectric charge-mode acceleration sensors Metra KD-91 (sensitivity $k = 0.5$ mV/(m/s^2)): one sensor was fixed on the drill-tip (position A) with the measurement axis aligned along the tool and the other sensor at concentrator-tip, near the place of drill bit mounting (position B) for measuring axial vibrations generated by the vibration tool itself. The recorded acceleration signal was converted and transmitted to the computer via digital oscilloscope Pico 3424. PicoScope software was used for data processing.

Measurement results provided in Fig. 2.37a indicate that at the excitation frequency of 12 kHz vibration amplitude of the drill bit reaches a moderate value and the amplitude at the concentrator-tip is relatively low. In contrast, at the excitation frequency of 16.6 kHz (Fig. 2.37b), the registered time responses of the drill-tip and concentrator-tip are of comparable amplitude. Moreover, these dynamic measurements reveal that at the excitation frequency of 16.6 kHz vibration responses of the drill bit and the vibration tool reach their peak values, i.e. maximal achievable amplitudes of axial vibrations are observed both at the drill-tip and concentrator-tip. This implies that at this particular frequency the piezoelectrically excited tool transfers the highest energy to the drill bit, which leads to the boost of axial vibration amplitude at the drill-tip. As a consequence, this creates the conditions to achieve the

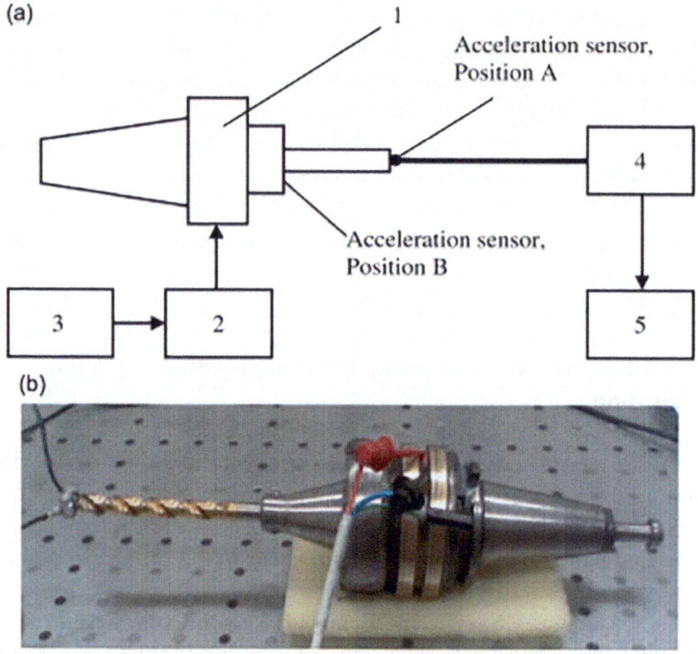

Fig. 2.36 **a** Experimental scheme for accelerometric measurements of transient vibrations generated at the drill-tip (Position A) and concentrator-tip (Position B): 1—vibration drilling tool, 2—power amplifier Piezo systems EPA-104, 3—signal generator Escort EGC-3235A, 4—analog–digital converter Pico 3424, and 5—computer and **b** vibration drilling tool with two acceleration sensors mounted on the drill-tip

largest positive impact of the superimposed high-frequency vibrations on the cutting process. In addition, the examination of vibration curves in Fig. 2.37b indicates that at 16.6 kHz there exists a 180° phase difference between the time responses at the concentrator-tip and the drill-tip. The observed out-of-phase character of vibration curves demonstrates that the drill-tip oscillates in one direction, while the concentrator-tip at the same time oscillates in the opposite direction, i.e. simultaneous extension and contraction of the tool in the direction parallel to its axis are induced at 16.6 kHz. Thus, the aforementioned amplitude amplification and presence of phase difference at 16.6 kHz between the two sensor signals confirm that the observed vibrations are attributed to the axial resonant vibrations of the vibration drilling tool itself.

Simulations analogous to the aforementioned vibration measurements were carried out in order to compare the results with the experimental findings. It is evident from the presented numerical results that the FE model of the drill bit reproduces the measured vibrational character of the tool fairly well: simulated time responses in Fig. 2.37c, d closely match the corresponding experimental responses in Fig. 2.37a, b in terms of generated amplitudes and phase differences between the excitation

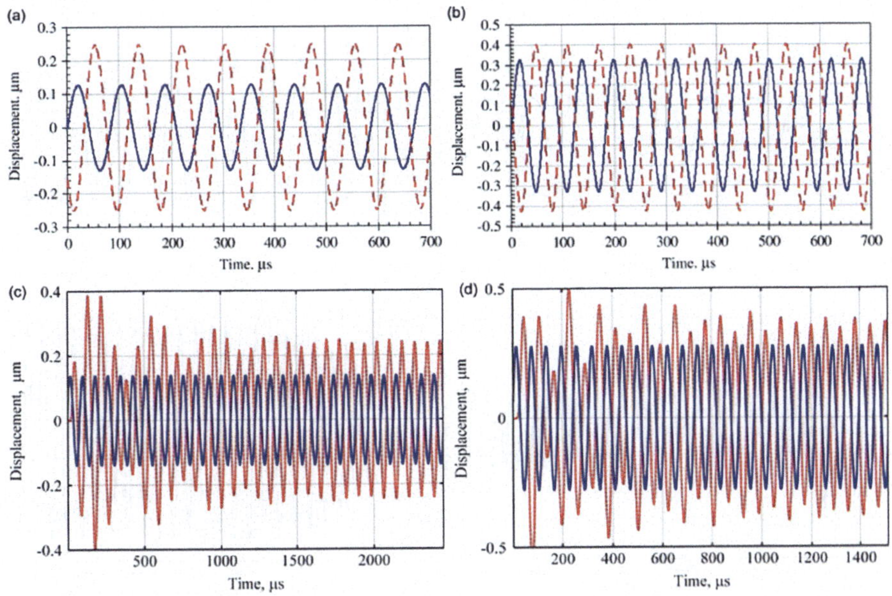

Fig. 2.37 Measured (**a, b**) and simulated (**c, d**) axial vibrations at the drill-tip (red-dashed line) and concentrator-tip (blue solid line) during harmonic excitation at 12 kHz (**a, c**) and 16.6 kHz (**b, d**)

and the drill-tip curves (experimental vibration curves for the concentrator-tip in Fig. 2.37a, b correspond to the applied kinematic excitation in Fig. 2.37c, d).

2.3.2.2 Time Responses to Harmonic Excitation—Study of Transverse and Torsional Vibrations

Another series of time response measurements were performed at the drill-tip in order to gain a more comprehensive view on the dynamic behavior of the vibration drilling tool at 16.6 kHz. The arrangement of acceleration sensors during the following measurements was as follows: two accelerometers were mounted on the drill-tip so as their measurement axes lie perpendicular to the axis of the drill bit (Fig. 2.36b). The sensors were fixed by gluing them on a small and thin plate, which was directly attached to the tip and thereby facilitated the repositioning of the sensors. Each sensor in this case measured the oscillations in the transverse direction. Furthermore, this particular arrangement of sensors enabled us to differentiate whether transverse or torsional vibrations are observed in the registered time responses. Figure 2.38a, b illustrates the fragments of time responses received during tool excitation at 12 kHz. It is obvious from Fig. 2.38a that signal curves are essentially concurrent in phase, thereby indicating purely transverse vibrations, which is consistent with the simulation results indicating that the fourth eigenmode of transverse vibrations is excited

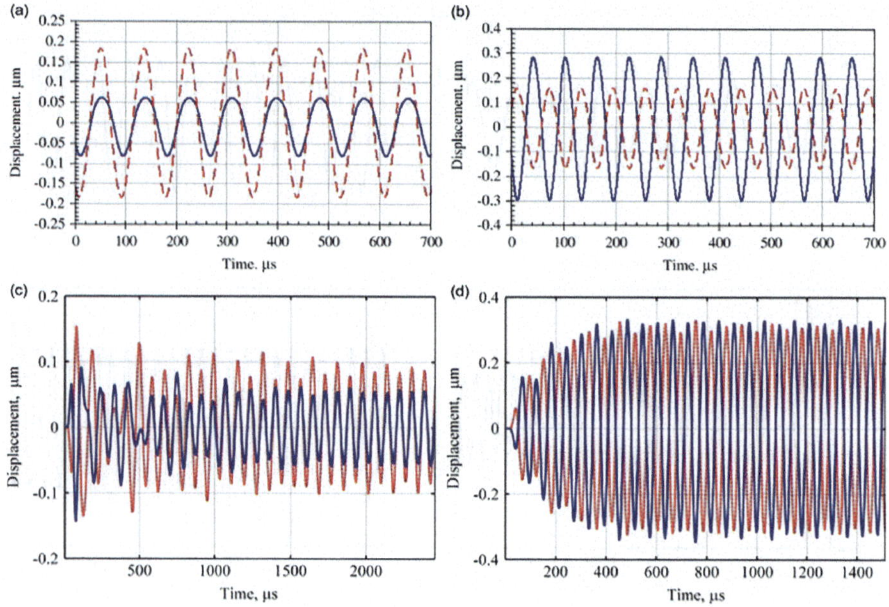

Fig. 2.38 Measured (**a, b**) and simulated (**c, d**) transverse vibrations of the drill-tip during harmonic excitation at 12 kHz (**a, c**) and 16.6 kHz (**b, d**). (For (**c, d**): solid and dashed lines correspond to time responses of the two points located at the opposite sides of the drill-tip cross-section)

at 12.229 kHz (Table 2.2). In addition, this conclusion is supported by the numerical time responses in Fig. 2.38c, which compare favorably with the corresponding experimental results in Fig. 2.38a.

Meanwhile, time responses in Fig. 2.38b, which were measured in the transverse direction at the excitation frequency of 16.6 kHz, are out of phase: the opposite ends of the drill-tip oscillate in different directions, which implies the presence of additional twisting/untwisting action that is generated simultaneously with the axial vibrations (Fig. 2.37). This experimental observation is confirmed by the simulations in Fig. 2.38d and is also in accordance with the results of numerical analysis of the drill bit, which clearly demonstrated the presence of coupling of the axial and torsional vibrations in the considered helical-shaped structure. Thus, it may be summarized that the dynamic behavior of the tool at 16.6 kHz is a combination of a prevalent axial mode of the vibration drilling tool itself accompanied by a parametrically excited twisting of the drill bit [18].

2.3.3 *Characterization of Vibration Drilling Process and Workpiece Surface Quality*

A sequence of conventional and vibration-assisted drilling experiments was carried out in order to characterize the effectiveness of the developed vibration drilling tool in terms of reduction of cutting force and torque as well as improvement of surface quality of the workpiece. Figure 2.39a illustrates the scheme of the experimental set-up, which was assembled in order to measure the axial cutting force and torque in the course of drilling operations. Four-component dynamometer platform Kistler 9272 was used for recording the magnitudes of the cutting force and torque generated during drilling of the cylindrical workpieces fabricated from steel XC48 (170 HB), which had a length of 40 mm and a diameter of 20 mm. The workpieces were mounted in the clamping device of the dynamometer, while the latter was installed on the desk (Fig. 2.39b). Axial cutting force and torque generated during drilling operations were registered and the signals were transmitted to the computer, where data analysis was performed by means of a special software.

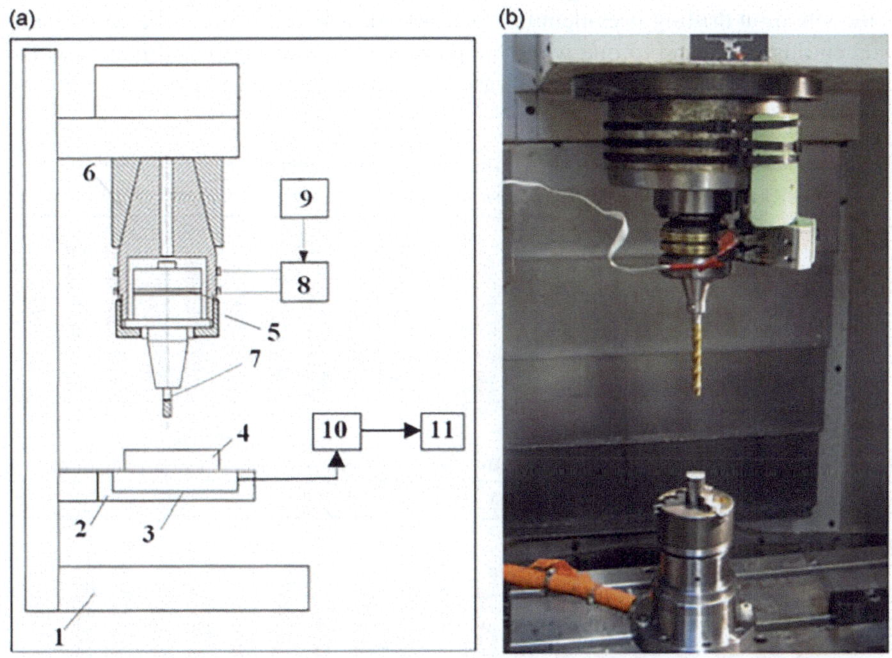

Fig. 2.39 **a** Scheme of the experimental set-up for cutting force and torque measurements during drilling operation: 1—machine tool construction, 2—desk, 3—Kistler dynamometer platform 9272, 4—workpiece, 5—vibration drilling tool, 6—standard tool holder DIN 6359, 7—drill bit, 8—power amplifier, 9—high-frequency generator Agilent 33220A, 10—controllers (Kistler amplifier), and 11—computer and **b** photo from the site of vibration drilling experiments

Vibration-assisted drilling experiments were conducted by driving the tool with two different excitation frequencies, which were found from dynamic measurements to be the dominant ones: 11.2 and 16.6 kHz. The following cutting regimes were applied: drilling depth 15 mm, feed rate 0.2–0.25 mm/rev, and drilling speed 600–900 rev/min. The following measurement procedure was used: two drilling holes and cutting force/torque measurements were performed for each different cutting condition, while three force/torque measurements were performed for each feed and cutting speed ratings. Obtained experimental results are summarized in Fig. 2.40. Variation of axial cutting force and torque illustrated in Fig. 2.40a, b provide the comparison between the values registered during conventional and vibration drilling processes at the excitation frequency of 11.2 kHz for the case of three different feed rates. Most of the measured values are consistent, indicating cutting force and torque increase with larger feed rates. These experimental findings do not exhibit meaningful differences between dynamometric measurements acquired during conventional and vibration drilling processes, particularly in terms of torque magnitudes. Furthermore, an inspection of results for axial cutting force at lower drilling speeds reveals a negative influence of the superimposed vibrations of 11.2 kHz because increased cutting force is observed during vibration drilling. In contrast, the positive influence of the vibration drilling is evident in Fig. 2.40c, d, which illustrates the variation of axial cutting force and torque when the piezoelectric transducer is driven with the excitation signal of 16.6 kHz. In this case, the cutting force during vibration drilling

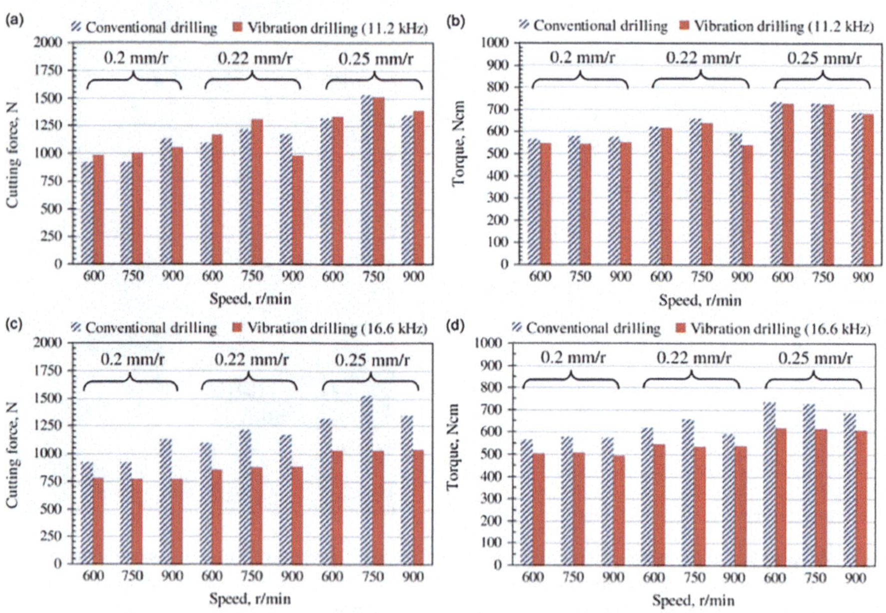

Fig. 2.40 Variation of axial cutting force and torque as a function of drilling speed for three different feed rates during conventional and vibration drilling processes when the tool is excited with the frequency of 11.2 kHz (**a**, **b**) and 16.6 **kHz** (**c**, **d**)

decreases by 12–46% with respect to conventional drilling. The corresponding torque reduction constitutes 13–20%. The most pronounced difference between force and torque magnitudes in conventional and vibration drilling is detected at the largest feed rate of 0.25 mm/rev, meanwhile the smallest difference is observed at the smallest feed rate of 0.2 mm/rev.

Reported results of cutting force and torque measurements point out the importance of tool mode control during the vibration drilling process. Excitation of the tool at 11.2 kHz, which corresponds to the first axial eigenmode of the considered drill bit, gave no conclusive answer regarding the positive impact of the mode on the cutting process in terms of generated cutting force and torque (detrimental influence of this particular mode was even observed in the case of specific cutting regimes). Meanwhile, the positive influence of superimposed high-frequency vibrations was intensified when the drilling tool was driven at the frequency of 16.6 kHz, which corresponds to the axial eigenmode of the tool itself.

After drilling experiments were finished, the machined cylindrical workpieces were subjected to roughness measurements by using roughness tester Time TR210 based on a high-precision inductive-type transducer. Workpiece roughness (R_a), obtained when drilling at an excitation frequency of 11.2 kHz, is lower approximately by 10% in comparison to the case of conventional drilling (reduction of R_a from 1.7 to 1.45 μm was observed). After vibration drilling with excitation of 16.6 kHz, the surface roughness of the workpieces decreased by approximately 25% (from 1.7 to 1.2–0.98 μm) with respect to conventional drilling. It is obvious that tool excitation at the first axial eigenmode of the drill bit insignificantly influences workpiece surface quality. The amplitude of induced axial vibrations is not sufficiently high in this case. At the excitation frequency of 16.6 kHz, the amplitudes of torsional and longitudinal vibrations at the drill-tip reach their maximum values, and the positive influence of the vibration drilling on the surface quality is evident.

In summary, results of the virtual–physical analysis of tool dynamics as well as data from drilling process characterization revealed that two axial vibration modes dominate in a frequency range of interest: the first axial eigenmode of the drill bit is generated during excitation at 11.2 kHz, meanwhile the axial mode of the tool itself is induced at 16.6 kHz. The influence of the superimposed high-frequency vibrations on process performance metrics at 11.2 kHz was found to be ambiguous: surface roughness of the workpieces was moderately reduced, while mixed results were obtained in the case of cutting forces and torques, which were shown to slightly fluctuate up and down in comparison to the magnitudes registered during the conventional drilling operation. In contrast, a definitely positive impact of vibration drilling was observed at 16.6 kHz, resulting in appreciable reductions in surface roughness as well as cutting forces and torques. The vibrational energy associated with the axial resonant mode of the tool at 16.6 kHz is considerably larger in comparison to the energy level of the axial eigenmode of the drill bit at 11.2 kHz, which, consequently, explains the observed differences in the positive impact of the vibration drilling on process characteristics at these excitation frequencies. Obtained results provide clear evidence that the performance of the drilling process is largely determined by the dynamics of the vibration drilling tool. Measurements also revealed that tool dynamics is not affected by the drill bit mounted in the tool. In addition, obtained results allow to claim that eigenmodes of a pre-twisted structure with "clamped-free" boundary conditions predetermine the dynamic response of the drill bit during the cutting process, when the boundary conditions of the pre-twisted structure are transformed to the "clamped-supported". Finally, research results suggest that it would be beneficial to tailor the design of the vibration drilling tool in such a way that the frequencies of axial vibration modes of the tool and the drill bit coincide. This could provide the conditions to further enhance tool effectiveness leading to improved performance metrics of the vibration drilling process.

2.3.4 Drilling Process Simulation Using Smoothed Particle Hydrodynamics Method

Ls-Dyna is a general-purpose FE program that can simulate complex real-world problems. There are many potential applications for Ls-Dyna and they can be applied to many areas. The Lagrangian Smoothed Particle Hydrodynamics (SPH) model is performed using Ls-Dyna software. SPH is a meshless method, so the large material distortions that result from the cutting problem are easy to control and the SPH contact control allows to naturally separate the workpiece and the chip. This method was used to create a virtual twin of the 3D drilling process. The solution was made on the assumption that the tool is considered to be a non-deformable interaction faced an impact when the cutting tool considered a non-deformable made of solid elements and the workpiece consists of microparticles rather than elements.

The research of the cutting process using virtual models can replace expensive experimentations for the microscopic level, and the selection of the research methods depends on the obtainable parameters and computer equipment capabilities. In the Langrangian approach, the numerical mesh moves and distorts with the physical material. So, the Lagrangian approach has artificial numerical tools to destroy the material as node separation, element deletion, and others. SPH is a modern effective technique to solve the problems of high deformation.

A proper SPH mesh must satisfy the following conditions: it must be as regular as possible and must not contain too large variations. The choice of the rectangular body for geometrical construction and particle generation is very convenient. Having the body which is called "volume" body (the case of the workpiece), the use of solid elements for SPH generation is very useful. In order to build SPH particles from meshed by solid bodies, drilling tool and workpiece were defined as "parts". In the next stage, SPH generation was performed from solid elements. Initially, a set of particles with two kinds of properties, physical and geometrical properties, were obtained. Physical properties were defined in automatic SPH generation by assigning physical properties to "solid nodes". After SPH part building, the workpiece part, composed from solid elements, has been deleted.

The drilling tool was built as a rigid body [19, 20]. Considering that the drilling tool is working with rotation and translation motion, the first numerical study was performed according to the bulge test. Drilling tool was used to perforate the copper plate (10 × 10 × 0.4) mm which was meshed by 160,000 Solid164 elements (8 nodes with three degrees of freedom at each node in *X*-, *Y*-, and *Z*-directions) [19]. The meshing of the drilling tool, shortened to 30 mm, was performed after construction in SolidWorks and importing in Ansys as a pre-processor. After SPH generation, the plate was composed of 202,005 particles (consistence up to 5000 particles per 1 mm^3). The tool velocity, by determining the time and displacement, was defined. After generating SPH particles, the boundary conditions were imposed, and all sides of the workpiece were fixed.

The kinematic deformation properties of the plate material were defined as follows: density 8.93 kg/mm^3, modulus of elasticity 110 GPa, Poisson's ratio 0.343, and yield strength 33.3 MPa. The tool material properties: density 13.300 kg/mm^3, modulus of elasticity 533 GPa, and Poison's ratio 0.24. The contact between tool and workpiece was modeled. Modeled drill and plate positions before and after contact interaction are shown in Fig. 2.41. The obtained results of the preparatory modeling with a specific drill allowed to more accurately select the number of SPH particles for drilling modeling.

For the workpiece, steel XC48 (14 × 14 × 10) mm 17,660 particles were generated from the solid part with 15,680 elements (Solid164). The tool motion—feed 0.25 mm/rev and speed 94.1 rad/s—was applied. The kinematic deformation properties of the steel material were defined as follows: density 7.8 kg/mm^3, modulus of elasticity 210 GPa, Poisson's ratio 0.3, and yield strength 53.3 MPa. The tool material properties: density 13.300 kg/mm^3, modulus of elasticity 533 GPa, and Poisson's

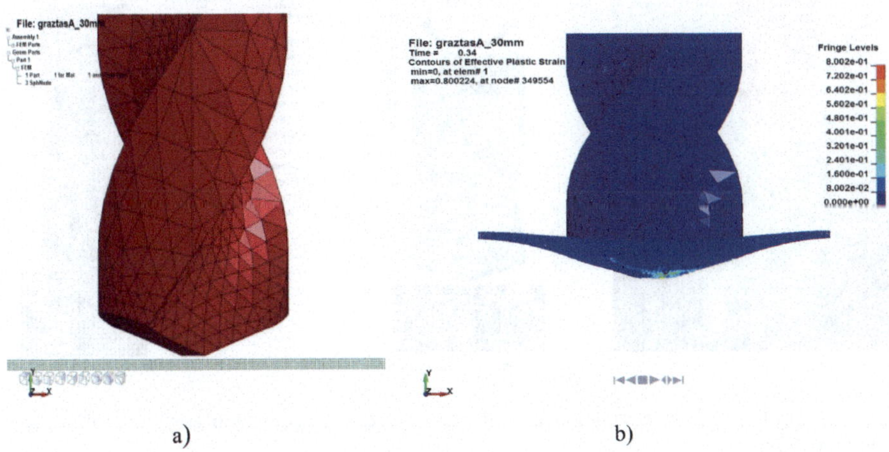

Fig. 2.41 **a** Drill and plate before contact and **b** plate perforation by the drilling tool

ratio 0.24. Modeling of the drilling process has shown (Fig. 2.42a, c) how the workpiece particles are separated and removed during cutting. Particle separation, drill indentation, and acting forces were recorded.

During the vibration drilling, not only the longitudinal (axial) velocity of movement, but also the high-frequency oscillations and their displacements in the axial direction are presented in Fig. 2.42b. The study of the drilling process demonstrated that using SPH method, the chip removing process is very "natural" and fast. As a result, the simulated cutting force difference from the experimentally measured cutting force is 1.7%. So the simulated drilling contact with the workpiece model is adequate for the real process. The simulated vibration drilling process shows the reduced thrust force compared with conventional drilling, and the same data is obtained during vibration drilling experiments. The cutting speed was chosen based on the best results of the vibration drilling effect obtained in the experimental studies of vibration cutting (Fig. 2.40). Axial force data during conventional drilling was compared to the vibration drilling data (Fig. 2.43).

The comparison of the results obtained when drilling in the conventional way and with excitation of the tool with high-frequency vibrations shows the difference between the acting drilling forces. At a cutting speed of 900 rpm, the axial force is reduced by an average of 5% compared to conventional drilling, but at a lower cutting speed of 750 rpm, a reduction in axial force of up to 5–25% is visible. When simulating the drilling process by excitation of the tool with high-frequency vibrations, the axial force is reduced by up to 25% compared to the conventional drilling model.

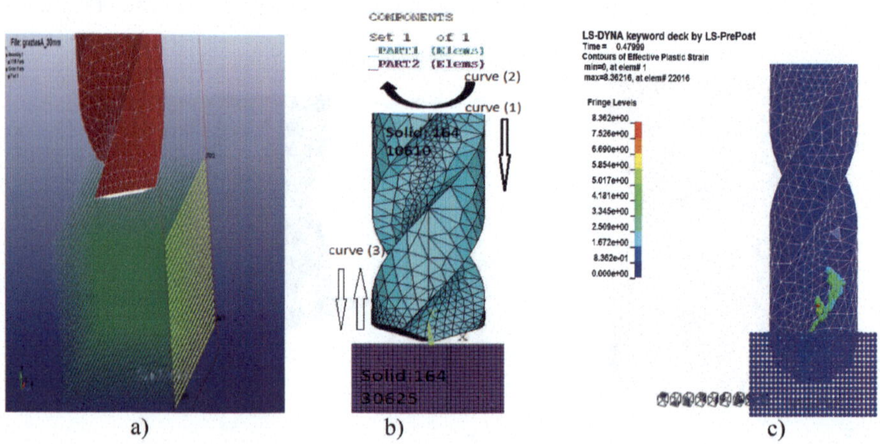

Fig. 2.42 Drilling process simulation. **a** Drill and the workpiece contact, **b** vibration drilling, and **c** chip formation

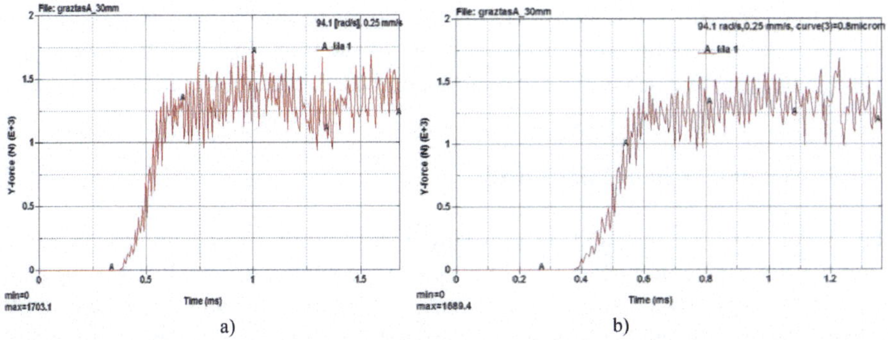

Fig. 2.43 Graph of axial forces acting on conventional **a** and vibration **b** drilling

2.3.5 Micro-drill Stiffness Amplification by Buckling Mode Control

Over the last two decades, the demands of drilling small holes (Ø 0.5 mm) at high rotational speed (80.0–180.0 rpm) are increasing due to the trend toward higher density circuits of computer parts and microelectronic packaging products. Several different micromachining technologies can be used for microholes manufacturing: microelectrochemical machining, laser beam machining, electron beam machining, micro-electro discharge machining, mechanical micro-cutting, etc. The main advantages of mechanical micro-drilling are as follows: less complicated equipment is necessary, the process is cheaper, the electrical properties of the workpiece do not influence the process, and machining time can be controlled easily. On the other hand, there are some disadvantages of mechanical micro-drilling: slender tools are less stiff and buckle easier so manufactured holes are less precise. When the tools with thicker web are used, it becomes more complicated to remove chips from the cutting zone. Also, it is difficult to remove broken drill bits and not to damage the manufactured piece.

The purpose of drilling printed circuit boards is to produce an opening through the board to provide the means for electrical interconnection between layers of the board and to permit through-the-board parts mounting with structural integrity and precision of location. Hence, the hole quality is very important for assuring the electrical connections of the board and the mounted components. During drilling, an excessive thrust force will cause the drill bit to buckle, possibly causing the drilled holes to widen and perhaps even deviate from a straight line. When the drill bit revolves with a rotational speed close to one of its flexural natural frequencies, the deflection of the drill bit will increase and may result in imperfect holes.

Buckling stiffness of micro-drill bit is an essential factor in order to secure the quality of micro-drilling process, and most of micro-drill bit failures happen because of buckling. The aim of this study is to investigate the possibilities to increase micro-drill stiffness by buckling it on a higher mode of the tool. This would allow to use higher cutting parameters and increase the efficiency of the micro-drilling process.

2.3.5.1 Virtual Twin of Micro-drill

The model of Ø 0.6 mm and 30 mm length micro-drill bit was created in order to predict the dynamic behavior of the tool. A geometrical model of the drill bit (Fig. 2.44a) was created in CAD software SolidWorks and then imported to FE software Ansys. In order to create a virtual twin that is efficient in terms of simulation time and results, only the micro-drill bit was designed and the influence of chuck was modelled by applying boundary conditions at the end of the micro-drill bit. The following material properties were used for creating the finite element model: Young's modulus $E = 2{\cdot}10^{11}$ Pa; Poisson's ratio $n = 0.28$; material density $\rho =$ 7800 kg/m^3. FE model (Fig. 2.44b) was created by using tetrahedral 10 nodes finite element Solid187 which is suitable for meshing complex geometry models created with CAD software. A density of FE mesh was chosen in order to properly adapt to the geometry of the cutting tool and to perform steady analysis results. FE model of micro-drill bit consists of 69,299 elements and 104,619 nodes. All nodes in three zones at the end of the micro-drill bit, where the tool is clamped in a chuck, were constrained in *x*-, *y*-, and *z*-directions.

The dynamics of the micro-drill bit is described by the equation of motion on a block form by considering that the base motion law is known and defined by nodal displacement vector $\{U_K\}$:

$$\begin{bmatrix} [M_{NN}] & [M_{NK}] \\ [M_{KN}] & [M_{KK}] \end{bmatrix} \begin{Bmatrix} \{\ddot{U}_N\} \\ \{\ddot{U}_K\} \end{Bmatrix} + \begin{bmatrix} [K_{NN}] & [K_{NK}] \\ [K_{KN}] & [K_{KK}] \end{bmatrix} \begin{Bmatrix} \{U_N\} \\ \{U_K\} \end{Bmatrix} = \begin{Bmatrix} \{0\} \\ \{R\} \end{Bmatrix}, \quad (2.37)$$

Fig. 2.44 **a** 3D model of micro-drill bit with chuck areas and **b** FE model of micro-drill bit

where $\{UN\}$, $\{UK\}$—modal displacement vectors representing displacements of free nodes and excited nodes; $[M]$, $[K]$—mass and damping matrices; $\{R\}$—vector of reaction forces of excited nodes.

In modal analysis, resonant frequencies of the micro-drill bit are found by solving the following equation:

$$\left([K] - \omega^2[M]\right)[\Phi] = \{0\}, \tag{2.38}$$

where ω^2—angular frequency and $[\Phi]$—mode shapes matrix.

The modal analysis of micro-drill bit constrained as pre-twisted cantilever was carried out in the range of 0–60 kHz by using the Block Lanczos mode extraction method. The results of this analysis are shown in Table 2.3. Fourteen eigenmodes of micro-drill bit vibrations were found in selected frequency range, twelve of which are transverse vibration eigenmodes, one torsional eigenmode at 32,084 Hz, and one axial vibration eigenmode at a frequency of 51,651 Hz. The first calculated transverse resonant frequency of the micro-drill bit is 691.83 Hz. Due to verification of the FE model adequacy to the physical one, damped free vibrations of the micro-drill bit were excited. The acceleration of micro-drill bit was measured by the accelerometer KD-91. Experimental data was collected by using oscilloscope PicoScope-4424 processed by PicoScope 6.0 software. Vibration above 1 kHz was filtered out and the vibration amplitude dependency on time is shown in Fig. 2.45. The first experimental eigenfrequency of the micro-drill bit is 677.8 Hz.

Table 2.3 Eigenmodes of micro-drill bit

No	Resonant frequency, Hz	Mode shape
1	691.83	
2	692.5	
3	3753.0	
4	3819.2	
5	9228.4	
6	9568.5	
7	18,134	
8	18,281	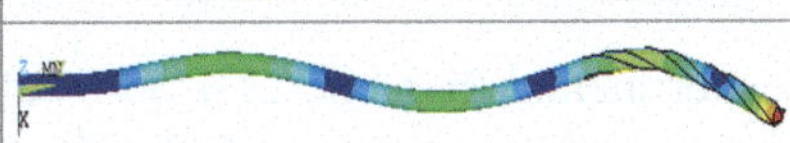

(continued)

Table 2.3 (continued)

No	Resonant frequency, Hz	Mode shape
9	30,523	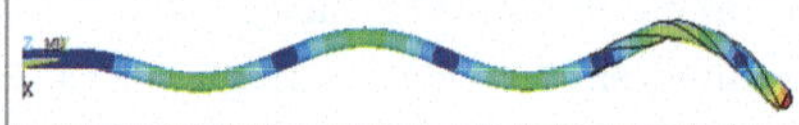
10	32,657	
11	32,084	
12	44,217	
13	45,620	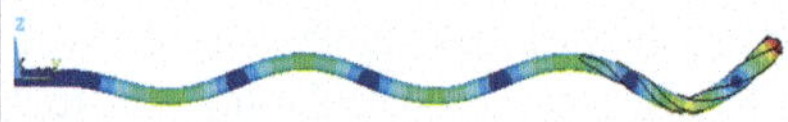
14	51,651	

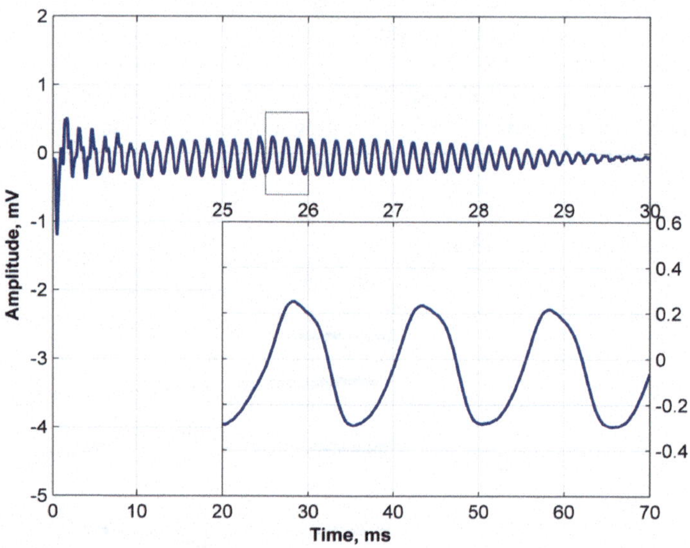

Fig. 2.45 Damped vibration amplitude dependence on time

The relative error of micro-drill bit model is calculated by the following formula:

$$\Delta = (1 - f/f_c) \cdot 100\%, \tag{2.39}$$

where f—experimental frequency, Hz; f_c—calculated frequency, Hz.

The calculated relative error is 3.41%. This confirms that the created FE model is adequate to physical.

Table 2.4 Critical buckling loads

Buckling mode no	Plane	Critical buckling load (N)
1	YZ	22.504
1	XY	30.375
2	YZ	68.772
2	XY	71.728

Micro-drill bit's free vibration characteristic is presented in Fig. 2.45.

2.3.5.2 Virtual Twin for Micro-drill Buckling Analysis

The eigenvalue buckling analysis was carried out in order to predict the critical buckling loads of micro-drill bit:

$$([K] + \lambda[\sigma])\{\psi\} = \{0\}, \tag{2.40}$$

where $[K]$—stiffness matrix; $[\sigma]$—stress stiffness matrix; λ—eigenvalue; $\{\psi\}$—eigenvector of displacements.

The boundary conditions remained the same as in the previous modal analysis: constrained zones that represent the micro-drill bit clamping in three-jaw chuck. In addition, a chisel edge of the micro-drill bit was fixed. This type of boundary condition simulates the most dangerous condition for micro-drill bit to buckle—when the unconstrained length of the tool is the biggest. The Block Lanczos method was selected for this analysis. The first two buckling eigenmodes were extracted. Calculated critical buckling loads are represented in Table 2.4 and the mode shapes are shown in Fig. 2.46. The first critical buckling load is 22.504 N (30,375 N, if micro-drill bit buckles in XY plane) and the second critical buckling load is 68.772 N (71,728 N, if micro-drill bit buckles in XY plane). So obtained second critical buckling load is 2.4–3 times bigger than the first critical buckling load.

2.3.5.3 Virtual Twin for Transient Analysis

The transient analysis of the beginning of micro-drilling process has been modeled with Comsol Multiphysics software. Pre-twisted 0.6 mm diameter and 30 mm length model of micro-drill bit was used in this analysis, which was divided into three stages. First of all, when the time t is between 0 and $t_1 = 0.0024$ s, the micro-drill bit is constrained as a cantilever: the end point of micro-drill bit is fixed and the tip point is left unconstrained (Fig. 2.47a). The micro-drill bit is kinematically

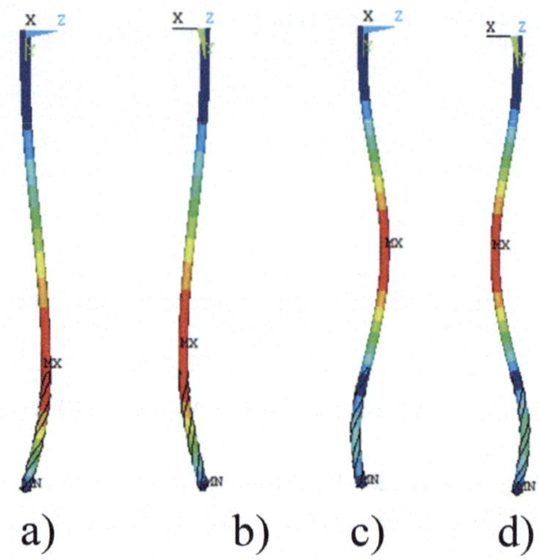

Fig. 2.46 Buckling modes shapes of micro-drill bit: **a** the first buckling mode in *YZ* plane, **b** the first buckling mode in *XY* plane, **c** the second buckling mode in *YZ* plane, and **d** the second buckling mode in *XY* plane

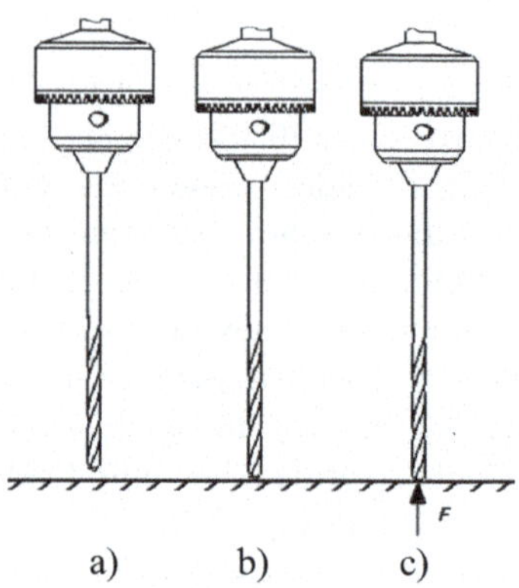

Fig. 2.47 Computational scheme of micro-drill bit buckling analysis: **a** micro-drill bit before touching the workpiece surface; **b** micro-drill bit touches the workpiece; **c** micro-drill bit goes deeper into the workpiece

excited by the sinusoidal force F at the end of the micro-drill bit (Fig. 2.47c) by the second eigenfrequency and vibrates according to the second transverse eigenmode (Fig. 2.48a). At this stage, the vibrations amplitude of the middle point of the micro-drill bit is negligible (Fig. 2.49) and the amplitude of micro-drill bit tip vibrations is much more significant (Fig. 2.50). At the second stage, when the time t is between $t_1 = 0.0024$ s and $t_2 = 0.006$ s, the tip of the micro-drill bit touches the workpiece and the micro-drill-tip is constrained in transverse directions but is able to rotate and no axial load is applied (Fig. 2.47b). The vibrations amplitude of the micro-drill-tip significantly reduces. At this moment, the transverse vibration mode of the micro-drill bit changes and it starts to vibrate according to the first buckling mode (Fig. 2.48b). This occurs because less energy is necessary to deform the micro-drill bit according to the first buckling mode. At the third stage, when the time t is more than $t_2 = 0.006$ s, the tip of the micro-drill bit is constrained and the axial load is applied to the micro-drill bit (Fig. 2.47c)—as during the cutting process.

Fig. 2.48 Transverse vibration mode shape: **a** micro-drill vibrates on the first eigenmode does not touching workpiece (**a**), on the first mode touching workpiece **b,** and on the second **c** eigenmode

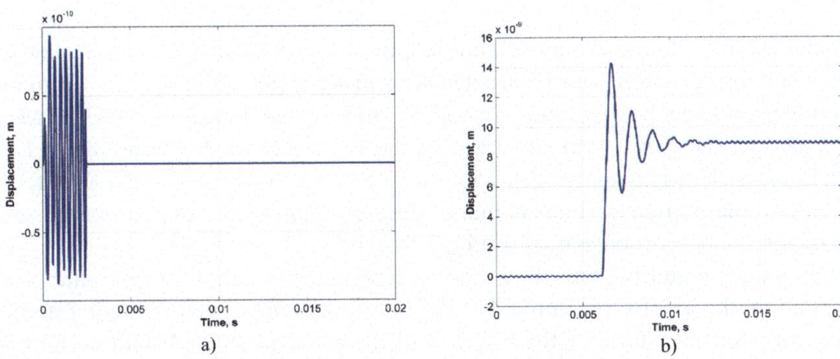

Fig. 2.49 Micro-drill bit middle point **a** and tip **b** transverse vibrations

After the transient state, the vibration amplitude of the micro-drill bit middle point significantly increases (Fig. 2.49a), but there are no changes in micro-drill-tip vibrations (Fig. 2.49b).

The reaction force occurs at the point of contact between the micro-drill bit and the workpiece (Fig. 2.50). This analysis shows that the second buckling mode of the micro-drill bit is unstable and changes to the first buckling mode when the micro-drill bit reaches the workpiece surface.

Fig. 2.50 Micro-drill bit tip reaction force

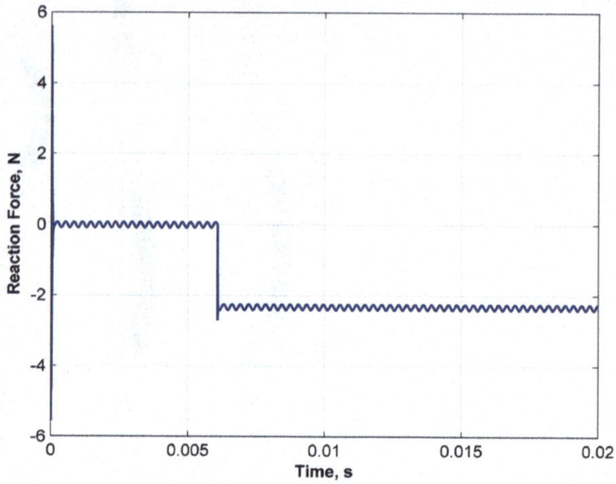

2.3.6 Experimental Study of Micro-drill Physical Twin

A slender 0.5 mm diameter and 60 mm length micro-drill bit was used for experiments. Such tool geometry has been selected in order to get better visual information about tool resonance. It was clamped into CNC milling machine tool DMU-35 M and was excited by using function generator Escort EGC-3235A, linear amplifier EPA-104, and piezoelectric actuator PSt 150/4/20 VS9. An axial force was measured by the Kistler dynamometer. The schematic representation of the experimental set-up and a picture of it are represented in Fig. 2.51a, b.

It was experimentally confirmed that the resonant frequencies of the micro-drill bit depend on the axial force applied to the micro-drill bit so that the nodal point of the second buckling mode of the micro-drill bit was constrained with nodal point constraining device, which is shown in Fig. 2.51b, to force the tool buckle according to the second buckling mode. The axial load was being applied to the micro-drill bit while the displacement of its end point reached 3 mm.

The experiment was carried out in three ways: non-rotating micro-drill bit with unconstrained nodal point of the second buckling mode was buckled, rotating micro-drill bit (1500 rpm) with unconstrained nodal point of the second buckling mode was buckled, and the rotating micro-drill bit with constrained nodal point of the second buckling mode was buckled. As it is seen in Fig. 2.52a, the buckling stiffness of the rotating micro-drill bit is less than that of the non-rotating one, when the nodal points of the second buckling mode are not constrained. It can be explained by centrifugal force which emerges when the micro-drill bit is rotating. Also, it was noticed that a bigger axial load is necessary to reach the same displacement of the micro-drill bit, when the nodal point of the second buckling mode is constrained: the buckling stiffness of rotating micro-drill bit with constrained nodal point of the second buckling mode (when micro-drill bit buckles according to second buckling

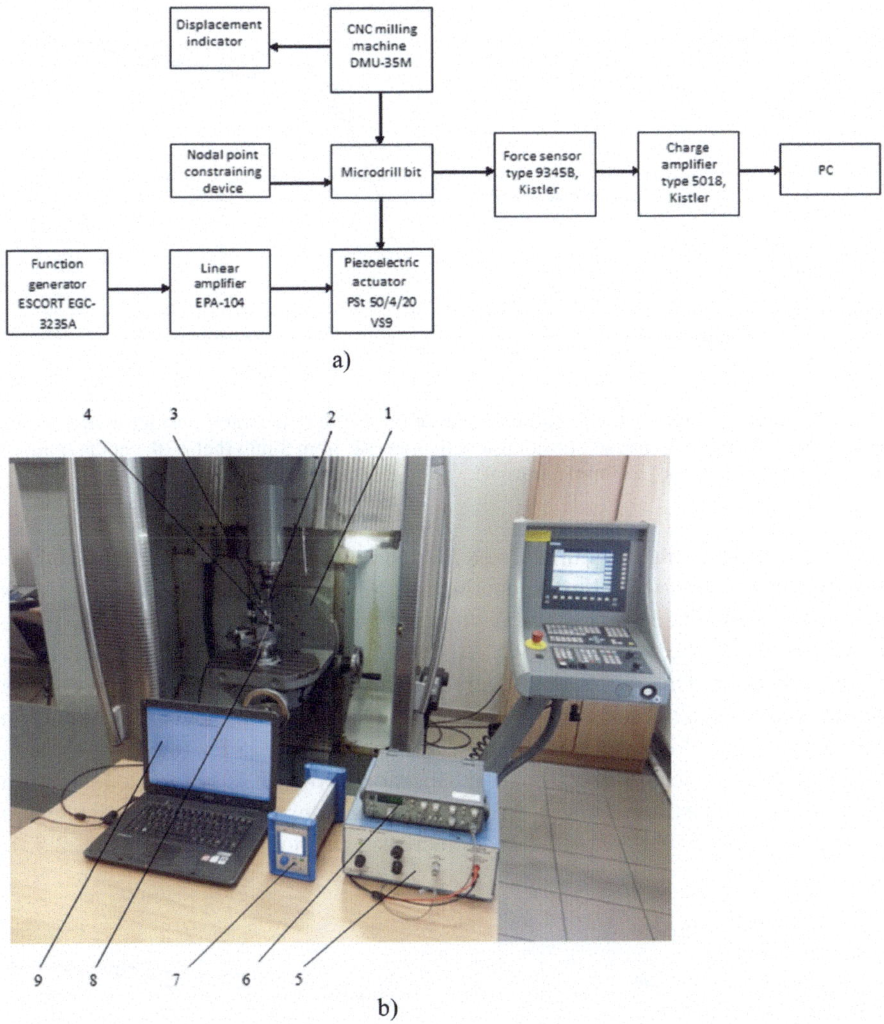

Fig. 2.51 **a** Schematic representation of the experimental set-up and **b** picture of the experimental set-up: 1—CNC milling machine DMU-35 M, 2—micro-drill bit, 3—nodal point constraining device, 4—piezoelectric actuator PSt 150/4/20, Piezomechanik, 5—linear amplifier EPA-104, 6—function generator Escort EGC-3235A, 7—charge amplifier type 5018, Kistler, 8—force sensor type 9345B, Kistler, and 9—PC

mode) is 2.5–3 times higher compared with the buckling stiffness of rotating micro-drill bit without constrained nodal point of the second buckling mode (when micro-drill bit buckles according to the first buckling mode).

Similar results were obtained by using the FE method: calculated critical load for the buckling according to the second mode is 2.4–3 times bigger than the load when micro-drill bit buckles according to the first mode.

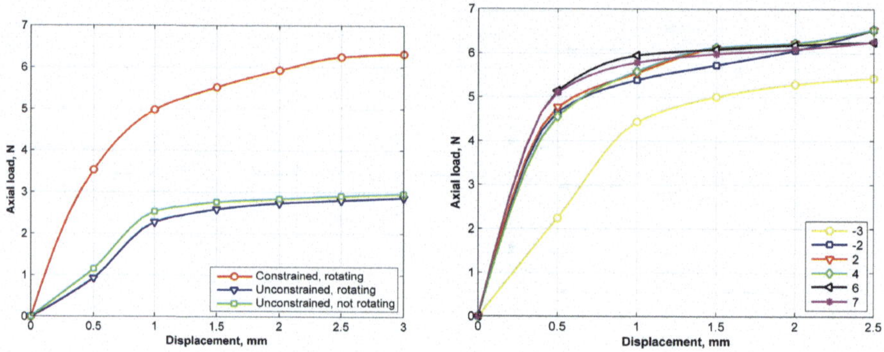

a) b)

Fig. 2.52 Micro-drill bit buckling load dependence on drill bit constraint position **a** and on the displacement of drill bit constraint position **b** in the interval from 3 mm (below the nodal point of the second buckling mode) to 7 mm (above it)

The constraint point was immobilized in such a way that the micro-drill bit gathers the shape of the second buckling mode. Figure 2.52b shows that the buckling stiffness of the micro-drill bit does not change significantly when the constraint is in the range from 2 mm below the nodal point of the second buckling mode of the micro-drill bit to 7 mm above it. This means that a very precise position of the constraint point is not necessary when micro-drilling process is carried out. For imitation of vibration cutting, the micro-drill bit was excited by a frequency of 225 Hz (the first resonant frequency of the micro-drill bit flexural vibrations, when it is not loaded). In both cases, when the constrained micro-drill bit and the unconstrained micro-drill bit were excited, the buckling stiffness of the tool increases but not significantly (Fig. 2.53).

Fig. 2.53 Vibrationally excited micro-drill bit buckling load dependence on the displacement of dill bit constraint

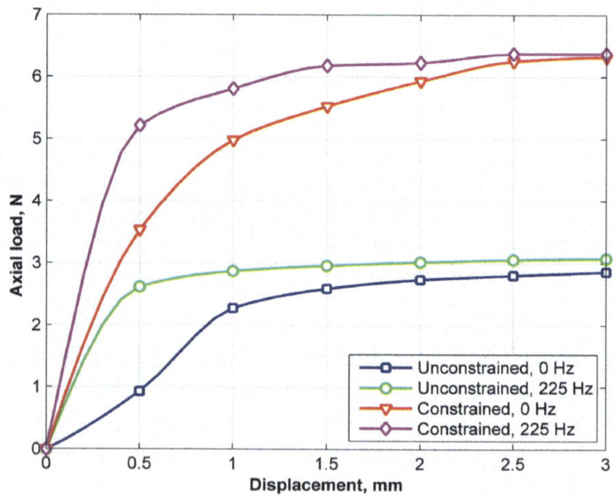

Performed experiments proved that the second buckling mode of the micro-drill bit is not constant and depends on the axial load applied to the tool. Because of that, additional means of exciting the second buckling mode have been used. The micro-drill bit can be excited in the second buckling mode by constraining the middle point of the tool. Experimental results show that the buckling stiffness can be increased by almost three times when the nodal point of the second buckling mode is constrained and that, of course, allow to use higher drilling regimes. This is possible to achieve by not using any complex equipment so can be easily applied practically.

2.4 Quality Improvement of Grinding Operations

Quality is the condition of the finished product, without defects, deficiencies, and significant variations. Quality could be improved by changes in the processing parameters of grinding machines, which are reflected in an increase in product acceptance in the market. Although research to improve grinding quality is mainly related to vibration reduction, here vibrations will be used to improve the smoothness of the grinded surface.

2.4.1 An Excitation Approach to Ultrasonically Assisted Cylindrical Grinding

Application of vibrational excitation to static parts (lathe tool, statically mounted workpiece, etc.) is rather straightforward. The modifications limit themselves to the mounting of an actuator to the said part. Application to moving parts (rotating tools and workpieces), however, is more complex, as it requires additional modules to be put in place to drive and control the actuators situated on the moving part. A novel, robust alternative for rotating component excitation that has the potential to be applied to any process involving a rotating tool or workpiece is proposed in [21]. Instead of mounting the actuator in the rotating interface, it is to be mounted immediately outside it, either through a collar or bracket link, thus, eliminating the need for additional modules. Such an approach has not been applied before. The main challenge is the development of an actuator that possesses reasonable power requirements, yet is still able to transfer the vibration between the moving and static interfaces.

2.4.1.1 Virtual Twin of Cylindrical Grinding

In order to maintain a sufficient level of control during the initial testing stages, a process with a relatively slowly rotating part had to be chosen. Such a part can be recognized as the workpiece in the cylindrical surface grinding process. When considering the specific set-up for the Supertec G32P-100NC grinding machine (Fig. 2.54), it can be seen that the piezoelectric transducer can be mounted via a clamp collar onto the sleeve of the cylindrical grinding machine tailstock; also, in the initial analysis stages, interaction between the workpiece and the grinding wheel was ignored. Two types of workpieces—a uniform and a stepped shaft—were used for experimentation.

The studies concerning the workpiece behavior were aimed at determining the most suitable excitation frequencies and required excitation force. These studies were tackled first, since their results served as inputs for further study and development of the actuator. The actuator development requires derivation and adjustment of ultrasonic horn dimensions, to suit operation to workpiece frequency. Upon completion of the design, the viability of the excitation approach was tested by determining the power requirements for producing expected workpiece excitation amplitudes.

Before studying the behavior of the workpiece any further, a set of target conditions were set forth:

- Since the tool approaches perpendicularly to the workpiece axis, the latter should vibrate in a transverse mode;
- The frequency of excitation must not be lower than 15 kHz as lower values would be too far away from the ultrasonic threshold;
- The target amplitude of excitation needs to be above 1 μm, as most encountered applications tend to employ amplitudes of such order of magnitude. The first two conditions can be tackled by employing modal analysis.

If the law of motion of the excited end of the workpiece is to be considered known and defined by a nodal displacement vector U_K, the block form of the equation of motion of the workpiece is written as follows:

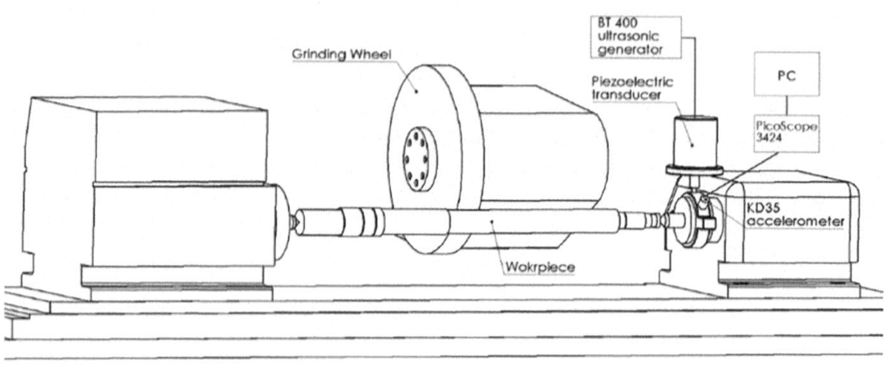

Fig. 2.54 Vibration grinding set-up for Supertec G32P-100NC grinding machine

$$\begin{bmatrix} M_{NN} & M_{NK} \\ M_{KN} & M_{KK} \end{bmatrix} \begin{Bmatrix} \ddot{U}_N \\ \ddot{U}_K \end{Bmatrix} + \begin{bmatrix} K_{NN} & K_{NK} \\ K_{KN} & K_{KK} \end{bmatrix} \begin{Bmatrix} U_N \\ U_K \end{Bmatrix} = \begin{Bmatrix} 0 \\ \overline{R} \end{Bmatrix}, \tag{2.41}$$

where N and K are indices representing nodes, whose displacements are unknown (free nodes) and known (excited or constrained nodes), respectively. Hence, U_N and U_K are nodal displacement vectors for free and constrained vectors, respectively, while pair combinations of these indices relay the position of these nodes in respective property matrices M (mass matrix) and K (stiffness matrix). R is the vector of unknown reaction forces of nodes under excitation.

Modal analysis of an undamped oscillatory motion finds resonant frequencies by solving the following equation:

$$\left(K - \omega^2 M\right)\hat{U} = \{0\}, \tag{2.42}$$

where ω is the angular frequency, while $\hat{U}$ is the mode shapes vector.

The modal analysis of the workpieces was performed employing the Ansys Workbench software package, modal analysis module. Ansys Workbench suite uses the Block Lanczos extraction method to obtain the eigenmodes in prescribed frequency ranges. Knowledge of appropriate frequencies of the workpieces allowed to apply a periodic forcing condition at the tailstock mount (Fig. 2.55). Transverse deformation amplitudes were observed as a result, and the most suitable force value and frequency were determined. In Ansys Workbench, a harmonic response module can be used for this purpose.

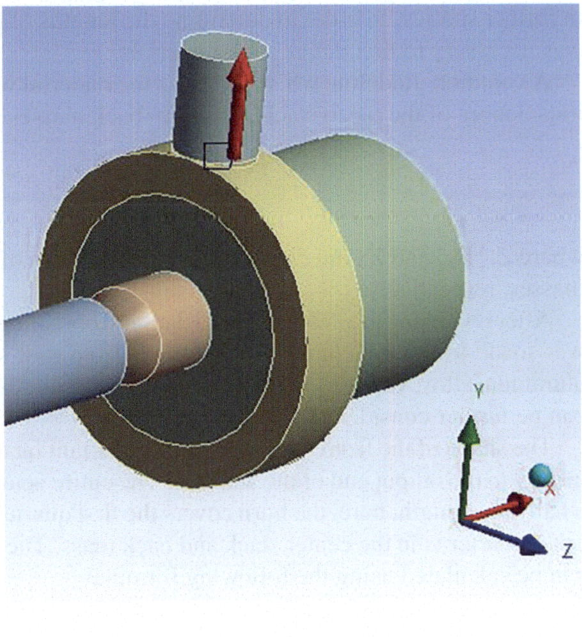

Fig. 2.55 Excitation force F application at the tailstock attachment

In the case of the two workpieces, both are constrained at the tailstock end and prestressed by the headstock. The mode search was performed between 15 and 25 kHz—resulting mode frequencies were then used to define excitation of the workpiece at an arbitrary magnitude of 1 N. The choice of the most suitable frequency has been simplified to fitting the three criteria in descending order of importance:

- Neighborhood of the frequency needed to be clear of other modes;
- The average peak-to-peak, workpiece displacement value—an average of absolute values of oscillation amplitudes across the length of the workpiece—was relatively high;
- The eigenmode frequency was in close proximity to a similar mode frequency in the other workpiece. Excitation frequencies at around 15.8 kHz have been chosen for both workpieces.

Carrying out a parametric sweep of the excitation force yields an average deformation amplitude of 1.5 μm at 2.1 N. Hence, the actuator will have to be able to be driven at 15.85 kHz and 15.794 kHz, respectively, and produced a reaction force of 2.1 N at the fixture.

The initial design of the actuator was based on pre-existing dimensional and material requirements, as well as wavelength considerations. The actuator was mounted on an M30 thread at the output end of the horn; hence, the variation of the output end diameter was limited (for safety, wall thickness was chosen as 4 mm, resulting in minimum output end diameter of 38 mm). The piezoelectric stack consisted of two piezo ceramic rings of dimensions (80 × 30 × 12) mm. This in turn defined the diameter of the input end of the horn and the backing mass to be 80 mm. For safe operation, the actuator was encased in a protective housing, which was mounted on an additionally formed flange, it was relatively light however and therefore neglected in further studies. In order to determine the lengths of the horn and the back mass, it was necessary to first define their materials.

A common guideline when choosing the material was to match the characteristic impedances of the piezo stack and front–back masses accordingly:

$$z_c = \sqrt{z_f z_b}, \tag{2.43}$$

where Z_c, Z_f, and Z_b are characteristic impedances of center stack, front, and back masses, respectively.

After reviewing available material data and obtaining impedance values, a decision was made to manufacture the front mass from tool steel and back mass from an aluminum alloy. With this knowledge in mind, longitudinal dimensions of the actuator can be further considered.

The shape of the front mass is also an important factor in the transfer of ultrasonic energy to the output end of the actuator. The entire actuator is expected to oscillate at a half wavelength; here, the horn covers the first quarter of the wavelength, while the other quarter is in the center stack and back mass. The wavelength for each material can be calculated using the following formula:

Table 2.5 Actuator material properties and physical dimensions

	Tool steel	PZT	Aluminum
Young's modulus [GPa]	170	73	78
Density [g cm^{-3}]	7.700	7.600	2.850
Sound velocity [ms^{-1}]	4698.715	3099.236	5231.484
Wavelength [m]	0.297	0.196	0.330
Length [m]	0.074	0.025	0.041

$$\lambda = \frac{c}{f}, \tag{2.44}$$

where c—the sound velocity in the medium, while f is the excitation frequency. The sound velocity can be determined accordingly:

$$c = \sqrt{\frac{E}{\rho}}, \tag{2.45}$$

where E—Young's modulus and ρ—the density of the material.

Wavelength values can be found in Table 2.5.

With all of the data obtained, the derivation of front-mass length was straightforward—a quarter of the wavelength for the tool steel. Since the value of the center stack length was already known, it was determined to take up an eight of its wavelength, making it possible to derive the back-mass value—an eighth of the wavelength for the aluminum alloy. These values were used in the initial design for modal analysis (Fig. 2.56). Here, a longitudinal eigenmode was expected to occur at 15.794 kHz. However, mode search at 15–25 kHz yielded frequency values 2 kHz above this value. A parametric sweep of the horn length was conducted and a longitudinal eigenmode was observed at 15.939 kHz when the horn length was 111 mm.

A newly designed actuator was tested in a harmonic response study; here, both planar faces of the piezo stack were assigned arbitrary forces of opposite signs in order to simulate excitation from the piezo stack. Force at the output end of the horn was measured, allowing to perform parametric sweep in order to determine the required input force, to produce 2.1 N at 15.794 kHz and 15.850 kHz on the

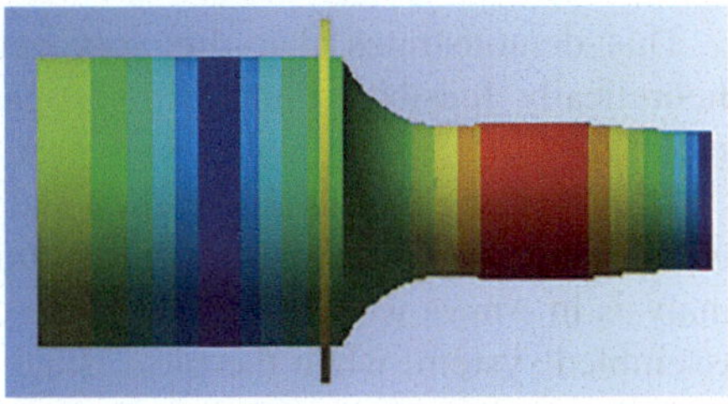

Fig. 2.56 Longitudinal mode of the initial actuator design

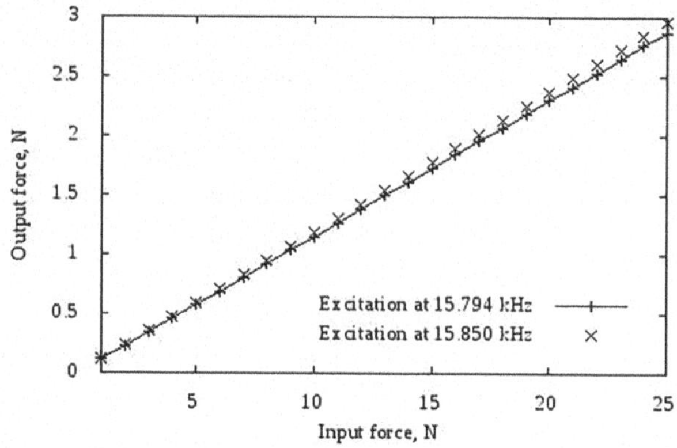

Fig. 2.57 Input–output force dependency for actuator at 15.794 and 15.850 kHz

output end, as required for excitation of the uniform cylinder and stepped shaft to 1.5 μm, respectively. The required input force was determined to be 19 and 20 N for 15.794 kHz and 15.850 kHz cases, respectively (Fig. 2.57).

Knowing the required force and velocity of vibration allowed to determine the power requirements:

$$P = Fv\cos\phi, \tag{2.46}$$

where φ is considered to be the phase difference between the velocity and force. The obtained results show the required power input of the transducer, required to achieve force F during excitation. Since accelerations at the piezo-material to the front-mass interface were determined to be 2070 mm/s^2 and 2000 mm/s^2 for 15.794 kHz and 15.850 kHz cases, respectively, peak velocities in both cases were found to be 20 mm/s. Assuming force and velocity to be in phase, Eq. 2.46 yields the following values:

$$P_{15.794} = Fv\cos\phi = 19 \times 20 = 380\ W \tag{2.47}$$

and

$$P_{15.855} = Fv\cos\phi = 20 \times 20 = 400\ W. \tag{2.48}$$

This demonstrates that the approach requires realistic power values and is theoretically feasible. Further investigations of the fully assembled system are necessary.

With working modes determined and the design of the actuator verified, a fully assembled system needs to be tested. Initial tests were performed using harmonic analysis in Ansys workbench; here, the displacement of the workpiece in the fully assembled system, when the piezo stack is excited at the predetermined force and frequency, is observed. Afterward, the excitation output was tested on a physical system. Lastly, the effectiveness of the approach was tested by observing surface grinding results with and without the operation of the actuator.

2.4.1.2 Harmonic Response of the Grinding System

The harmonic analysis data for both workpiece configurations is presented in Table 2.6.

The workpiece displacement graphs (Fig. 2.58) show a noticeable reduction in amplitude (Fig. 2.58).

Additional modal analysis of the entire assembly was performed, demonstrating that the addition of the actuator to the system shifted the transverse mode frequency to 17.349 kHz and 17.722 kHz for uniform cylinder and stepped shaft workpieces, respectively (Fig. 2.59), which was the cause for a decrease in displacement.

Table 2.6 Workpiece excitation parameters

Workpiece	Excitation force, N	Frequency, kHz
Uniform cylinder	20	15.850
Stepped shaft	19	15.794

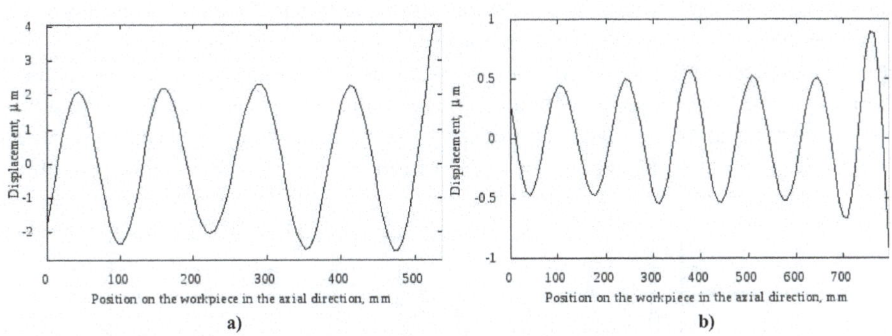

Fig. 2.58 Response of the uniform cylinder workpiece, being driven by 20 N at **a** 15.85 kHz and **b** stepped shaft workpiece 15.794 kHz frequency, with the actuator attached

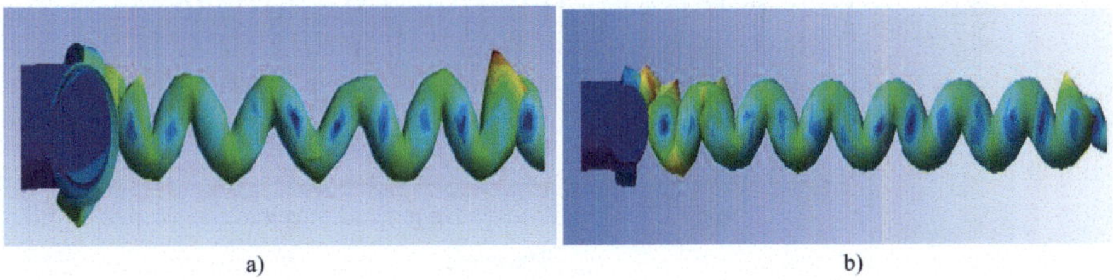

Fig. 2.59 Shifted transverse mode for **a** uniform cylinder at 17.349 kHz and **b** for stepped shaft at 17.722 kHz

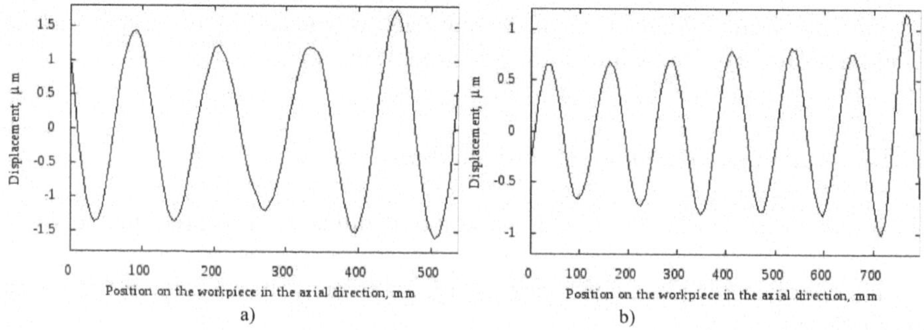

Fig. 2.60 Response of the uniform cylinder workpiece, being driven by 19 N at 17.35 kHz frequency **a** and of the stepped shaft workpiece, being driven by 19 N at 17.72 kHz frequency, **b** with the actuator attached

Driving the systems at 17.349 and 17.722 kHz yielded more appropriate responses in the order of micrometers (Fig. 2.60):

Similar changes are expected to occur in the physical system as well; therefore, experimental vibration analysis is essential in this case.

The physical twin consisted of Supertec G32D 100 NC cylindrical grinding machine with an actuator mounted at the tailstock. The actuator was driven by Sensotronica BT400 ultrasonic generator; from the tailstock, the oscillations were expected to be transferred to the workpiece at 17.35 kHz for a uniform cylindrical workpiece and 17.72 kHz for a stepped shaft workpiece. For measurement, PicoScope-3424 oscilloscope with a KD-91 accelerometer was used. With the ultrasonic generator at 40% power, amplitudes of 0.368 μm were observed to occur at 16.82 kHz for both workpieces. The acceleration (Fig. 2.61) was in close proximity

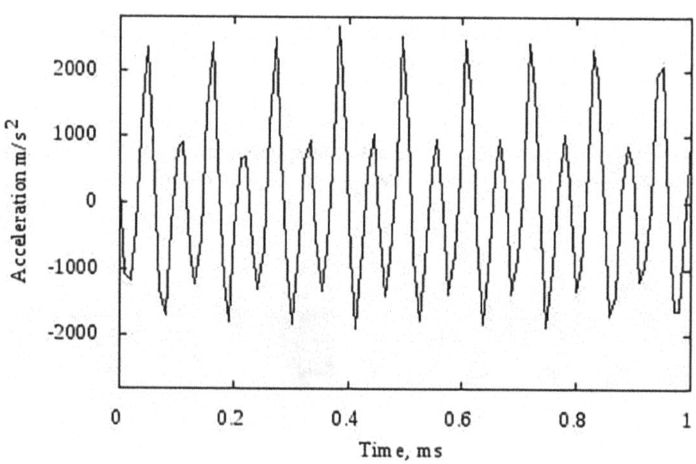

Fig. 2.61 Acceleration at 16.82 kHz

to the values obtained during the simulation. This change corresponds with observations, regarding the shift of eigenmode frequency, due to the addition of the actuator. These conditions are to be further used in vibration-assisted grinding experiments.

2.4.1.3 Physical Twin of Cylindrical Grinding

In order to test the effectiveness of the system, a grinding trial was conducted. Experiments were carried out with and without vibrations for wet (Fig. 2.62) and dry grinding. The first experiment was carried out on a C40E/1.1186 steel shaft for the conventional (without ultrasonic vibrations) grinding.

After each grinding run, surface quality was measured using a surface roughness tester Mitutoyo Surftest SJ-210. Each measurement was performed five times, at sampling and cut-off lengths of 0.8 mm. Settings of machining parameters for the experiment are summarized in Tables 2.7 and 2.8. Ultrasonically assisted grinding zones are shown in Fig. 2.63.

Conventional (without ultrasonic vibrations) and ultrasonic-assisted grinding zones are shown in Fig. 2.63a. During the cylindrical grinding process, the surface being ground was divided into two equal zones for surface roughness measurement, and the obtained data was subsequently processed.

Fig. 2.62 Experimental set-up of ultrasonic-assisted grinding. 1—Workpiece (steel bar); 2—accelerometer (KD35); 3—piezoelectric transducer; 4—PicoScope (3424); 5—tailstock of a cylindrical grinding machine (Supertec G32P-100NC)

Table 2.7 Experimental study consumable list

Grinding wheel	Vitrified bond *ZrA* grinding wheel, grain size 80, 405/210/50
Workpiece 1	1.1186 medium carbon steel, Vickers hardness 280 (L = 814 mm, Ø = 50 mm)
Workpiece 2	1.7225 42CrMo4, Vickers hardness 207 (HBW) (L = 500 mm, Ø = 35 mm)

Table 2.8 Major machining parameters

Grinding process	Dry/wet cylindrical grinding
Conventional grinding conditions	Feed speed $v_f = 5$ m/min; depth of cut a = 0.01 mm; cutting speed $v_c = 34$ m/s
Ultrasonic vibration assistance conditions	Frequency f = 16.81 kHz; power P = 40%; amplitude A = 0.386 μm

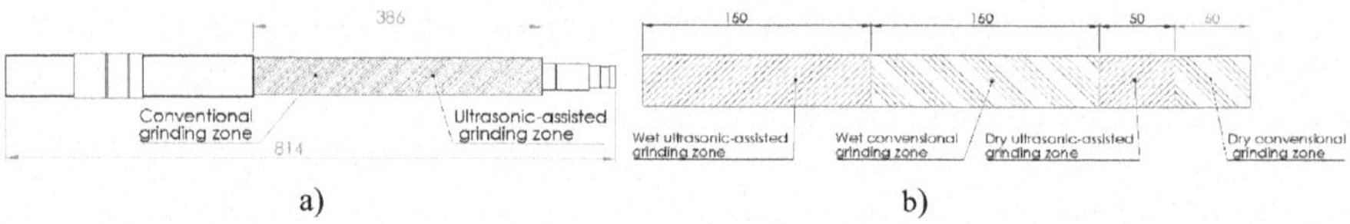

Fig. 2.63 Grinding zones during the cylindrical grinding of a shaft **a** and of a bar **b**

The second experiment was carried out on a 42CrMo4/1.7225 steel bar. The grinding zones of the experiment are shown in Fig. 2.63b. After the experiments, the roughness of the shaft and bar was measured with a surface roughness tester at defined zones for conventional and ultrasonically assisted grinding with and without cooling fluid. Results of surface roughness are summarized in Table 2.9 and shown in Fig. 2.64. Here, *UG* and *CG* refer to wet grinding with and without ultrasound, when applied to the stepped shaft, while UDG and CDG stand for ultrasonic and conventional dry grinding with UWG and CWG applying for wet grinding, respectively, when dealing with the uniform steel bar. The results in the case of dry grinding are not much different between ultrasonic and conventional counterparts. Noticeable improvements were observed in ultrasonic wet grinding of a stepped shaft (UG and CG in Fig. 2.64b) when compared to its conventional counterpart. This may suggest

Table 2.9 Surface roughness values

Surface Roughness parameter, [μm]	1.7225 steel bar			
	Dry grinding		Wet grinding	
	UDG	CDG	UG	CG
Ra	0.442	0.491	0.418	0.432
Rq	0.556	0.611	0.516	0.545
Rz	3.323	3.315	2.818	2.843
Roughness parameter, [μm]	1.1186 shaft			
	Wet grinding			
	UWG		CWG	
Ra	0.251		0.298	
Rq	0.334		0.408	
Rz	2.291		2.939	

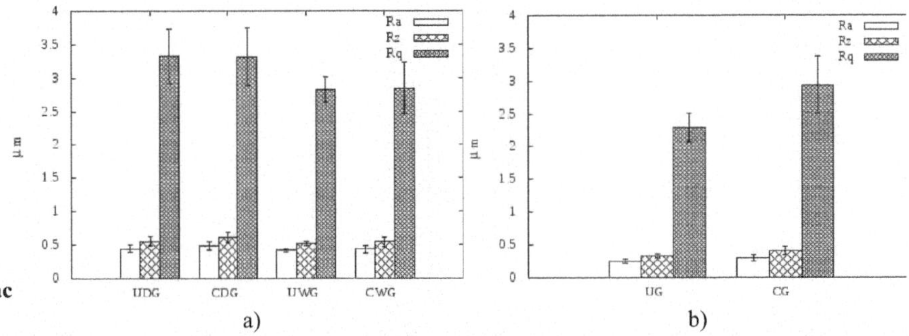

Fig. 2.64 Surface roughness values for grinding trials on a uniform cylinder **a** and stepped shaft, **b** workpiece

that high amplitudes may exert a destructive effect on the surface. However, the lack of noticeable improvement in dry grinding also suggests that vibrational assistance should best be applied to stationary interfaces and that ultrasound enables a better lubrication mechanism, improving the surface quality as a result.

2.4.2 Development of Actuator for Back Grinding

The back grinding process was chosen as testing grounds for the different vibration excitation approaches [22]. Tool excitation, whereby vibrations are transferred to the tool–workpiece interface through an ultrasonic tool holder, is the most common approach to vibration-assisted grinding. Tool excitation approach, however, is generally limited in terms of low excitation frequency, low spindle speed, and limited applicability of the method for difficult to manufacture materials treatment. The development of ultrasonic excitation equipment is crucial in order to successfully carry out the experimental studies. The development of each actuator follows a similar sequence. Previously outlined experimental studies were used for the specification of design parameters of each actuator. According to the specification parameters, the initial dimensions of the actuators were approximated by using analytical relations outlined previously. The designs were validated and adjusted by employing numerical techniques. All the actuators developed in this book are of the Langevin, sandwiched piezo transducer type. These actuators consist (Fig. 2.65) of a piezoceramic element stack along with electrodes 8 placed in between two masses 7 and 9 that are connected with a bolt 6 in such a way that as the bolt gets tightened, the piezoceramic stack is compressed. The mass commonly used for transmitting the vibrations to the working end of the device is commonly termed the front mass, whereas the piezoceramic stack is commonly termed as the center mass, and the counterbalance mass is termed as the back mass.

Actuator's cross-sections may vary along with the range of different shapes, as does the symmetry axis and the conditions. In the case of rotary tool applications, in

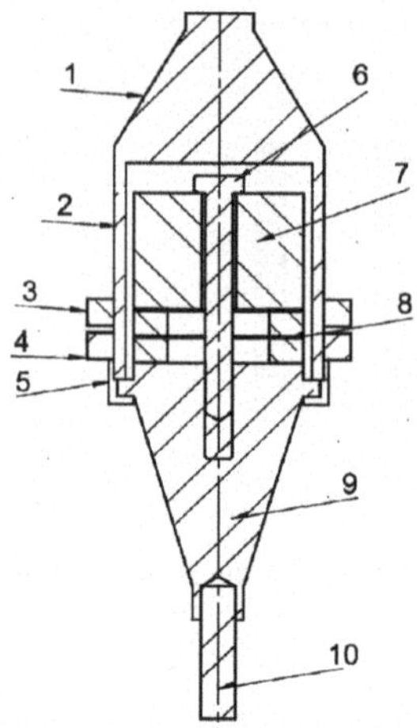

Fig. 2.65 Schematics of a Langevin-type actuator in a tool holder configuration: 1—tool holder interface, 2—housing cylinder, ¾—commutation rings, 5—retention nut, 6—retention bolt, 7—back mass, 8—piezoceramic stack, 9—front mass, and 10—grinding burr

order to reduce the effects of the eccentric load, all the parts are best kept in concentric assembly with each part maintaining the circular cross-section. This design condition was maintained for all the three cases of front-mass profile shapes (Fig. 2.66a). The front-mass geometry can either stay uniform or have a widening or a narrowing profile in order to disperse or concentrate the vibrations, respectively. The generally accepted front-mass profile shapes are cylindrical (stepped), conical, or catenoidal.

The general practice for constraining the actuators is via a flange along the nodal plane (Fig. 2.66b). This method is employed in the tool excitation approach; however, in the static workpiece study and in the rotating workpiece study, the constraint is placed at the exposed back-mass and at front-mass faces, respectively. The first

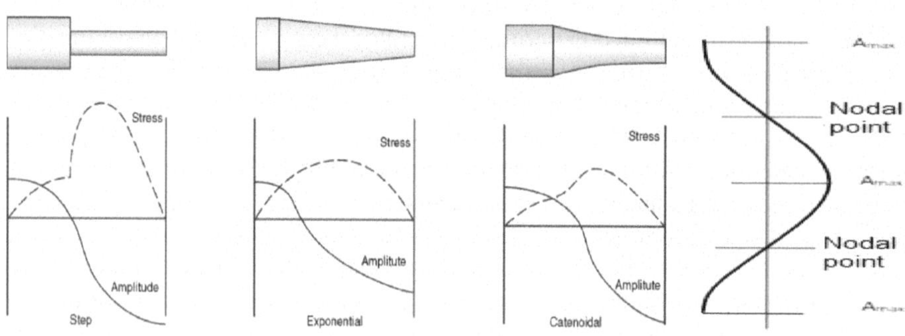

Fig. 2.66 Different front-mass profile shapes (**a**) and nodal point and displacement diagram in a Langevin transducer (**b**)

limitation is a result of the general dimensional constraints of a tool holder assembly. To function as a tool holder, a certain amount of actuator front-mass length would have to be dedicated for tool holder features. Tool dimensions have been estimated for 20 kHz, namely due to this constraint. The latter two limitations arise as a result of general challenges to operating an ultrasonic actuator in a rotating interface. The piezoceramic rings require a supply of an electric signal to operate. To transfer the signal to the actuator, it is necessary to employ a commutation approach that would not impede on the rotational motion of the spindle. For this, additional constructions are required to support either a carbon brush or a liquid contact connection.

2.4.2.1 Physical Twin of Back Grinding

The grinding runs were performed using a five-axis Starrag Heckert LX 151 CNC milling machine (Fig. 2.67). In practice, back grinding of tungsten carbide generally employs CBN grinding burrs. In this case, because the workpiece 1 (Fig. 2.67) surfaces were pre-polished and because the nature of the tests did not focus on roughing stages of the process, but rather finishing, a galvanically bonded CBN grinding burr 2 with a relatively fine grit size of D 126 (Diamond CBN ISO 6106) diameter of 10 mm and length of 10 mm was chosen.

Outlined in Fig. 2.67 is the main set-up for grinding of tungsten carbide with ultrasonic tool excitation assistance. Here, the tungsten carbide workpiece 1 is machined using a grinding burr 2 connected to the tool actuator 3, 4, 6, 5, and 9, which is housed in the spindle of the head of Starrag Heckert LX 151 5 axis CNC milling machine 10, and exhibits ultrasonic vibrations when supplied with an electric signal

Fig. 2.67 Tool excitation grinding set-up. 1—workpiece; 2—grinding burr; 3—tool holder horn; 4—retention nut; 5-carbon brush connection; 6—commutation rings; 7—carbon brush holder structure; 8—actuator housing cup; 9—HSK 65A spindle interface; 10—Starrag Heckert LX 151 5 axis CNC milling machine

Table 2.10 Properties of materials used for tool actuator part production

Material	Young's modulus	Density
	E(GPa)	ρ(kg/m^3)
40 × steel	211	7800
PZT4 piezoceramic	78	7600

through carbon brushes located in the carbon brush assembly 7 and 5, which is also attached to 10. The tool actuator at its core is a simple half-wave Langevin, sandwiched piezo transducer. To hold the tool and guide longitudinal vibrations to its tip, the front mass of the transducer was designed as an exponential horn with a tool holder feature which is shown in Fig. 2.66a. As seen in Fig. 2.67, the entire transducer assembly is retained inside the actuator housing cup 8 by a retaining nut 4. The actuator housing cup 8 attaches to the spindle through the HSK-65 spindle interface 9. To operate the transducer, an electric signal is transferred through a carbon brush connection 5 into the commutation rings 6 attached to the actuator housing cup 8. Since the core part of the tool actuator design is the Langevin, sandwiched piezo transducer, determination of correct design features and dimensions is of paramount importance for the correct function of the tool actuator. Therefore, the transducer dimensions were chosen in accordance to initial conditions and then validated and adjusted using the FE method. The horn and back-mass parts of the actuator are commonly produced from high stiffness alloys, to ensure efficient vibration transfer to the output end of the actuator. Because the actuator is constrained at the nodal point, it is necessary to ensure that the impedance of the back mass is higher than that of the front mass. Therefore, in this study, the horn and back-mass parts of the tool actuator were produced using 40 × steel, while PZT4 piezoceramic rings were used for the piezoceramic stack. The relevant properties of these materials are outlined in Table 2.10.

The surface quality was evaluated using white light interferometry measurements on Polytec MSA 500 (Fig. 2.68a). The actuators were excited using EPA-104 signal amplifier and Agilent 33220A signal generator; their modes of vibration were validated by Polytec PSV-500 3D scanning vibrometer (Fig. 2.68b).

The three commonly used front-mass shapes are uniform cylindrical shape, variable width conical, and exponential horns (Fig. 2.66a). In this case, an exponential profile with a 2:1 input to output end ratio was chosen for the design of the ultrasonic horn. It is important to avoid operating the actuator in the same frequency as lateral vibrations of the tool. In order to identify this frequency, an impact test was conducted using a Polytec OFV-505 sensor head and Polytec OFV-5000 vibrometer with a Picotech digital oscilloscope PicoScope connected to a PC for data acquisition. The impact test result of the tool shows (Fig. 2.69) that the lateral vibration of its shank is 5 kHz. To prevent the actuator from inducing tool resonance, it should be driven at a frequency higher than the one observed in the impact test. Therefore, to maintain stable operation while still being able to test the effects of actuation on the low end of the frequency spectrum, the actuator was designed to work at 15 kHz.

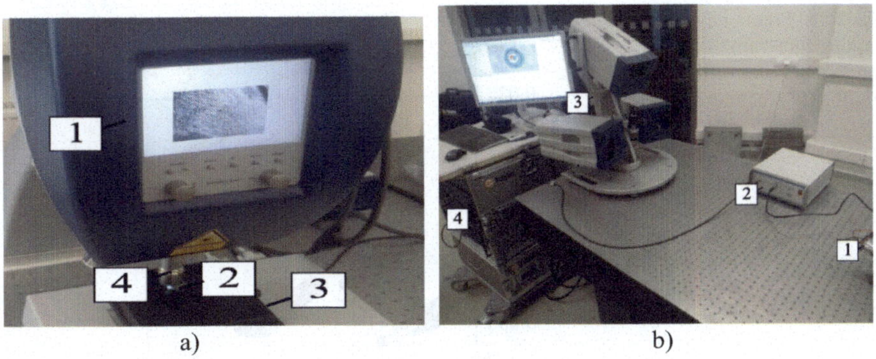

Fig. 2.68 **a** Polytec MSA 500 microsystem analyzer set-up used for white light interferometry measurements: 1—analyzer head; 2—sample surface; 3—*XY* positioning table; 4—interferometric lens and **b** Polytec PSV-500 3D scanning laser vibrometer set-up: 1—actuator; 2—signal amplifier; 3—scanning laser head

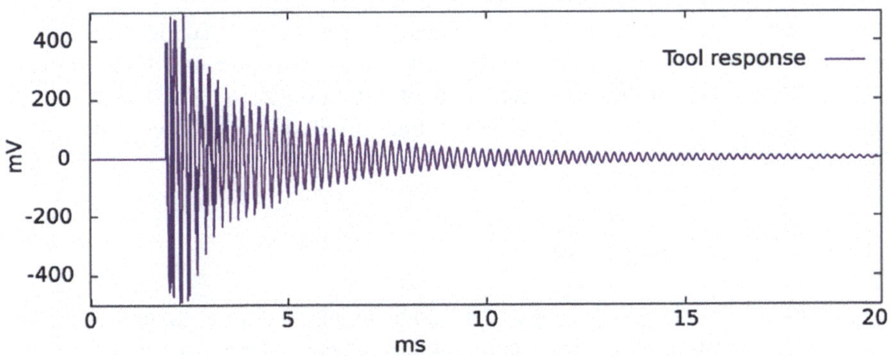

Fig. 2.69 Tool impact test graph at 75 mm protrusion, response frequency of 5 kHz

To ensure the successful operation of the actuator, produced according to previously outlined design parameters, it is necessary to verify the designs through numerical simulation. In addition, a validated model serves as a basis for rapid design modifications. The actuator was modeled using FE, based on the design parameters outlined earlier. The model was produced in the SalomeMeca suite, using hexahedral elements. The analysis was carried out in Code_Aster—contact between separate parts was defined using Lire_mailage keyword, and flange faces of the tool actuator were constrained using Enforce_dof keyword. For verification, modal analysis was carried out using the equation of free vibration:

$$[M]\{\ddot{u}\} + [K]\{u\} = \{0\}, \tag{2.49}$$

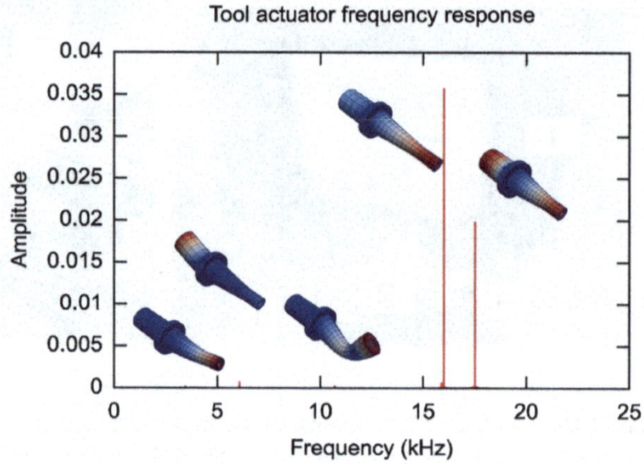

Fig. 2.70 Frequency response diagram and dominant modes of the grinding tool actuator

where $[M]$ is the mass matrix, $[K]$ is the structure stiffness matrix, $\{\ddot{u}\}$ is nodal acceleration vector, and $\{u\}$ is the nodal displacement vector.

The analysis yielded the longitudinal eigenmode at 14.060 kHz. However, more intense displacements were observed on the back-mass portion of the actuator (Fig. 2.70). This is due to poor execution of condition $Z_b > Z_f$. Therefore, the back-mass length was increased by 10%, which resulted in a displacement shift to the front with the longitudinal eigenmode at 16.3 kHz.

2.4.2.2 Back Grinding Results

In order to observe the morphology of the chips obtained during the tool excitation study, wolfram carbide (WC) dust samples were collected on pieces of clear tape after each test run while making sure to eliminate any residue before proceeding with the subsequent test run (Table 2.11). The samples were then observed and photographed on a Microscope at 100 × magnification (frame size 100 × 100 μm). To better observe the morphology of the dust, a conscious effort was made to focus on the parts of the sample where a significant amount of background is not covered by the dust.

Provided in Fig. 2.71 are samples from machining tests at 25 μm cutting depth

Table 2.11 Experimental grinding runs for tool actuator study

Depth	Excitation frequency	Spindle speed	Feed rate
(μm)	(kHz)	(rpm)	(mm/min)
25	0	5000	5
	14.5		
50	0		
	14.5		

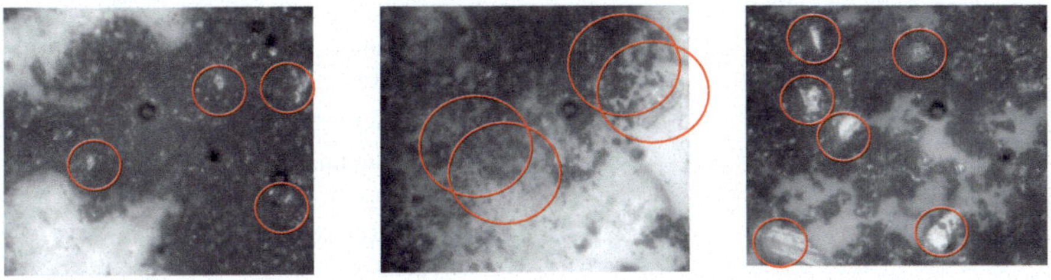

a) b) c)

Fig. 2.71 WC dust obtained at 25 μm cutting depth without excitation (**a**), with longitudinal excitation (**b**), and with transversal excitation (**c**)

without ultrasonic excitation, with longitudinal excitation, and transversal excitation in the respective order. It can be seen in all of the pictures that the dust is of irregular size in all the three instances; however, the most consistent size of dust particles can be observed in the sample obtained from longitudinal excitation, followed by the no excitation sample, thus, making the transversal excitation sample the least consistent as some particles measure up to 10 μm. Such irregularity may have an effect on the obtained surface's roughness of the sample and may indicate the occurrence of destructive processes—which is most likely the case in the transversal excitation sample. The grinding dust sample in the no excitation case features larger particles measuring up to 3.5 μm (particles circled in red). These particles are of the spherical nature and are indicative of thermally induced plastic deformation. Meanwhile, the longitudinal excitation case produced small 1.5–2 μm chips which were more common in the ductile turning induced by large hydrostatic pressures. The transverse excitation case provided large flat-faced chips which are indicative of a brittle fracture.

Data resulting from a repeated test run at the cutting depth of 50 μm shows sample photos for all the three conditions in the previously mentioned order (Fig. 2.72). In this case, the transversal excitation sample demonstrates the same behavior (particles circled in red), yet the control condition sample seems to exhibit similar signs of irregular fracture (circled in red in Fig. 2.71) as it also features irregular size particles

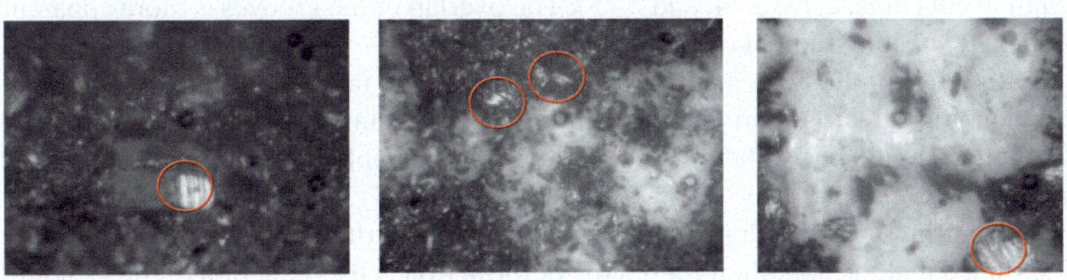

a) b) c)

Fig. 2.72 WC dust obtained at 50 μm cutting depth without excitation (**a**), with longitudinal excitation (**b**), and with transversal excitation (**c**)

Table 2.12 Surface roughness parameters of WC samples

Set-up	Excitation condition	Ra (μm)	Rz (μm)
25 μm cutting depth	No excitation	0.198	1.8448
	Longitudinal excitation	0.1536	1.0699
	Transversal excitation	0.2696	1.8059
50 μm cutting depth	No excitation	0.408	2.4072
	Longitudinal excitation	0.2471	1.5807
	Transversal excitation	0.5651	3.9809

in the image. The consistency of the grain size changes includes particles indicative of thermally induced ductile mode deformation (circled in red in Fig. 2.72).

Surface profilometry provides an adequate quantitative picture of the surface quality, and hence, it was used to further evaluate the different machining techniques tested in the experiments. The table below (Table 2.12) provides the surface roughness parameters (Ra and Rz) of both machining set-ups (at 25 and 50 μm cutting depths).

From the obtained surface profilometry results, the transversal excitation resulted in the worst surface finish at both depths. By a small margin, the longitudinal excitation resulted in a better surface finish than the conventional method in both cases. However, the margin is not high enough to claim the superiority of the method. Even though white light interferometry may serve as a quantitative guide to the surface quality (as it provides calculated surface roughness values), it could be used as a qualitative measure to investigate surface morphology changes at different machining conditions. When removing material, the tool leaves behind grooves characteristic of the geometry of the cutting edge. As the cutting edge in the grinding process is not clearly defined, it is expected to be characteristic of the galvanically bonded grain size of the tool. The expected output from excitation is the blending of these grooves. Provided below are white light interferometry pictures of the samples at a 25 μm depth of cut (Figs. 2.73, 2.74, and 2.75). The overlap of the grooves is more apparent in the case of longitudinal excitation.

In the case of the 50 μm cutting depth, as provided in Figs. 2.76, 2.77, and 2.78, the blending effect is more apparent in the case of transversal excitation. A better distinction should be available at a higher excitation amplitude and an increased feed rate.

The cutting force generally serves as a predictor of ductile mode deformation. It is assumed that the condition for ductile deformation is the reduction of the grinding force. However, depending on the ductile deformation mechanism (thermally induced or high-pressure-induced plasticity), the reduction may not be as apparent. The force

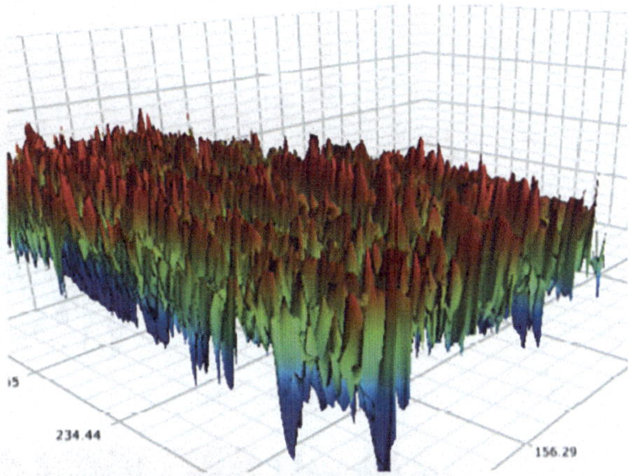

Fig. 2.73 White light interferometry of the surface obtained without excitation at 25 μm

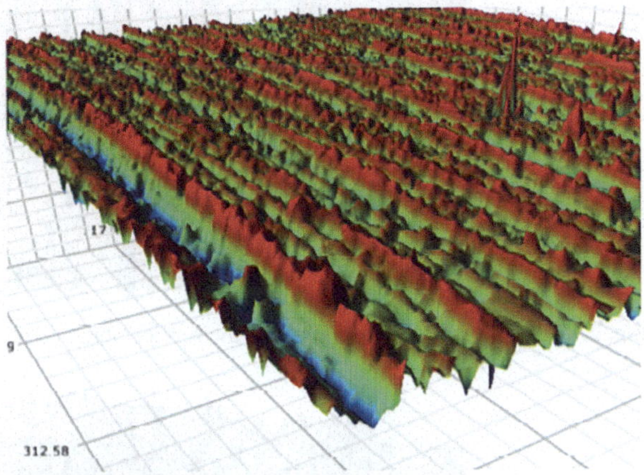

Fig. 2.74 White light interferometry of the surface obtained with longitudinal excitation at 25 μm

Fig. 2.75 White light interferometry of the surface obtained with transversal excitation at 25 μm

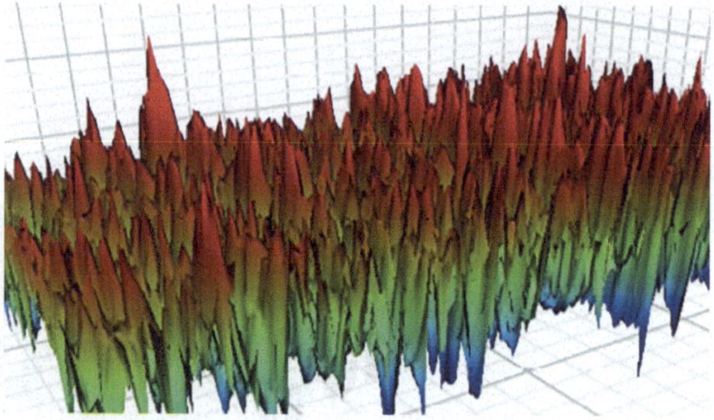

Fig. 2.76 White light interferometry of the surface obtained without excitation at 50 μm

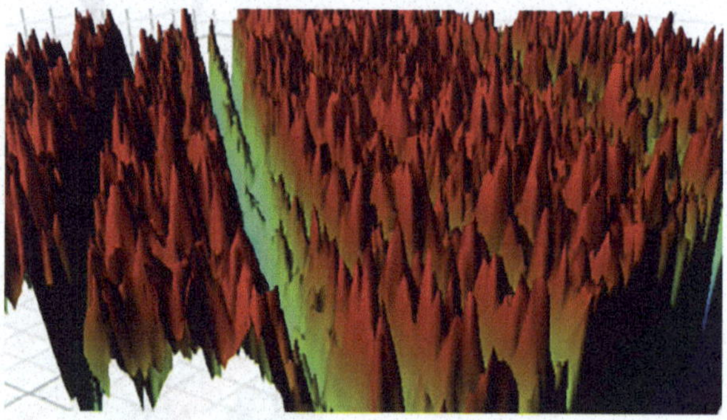

Fig. 2.77 White light interferometry of the surface obtained with longitudinal excitation at 50 μm

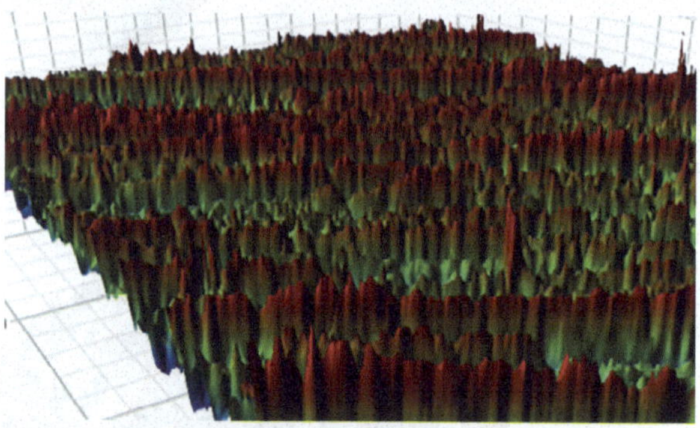

Fig. 2.78 White light interferometry of the surface obtained with transversal excitation at 50 μm

component along the Z-axis allows determining the load applied to the bottom of the grinding path by the bottom face of the tool as a result of tool deflection and other effects. It can be seen from the graphs below that the depth change does not noticeably influence the force along the Z-axis when grinding without ultrasonic excitation. While the average value remains unchanged, the force more noticeably fluctuates when the depth is increased.

Upon application of ultrasonic excitation along the longitudinal direction, a noticeable decrease (~50%) of the grinding force and fluctuations was observed at 25 μm grinding depth (Fig. 2.79). At 50 μm grinding depth (Fig. 2.80), the fluctuations became more apparent than in the case without ultrasonic excitation, while the average force value increased. Compared to the case of longitudinal vibration, when grinding at 25 μm, the analogous case with lateral vibration produces a larger force (~20%). At 50 μm, the fluctuations are not as apparent; however, the average force

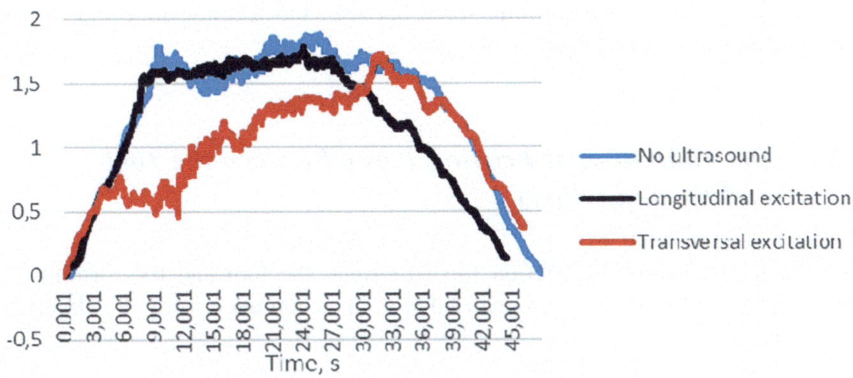

Fig. 2.79 Normal grinding force at 25 μm depth with and without ultrasonic excitation

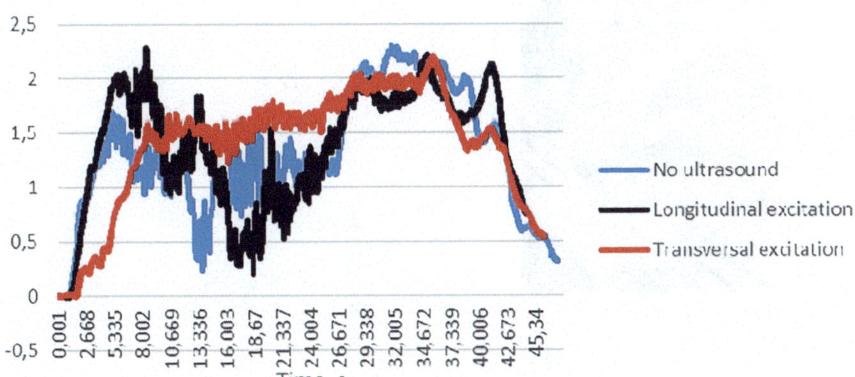

Fig. 2.80 Normal grinding force at 50 μm depth with and without ultrasonic excitation

value is about the same as in the longitudinal case, which is higher than in the conventional case.

2.5 Artificial Neural Networks Approaches for Quality Prediction in Robotized Incremental Sheet Forming

Single Point Incremental Forming (SPIF) is a recent sheet forming technology in several decades. Nowadays, SPIF technology continues to be researched, applied, and improved in the sheet production industry. Several variables related to SPIF technological parameters, such as material conformability and friction phenomena, have not been sufficiently studied. In this book, numerical, experimental, and artificial intelligence methods are used to analyze the deformation forces. A few innovative methods for reducing the forming force between the forming tool and the sheet surface are proposed and developed.

2.5.1 Determination of Friction Force Between the Tool and Forming Sheet

In order to determine the friction force between the tool and the forming sheet (Fig. 2.81a), calculations were carried out in accordance with the diagram in Fig. 2.81b.

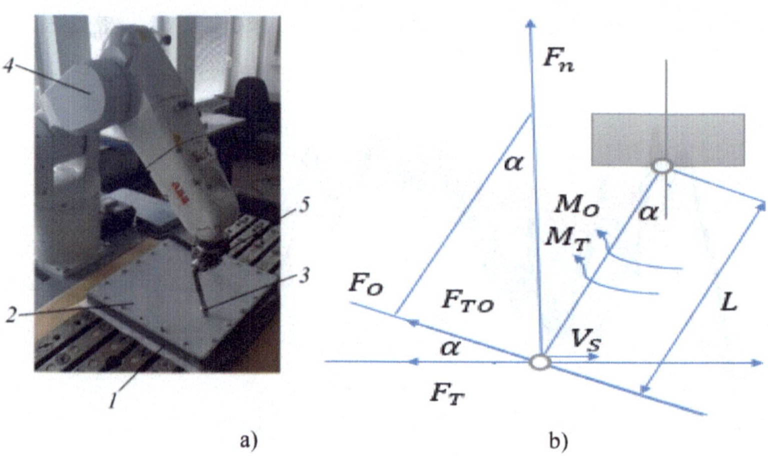

Fig. 2.81 Determination of friction force: **a** interaction and **b** calculation scheme of single point tool and sheet: 1—frame for sheet, 2—sheet, 3—forming tool, 4—Robot ABB IRB1200, and 5—mechanical (friction) torque meter STJ100

Since the roughness momentum gauge was placed with an angle of 30°, both pressure and friction torque could be obtained [23]. The friction force can be calculated as follows:

$$F_T = k_T F_N. \quad (2.50)$$

$$F_{TO} = F_T \sin\alpha. \quad (2.51)$$

$$M_O = F_O L. \quad (2.52)$$

$$F_N = \frac{F_O}{\sin\alpha}. \quad (2.53)$$

$$M_O + M_T = (F_O + F_{TO}) \cdot L. \quad (2.54)$$

Here, L—the length of the tool link connecting sphere and moment gauge (150 mm), α—an angle between the sample sheet surface and the tool link (30°), M_O, M_T—tool pressing force and tool friction force generated angular momentums, F_O, F_{TO}—tool pressure force and the friction force projection to tools link, F_N, F_T—tool's sphere pressure and friction force, k_T—friction force coefficient between the sample sheet and the steel tool sphere, and v_s—steel tool's sphere speed on the surface of the sample sheet (mm/s).

2.5.2 Evaluation Methodology of Metal Sheet Forming Process

Metallic materials used in the deforming process, such as SPIF, visibly deform on contact with the tool area. The key challenge of the SPIF process is to effectively assess the shaping forces. The forces generated during SPIF may be controlled by varying the coefficient of friction between the tool and the workpiece. Lubrication is an important factor in the SPIF process, reducing friction in the tool–workpiece contact area, but the use of grease is associated with environmental problems. Therefore, it is necessary to find other ways to reduce the shaping forces associated with the process dynamics. This requires, in particular, the development of a physically adequate mathematical model of the process, whose mechanical parameters can be adjusted experimentally using the 3D scanning device shown in Fig. 2.82.

An 300 × 300 × 0.5 aluminum alloy AW1050 sheet embedded in a 30 × 30 mm welded steel frame was formed. The frequencies of the resonant modes were determined using the Comsol multiphysics FE method. In an attempt to change the lubrication of the contact surface between the forming tool and the sheet metal, an attempt was made to excite the sheet metal by vibrations. For this purpose, equipment, which

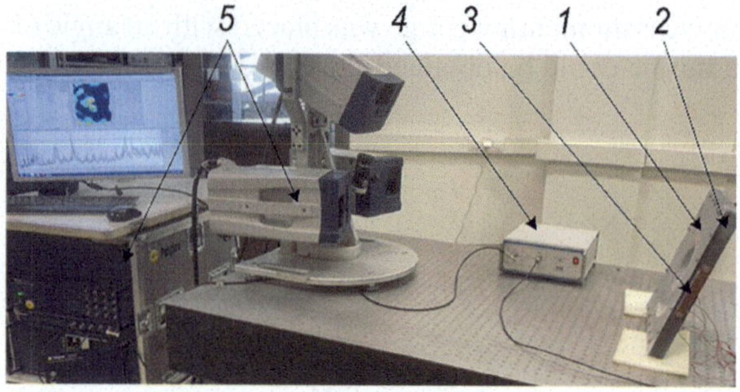

Fig. 2.82 3D scanning set-up: 1—an experimental body—aluminum alloy sheet, 2—steel base frame, 3—piezoelectric actuator, 4—liner amplifier P200 (FLC Electronics AB, Sweden), and 5—3D scanning vibrometer PSV-500-3D-HV (Polytec GmbH, Germany)

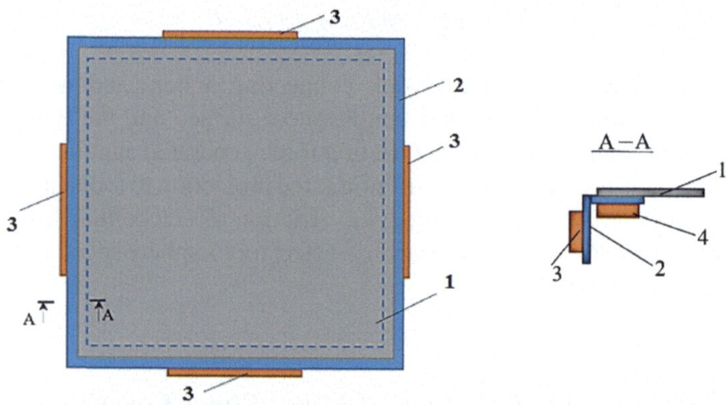

Fig. 2.83 Scheme of excitation of ultrasonic 3D vibrations in the sheet: 1—metal sheet; 2—frame; 3 and 4—bimorph type piezoelectric actuators

scheme is presented in Fig. 2.83, was developed to excite 3D ultrasonic frequency vibrations in SPIF sheet metal. This equipment consists of an angled profile metal frame 2, to which piezo ceramic elements 3 and 4 are glued in different planes, to excite bending oscillations of the higher harmonics in the frame. Since the sheet metal is rigidly attached to the frame by its outer contour, the vibrations of the frame also excite the sheet metal. This principle of ultrasonic 3D vibrations excitation in the sheet metal makes it possible to reduce the friction between the sheet metal and the tool during SPIF, to improve the lubrication conditions at the tool–workpiece contact pair, and in some cases, to completely eliminate the lubrication. From 2 to 8 piezo ceramic elements in the rectangular can be used to excite the 3D vibrations in the sheet metal according to the given scheme (Fig. 2.84) by alimenting them with harmonic ultrasonic frequency electrical signals. Investigated sheet 1 is clamped to the frame and it is excited with two piezoelectric transducers. Power amplifier 2 is used to generate the vibrations for piezoelectric transducer. While the robot hand moves the

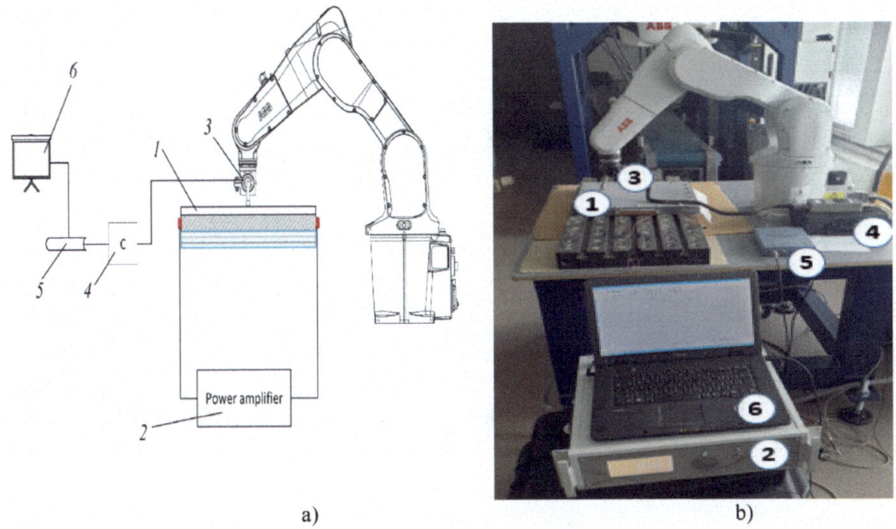

Fig. 2.84 Experimental set-up scheme (**a**) and photo (**b**) for sample excitation: 1—aluminum alloy sheet, 2—power amplifier, 3—hemispheral forming tool in robot hand, 4—sensor controller Keyence LK G3001P, 5—oscilloscope PicoScope-3424, and 6—PC

tool with the sphere attached to it incrementally forms the sheet. The tool is specifically attached with a 30° angle, so that the Mechanical momentum gauge STJ100 (Series BGI Digital Force Gauge with torque sensor STJ100, Mark-10 Corp) 3 could measure both pressing and friction momentum. The gauge is connected to Sensor Controler Keyence LK G3001P 4 which is connected to Oscilloscope (PicoScope-3424) 5 which writes the results received from the Mechanical moment sensor to the computer 6.

Experimental studies on a 0.5 mm thick aluminum plate show that the maximum reduction in friction forces at the tool–workpiece contact pair was observed when the piezoelectric actuators were excited in the 25–35 kHz frequency range. The sheet metal vibrational analysis with the Polytec scanning vibrometer PSV-500-3D-HV shows that in this frequency band the workpiece is dominated by planar (*XY*) higher frequency harmonic vibrations, which are significantly less suppressed in the tool–sheet metal pair friction contact than in the perpendicular to the metal sheet in the *Z*-direction (Fig. 2.85). These results show that when two piezoelectric actuators in different planes are excited, the in-plane (*XY*) oscillations in the 30–33 kHz band are greatly enhanced [24].

During the SPIF process, the tool moves by deforming the aluminum alloy sheet, so the amplitudes and frequencies of the eigenmodes of this sheet are influenced by the position of the tool and forming depth. Experimental studies were performed to evaluate the influence of the forming tool contact locations (1 and 2 in Fig. 2.86) on the dynamics of the ultrasonically activated sheet.

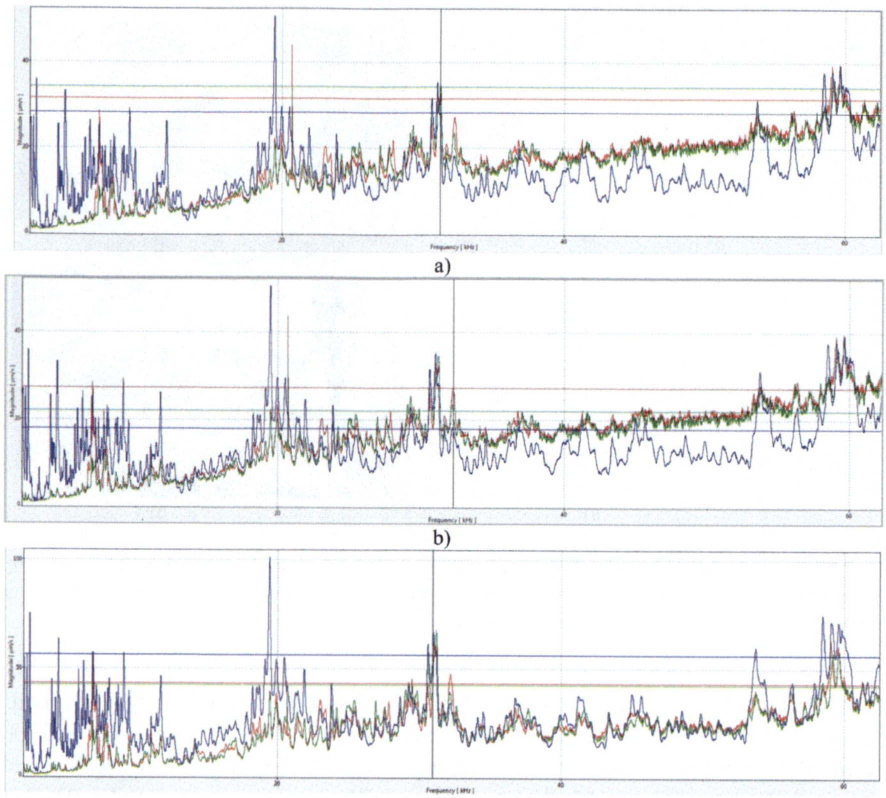

Fig. 2.85 The amplitude–frequency characteristics of the sheet metal measured with a Polytec 3D scanning vibrometer in the frequency range from 2 to 60 kHz after excitation of two differently arranged piezoelectric actuators (according to Fig. 2.84): **a** actuator 4 is excited, **b** actuator 3 is excited, and **c** both actuators are excited, where *X* (red) and *Y* (green) are lateral vibrations, and *Z* (blue) are vibrations in the normal direction to a sheet, respectively

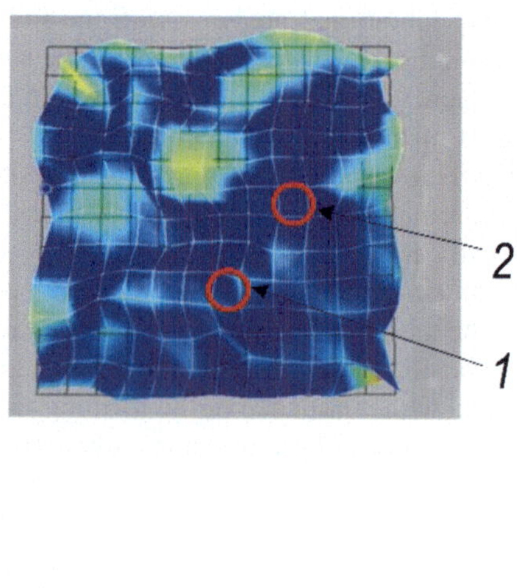

Fig. 2.86 An aluminum alloy sheet vibrations at an excitation frequency of 30 kHz and forming tool contact with a sheet location: 1—bottom; 2—on the side

Table 2.13 Mechanical properties and chemical composition of aluminum alloy AW1050 used in experiments

Parameter	Value	Unit
Proof stress	85 min	MPa
Tensile strength	105–145	MPa
Hardness Brinell	34	HB
Elongation A	12 min	%
Density	2.71	kg/m^3
Melting point	650	°C
Modulus of elasticity	71	GPa
Electrical Resistivity	0.282×10^{-6}	Ω m
Thermal conductivity	222	W/m K
Thermal expansion	24×10^{-6}	/K

An aluminum alloy AW1050 (mechanical properties and chemical composition of aluminum alloy presented in Table 2.13) sheet with dimensions of 350 × 350 × 0.5 mm is rigidly attached to the frame and one side is facing three laser scanning heads (Fig. 2.82). On the other side of the metal sheet, a forming tool with a steel sphere of 8.5 mm radius is in contact with a sheet at locations 2 presented in Fig. 2.86. Some vibrations of an aluminum alloy sheet are excited in the frequency range from 0.5 to 60 kHz with the piezoceramic transducers attached to a frame. Thereby, the 3D frequency response and deformations of an aluminum alloy sheet are obtained.

The presented vibrograms (Fig. 2.87) show that regardless of the location of the tool in contact with the metal sheet and the forming depth, the amplitudes of the *Z* eigenmodes perpendicular to the sheet surface are significant, while the influence of the tool location on the amplitudes of the *X* and *Y* natural oscillation modes in the 2D plane is insignificant.

The experimental results have shown that the SPIF process is most effective in the frequency range from 28 to 36 kHz, where the lateral vibrations in the *XY*-direction of the sheet dominate over the normal Z-direction vibrations of the sheet. The effect of ultrasonic vibration on the sliding friction of aluminum alloy specimens in sliding tool steel has been studied in [25]. A significant reduction in sliding friction (up to > 80%) was observed and good agreement was found between the measured values and the predictions of two simple models for the effects of longitudinal and transverse vibrations. Ultrasound not only reduces friction between the tool and the workpiece but in our developed technology, ultrasound also makes it easier for the tool to slide on the sheet surface. Ultrasonic-assisted friction reduction is well known in the field of metal-to-metal contacts. Due to the vibration, the stick phase in the contact phase vanishes and only sliding occurs. As long as the macroscopic relative velocity of the contact partners is much lower than the vibration velocity, the force required to move the parts tends to (nearly) zero. This finding further reaffirms that ultrasonic excitation of the sheet in the 2D plane can have a significant impact on the efficiency of the SPIF process.

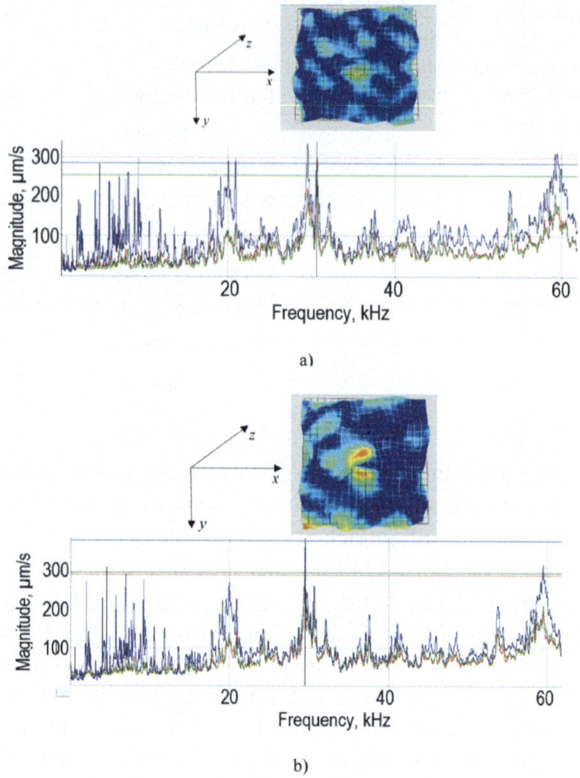

Fig. 2.87 3D frequency response and deformations of an aluminum alloy sheet in the frequency range from 0.5 to 60 kHz measured with a Polytec PSV-500 3D laser Dopler vibrometer scanner: *X* (red) and *Y* (green) are lateral vibrations, *Z* (blue) in the normal direction, respectively: influence of tool position and forming depth on the eigenmodes of the sheet at tool position 1, forming depth: **a** tool position 1, forming depth 15 mm; **b** tool position 2, forming depth 30 mm

The mechanical (friction) torque meter STJ100 (BGI series digital force gauge with torque sensor STJ100, Mark-10 Corp., USA) measured that the sensitivity factor was 6 Nm/V during the experiment. The results of the tool friction coefficient and friction force measurements on dry, lubricated, and ultrasonically excited surfaces in the 30–33 kHz frequency range, which reduces the friction force at the tool–sheet metal contact pair, are shown in Table 2.14.

Table 2.14 shows that the coefficient of friction between the steel tool and the aluminum alloy sheet is close to that of the lubricated surfaces when subjected to ultrasonic vibrations. This makes it possible to solve environmental problems, and the surface of the manufactured part does not need to be cleaned, while at the same

Table 2.14 Friction coefficient and friction force measurement results

Method	Friction force (N)	Friction coefficients
Without lubrication and vibration	3,2	0,5
With lubrication	1,6	0,1
With vibration	1,9	0,12

time its roughness is reduced by half. This revealed phenomenon has been patented by the authors [26].

2.5.2.1 Development of Metal Sheet SPIF Virtual Twin

A 0.5 mm thick sheet of aluminum alloy EN AW1050A H24 was used for the experimental and numerical analysis. Material properties taken from supplier datasheet are presented in Table 2.15.

As the failure strain is a stress state-dependent parameter [27, 28], and the value obtained in the uniaxial test may differ from the failure strain in the SPIF process, the simple cupping was selected for the calibration of the material model. In the case of the cupping test, the biaxial tension is dominant and the failure strain is larger than in the uniaxial tensile test. During the incremental forming process, the strain path is nonlinear [29]; therefore, not only the uniaxial failure strain but also the conventional forming limit diagram criterion can cause inaccurate failure prediction.

The stress–strain relationship needed for the numerical analysis was obtained by reverse engineering approach using data from physical cupping test. It was chosen to describe the material behavior by the Power-Law material model (Fig. 2.88) to decrease the number of material model constants needed to calibrate. Using the range of material properties from the material datasheet (yield stresses, ultimate strength range, and failure strain), the Power-Law parameters such as the strength coefficient K, the hardening exponent n, and the failure strain were calibrated using LS-Opt system. The input range of the stress–strain curves for the Power-Law models used in LS-Opt is presented in Fig. 2.88, where the solid line represents the calibrated Power-Law material curve. The Power-Law equation is provided below:

$$\sigma = K\varepsilon^n = 143\varepsilon^{0.097}. \tag{2.55}$$

Table 2.15 Mechanical properties of aluminum alloy EN AW1050A H24 from the datasheet

	Modulus of elasticity, GPa	Proof stress, MPa	Tensile strength, MPa	Elongation A, %	Strength coefficient K, MPa	Strain hardening coefficient n
Datasheet Values	71	>85	105–145	3–8		
Range for curve fitting				2–40	140–200	0.05–0.2
Calibrated values LS-Opt & LS-Dyna				37.9	143	0.097

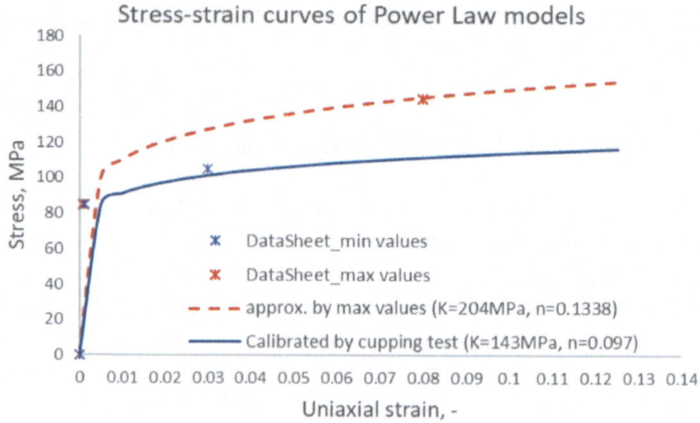

Fig. 2.88 Stress–strain curves of the Power-Law model

2.5.3 Cupping Test for the Material Model Calibration

In order to develop a validated numerical model of the SPIF process, several sets of experimental tests were performed. Firstly, the material model was calibrated by reverse engineering using experimental results of the cupping test. Subsequently, the same material model was used for the SPIF simulation. In order to characterize the plastic behavior and ductile fracture of the aluminum alloy sheet, the Erichsen cupping tests on square aluminum plates were performed in accordance with ISO 20482: 2013 (Metallic materials—Sheets and strips—Erichsen tensile test). The Erichsen standard provides information on fracture under the equibiaxial state of stress. The results are influenced by the sheet thickness and the friction between the sheet and the tool surface. For the Erichsen compression test, the specimen was attached to a 55 mm die with a heavy flange and bolts tightened to achieve a force of approximately 10 kN on the workpiece holder (Fig. 2.89). The radius of the hemispherical punch was 10 mm and the test speed was 5 mm/min. The experiment was carried out by applying a hemispherical punch to a sheet of metal until a crack appeared.

The force–displacement response was obtained by plotting the reaction force on the punch versus the displacement of the punch (Fig. 2.90).

The maximum load for all 5 cupping tests was $F_{\max} = 1160$ N, and the Erichsen index IE = 5.36 mm. The force versus displacement curve was chosen as the validation criterion for the numerical model.

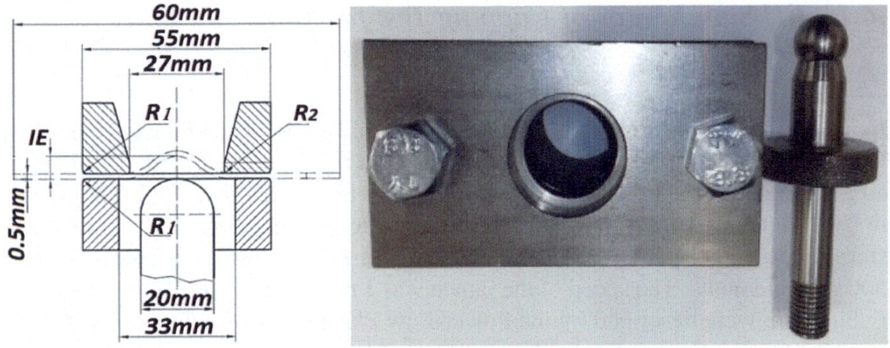

a) b)

Fig. 2.89 Erichsen cupping test: **a** scheme and **b** test die

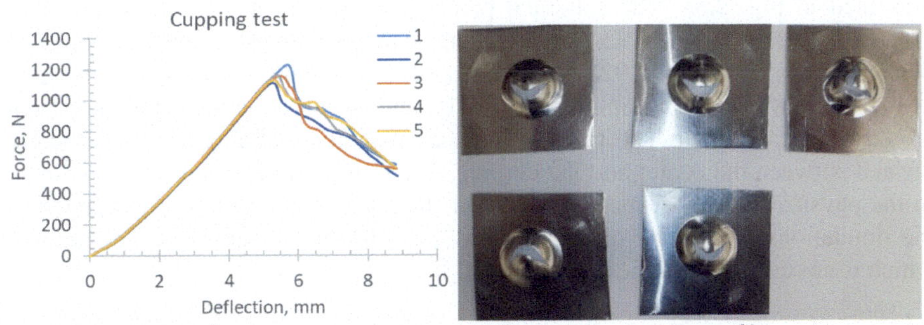

a) b)

Fig. 2.90 Erichsen cupping test results: **a** cupping curves for specimens 1–5 and **b** nature of the failure of the specimens 1–5 after the cupping test

2.6 FE Simulations

The strain path, the stress state, and the prediction of the formability of the SPIF were evaluated by FE analysis. In all subsequent simulations, the explicit FE code LS-Dyna was used. The aluminum sheet was modeled using fully integrated shell elements with thickness stretch allowing. Through the thickness a linear variation of strain was evaluated by 5 integration points and a shear correction factor of 0.833 was used. The contact between the punch/tool and the aluminum sheet was described using the keyword: *CONTACT_FORMING_ONE_WAY_SURFACE_TO_SURFACE. There were no damage models involved, only a simple failure model by maximum failure effective strain criteria has been used.

2.6.1 FE Simulations of Cupping Test for Material Model Calibration

To simulate the cupping test and calibrate the material model. The FE model consisted of four parts: a deformable aluminum sheet and a rigid spherical punch, a bottom, and a top holder. The holders were compressed with a force of 10 kN. The size of the shell elements varied between 0.5 at the center/contact zone and 1 mm at the external side of the specimen. The FE model (Fig. 2.91a) consists of approximately 8500 shell elements. The speed of the punch was 1 m/s. The elasto-plastic properties of aluminum were described by the Power-Law Plasticity material model and three parameters were chosen for calibration: strength coefficient K, Strain hardening coefficient n, and failure strain ε_u. The simulation results are presented in Fig. 2.91b.

After curve fitting, the experimental force–displacement relation and the FE simulation prediction (Fig. 2.91b) correlated well. Calibrated material model curve is presented in Fig. 2.88. The spherical punch penetration into an aluminum alloy sheet is illustrated in Fig. 2.92. Comparing the fracture paths also reveals a common behavior (Figs. 2.89b and 2.92). The most highly stressed point is located at the apex of the dome where the components of radial and circumferential stress and strain are equal to each other. The fracture lines are diagonal and almost symmetrically located between the center and the corners. Three diagonal fracture lines dominate in the physical test experiments while four diagonal fracture lines are obtained in the simulations. One of the reasons for this could be the anisotropy of the material, which is not evaluated in the FE model.

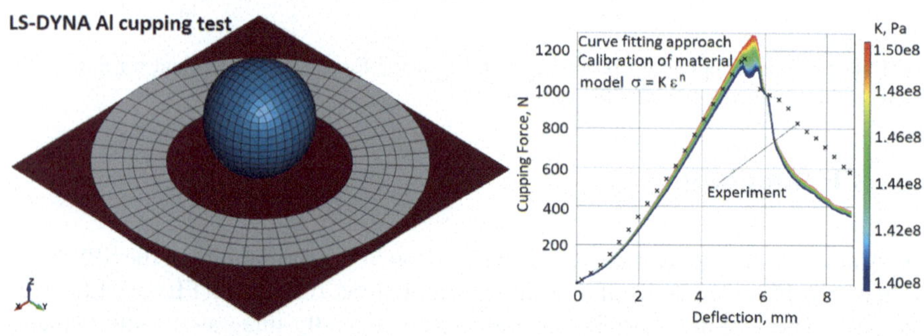

Fig. 2.91 Modeling of the cupping test: **a** numerical model of the cupping test, **b** force–displacement plot of the cupping experiment and sensitivity of the simulation results to the strength coefficient K of calibrated Power-Law material model (strain hardening coefficient $n = 0.097$ and failure strain $\varepsilon_u = 0.38$)

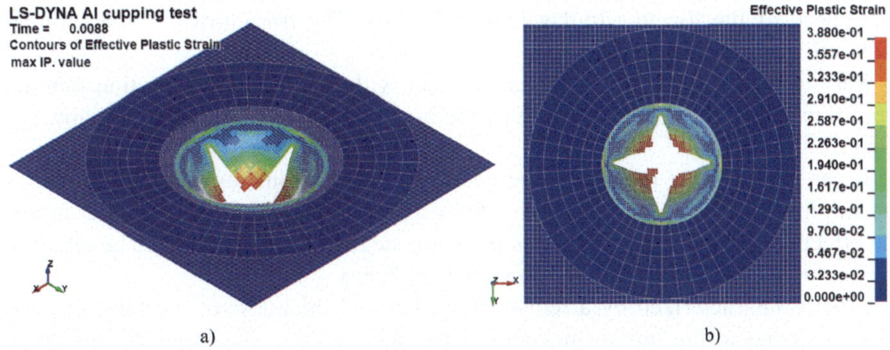

Fig. 2.92 Simulated images of a spherical punch piercing an aluminum alloy sheet during the Erichsen test: **a** isometric view and **b** at the bottom of the sheet

2.6.2 Numerical Simulations of SPIF Process

The SPIF model consisted of two parts: an aluminum sheet and a forming tool (Fig. 2.93). The aluminum sheet was modeled in the same way as for the cupping test. The size of the shell elements varied between 1 at the center/contact zone and 5 mm at the external side of the specimen. The FE model consisted of about 20,000 shell elements. The main features of the FE model were copied from the calibrated cupping test model. The elasto-plastic properties of aluminum were described by the Power-Law Plasticity material model calibrated with cupping test (Fig. 2.89). The forming tool was modeled as a rigid body using shell elements. The aluminum sheet holder was simulated with fixed nodes (25 mm wide around all 4 edges). The major diameter of the helix was 140 mm.

The load was applied by three functions of tool displacements: *f*(*x*), *f*(*y*), and *f*(*z*). All functions were continuous and varied so that the horizontal displacement per revolution was 0.5 mm and the vertical displacement per revolution was 0.5 mm.

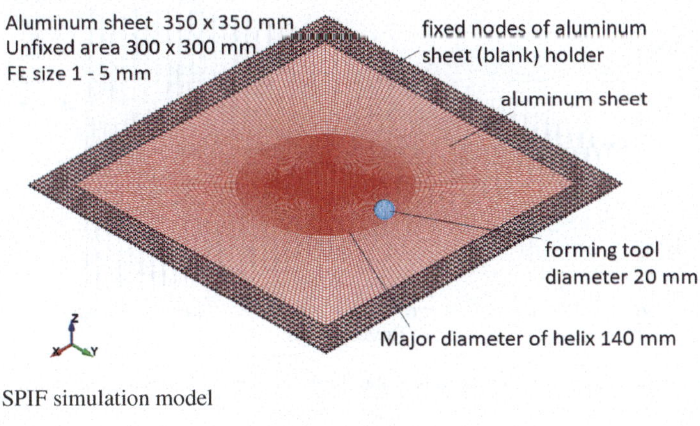

Fig. 2.93 SPIF simulation model

The speed of the forming tool was about 2 m/s. The trajectories of the forming tool-path are shown in Fig. 2.94.

Various simulations have been carried out with coefficients of friction ranging from 0 to 0.5. The effect of friction on the horizontal force component is shown in Fig. 2.95.

It can be seen that increasing the coefficient of friction from 0 to 0.5 increases the amplitudes of the X force component by a factor of approximately 2. In contrast, the effect on the vertical Z force component is negligible (Fig. 2.96a). The effect of friction on the resultant force is shown in Fig. 2.96b.

SPIF is characterized by a reduction in the wall thickness of the final finished part compared to the initial thickness of the sheet metal. The excessive thickness reduction ratio in the deformation zones when the sheet metal is formed separately has a significant effect on the forming limit. The prediction of the thickness of the deformation zone is an important approach to control the thinning ratio. With regard to the object of study, aluminum alloy, the principle of thickness deformation in the

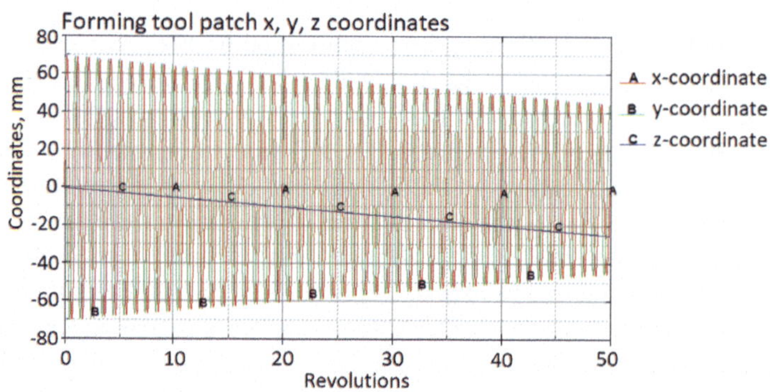

Fig. 2.94 Loading curves displacements in XYZ-directions

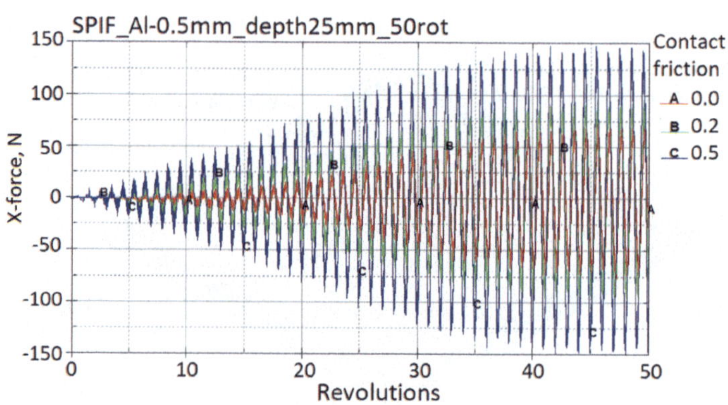

Fig. 2.95 X-contact force versus rotations at the different contact friction coefficients

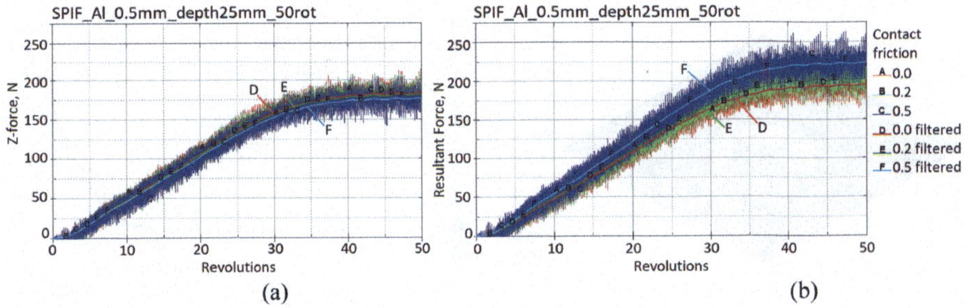

Fig. 2.96 Contact force versus rotations for different coefficients of contact friction: **a** z-force (vertical force) and **b** resultant force

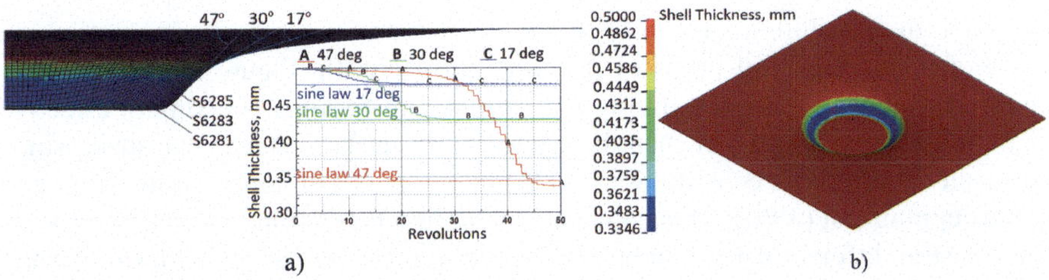

Fig. 2.97 Shell thickness reduction: **a** variation at elements which after the forming depth of 25 mm has the drawing angles of 47°, 30°, and 17°, **b** shell thickness distribution after the forming depth of 25 mm

SPIF process is presented in Fig. 2.97. The relationship between the wall thickness and the drawing angle α can be expressed by the sine law used in SPIF:

$$t_f = t_0 \sin(90 - \alpha), \tag{2.56}$$

where t_0—is the initial thickness, t_f—is the final thickness, and α—is the drawing angle between the initial flat surface and the deformed surface.

The drawing angle obtained in simulation varies from 47°, therefore, according to the sine law, this would lead the reduced thickness up to 0.34 mm. In contrast, the minimum thickness in the simulation was 0.335 mm. The results obtained from the SPIF simulation have a good correlation with the sine law. The strain distribution after the forming depth of 25 mm is presented in Fig. 2.98a. The shell element 6285 of the aluminum alloy sheet gets in contact with forming tool at approximately 35 revolutions. In Fig. 2.98b, it is shown that a significant increase in effective deformations begins after contact with the tool and increases further tool rotation. Figure 2.98b shows that a significant increase in effective strains starts after contact with the tool and increases with further rotation of the tool.

Figures 2.97 and 2.98 show that the decrease in sheet thickness during SPIF correlates with effective plastic strain changes. The thinning of the forming sheet is

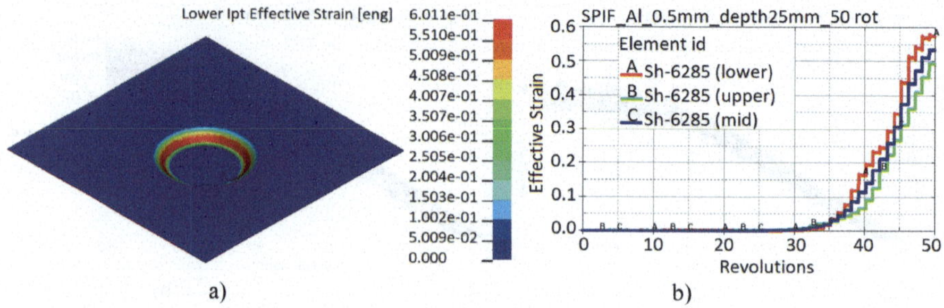

Fig. 2.98 Effective strain distribution at the forming depth of 25 mm (**a**) and its evolution in shell element 6285 during the SPIF simulation process (**b**)

one of the main failure modes in SPIF and is related to the drawing angle α which is one of the geometrical limits of the SPIF process [30]. Figure 2.98 shows that the effective strains achieved during the incremental forming (maximum effective strain 60%) are much higher than the values given in the material datasheet, since the tensile failure strain constitutes (3–8) % and the values of failure strain calibrated by the cupping test (38%). This confirms that failure prediction in the SPIF process is not so straightforward and that other failure criteria should be applied. Comparing the evolution of stress triaxialities during the cupping test and SPIF (Fig. 2.99), it is evident that during the cupping test, the stress triaxiality is stable and equal to the 2/3, which corresponds to the equibiaxial tension stress state, while the stress triaxiality during the SPIF process varies from -0.6 up to 0.6. It confirms that the stress state during the SPIF process varies from plane strain compression up to plane strain tension. While at plane strain tension state, the failure strain of material has a smaller value compared with failure strains at the other stress states, higher formability can be achieved during the SPIF process.

Deformation phenomena during SPIF are still under discussion. In the absence of direct experimental evidence, some researchers have claimed that deformation is caused by stretching, while the others state that it is caused by through-thickness shear [31, 32]. If out-of-plane shear dominates, the principal strains are not in the plane [33];

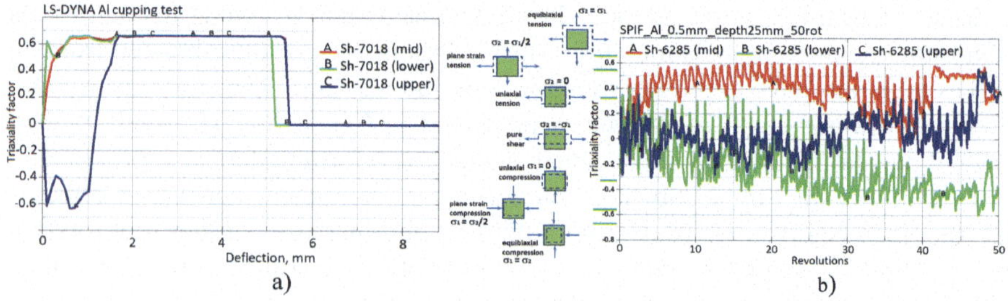

Fig. 2.99 The stress triaxiality evolves in certain elements that come into contact with the forming tool during: **a** cupping test and **b** SPIF

therefore, the use of shell elements can be limited, as shell elements typically do not allow capturing through-thickness shear properties in simulation. Fully integrated shell elements with a thickness stretch in LS-Dyna allow the capture of through-thickness shear stresses and shear strains. Three elements (S2681, S2683, S2685 see Fig. 2.15a) along the x-axis were selected to analyze the stresses and strains during the simulation of the SPIF process. The strain distribution through the thickness is shown in Fig. 2.100.

The results show that the through-thickness shear strains have significant values compared to the in-plane (*xy*) shear strains. In particular, the transverse through-thickness deformation of element No. 6285 accounts for about 50% of the in-plane shear deformation.

The simulation of vibrational excitation of the aluminum alloy sheet sample was performed by piezo actuators attached to the steel frame using Comsol multiphysics software. The results of experimental and theoretical studies for the first embedded sheet in steel frame mode, presented in Fig. 2.101, show a good correlation between

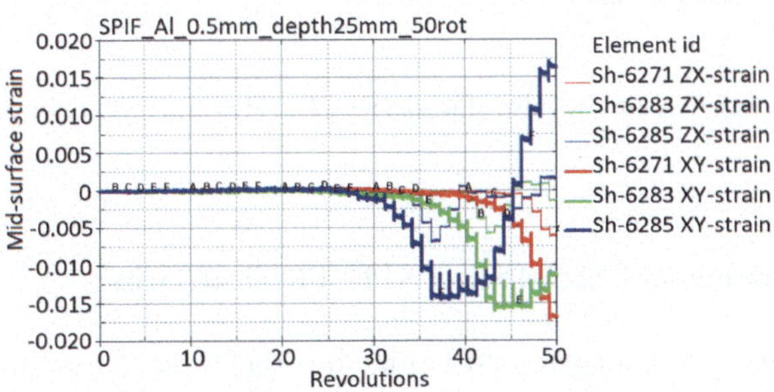

Fig. 2.100 Variation of in-plane and out-of-plane shear strains during the SPIF process

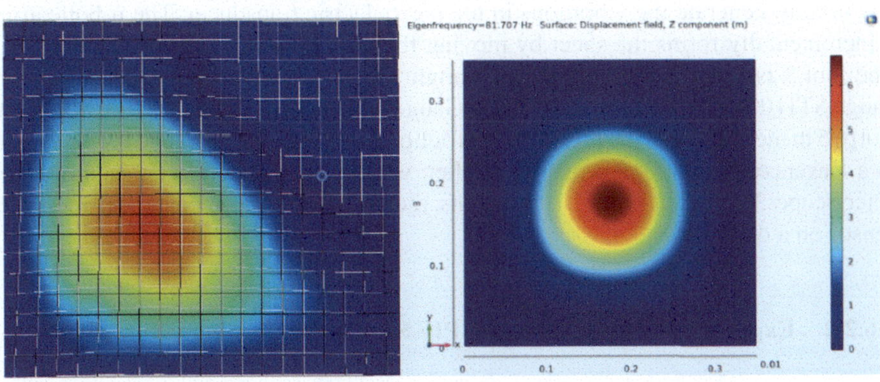

a) b)

Fig. 2.101 The first modes of transverse vibrations of an aluminum alloy sheet fixed in a steel frame: **a** measured at a frequency of 78.8 Hz and **b** calculated at a frequency of 81,707 Hz

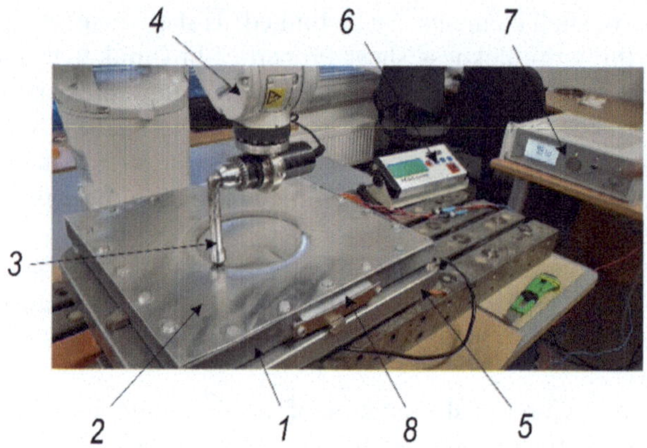

Fig. 2.102 Experimental set-up of incremental aluminum alloy sheet forming: 1—frame for aluminum sheet, 2—aluminum alloy sheet, 3—forming tool, 4—robot ABB IRB1200, 5—pressing force sensor FCS-4035-150, 6—controller of pressing force sensor, 7—power amplifier, and 8—piezoelectric actuators for ultrasonic excitation

the dynamic properties of the simulated (81.707 Hz) and experimental (78.8 Hz) models.

2.6.2.1 Development of Metal Sheet SPIF Physical Twin

In order to find a more efficient control of the forming force, an attempt was made to excite the aluminum alloy sample by ultrasonic vibrations. The experimental set-up created for this purpose is shown in Fig. 2.102. An investigated sheet 2 is clamped into frame *1* and is excited with two piezoelectric transducers 8. A power amplifier 7 is used to generate the vibrations in the piezoelectric transducer. The robotic arm 4 incrementally forms the sheet by moving the tool with the sphere attached to it. The tool 3 is specially mounted at 30° angle to allow the Mechanical momentum gauge STJ100 (Series BGI Digital Force Gauge with torque sensor STJ100, Mark-10 Corp) 5 to measure both the pressing and friction momentum. The gauge is connected to a Keyence LK G3001P sensor controller, which is connected to an oscilloscope (PicoScope-3424) that records the results received from the mechanical moment sensor on a display *6*.

2.6.2.2 Experimental Validation of SPIF Simulation Results

The SPIF experiment was performed under dry, lubricated, and vibration-excited friction at the tool–sheet metal contact surfaces. Vertical force dependences on the process conditions are graphically presented in Fig. 2.103.

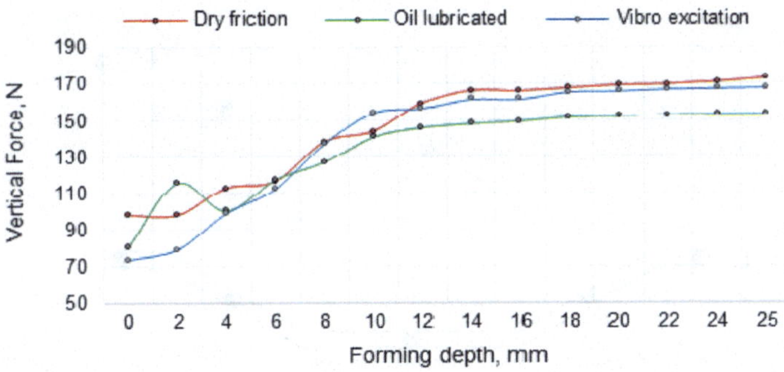

Fig. 2.103 Graphical representation of vertical force dependency from different friction when sheet thickness is 0.5 mm

Fig. 2.104 Graphical representation of the puncture force of an aluminum sheet

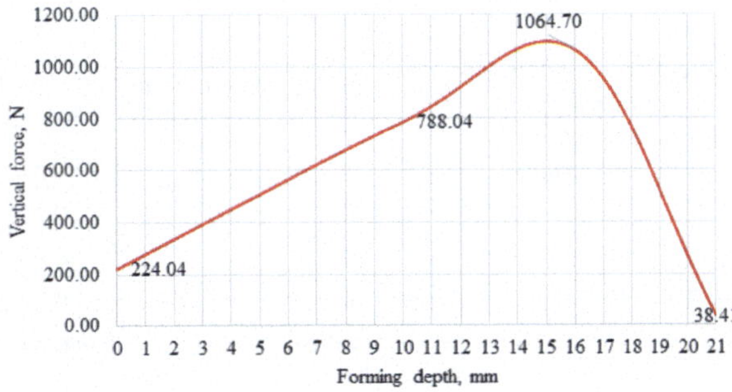

Accordingly, to verify modeling results, presented in Fig. 2.91b, the puncture forces of the sheet metal during the experimental investigation are shown in Fig. 2.104. It has been observed that at a depth of 0 mm, the vertical force is equal to 224.04 N; at a depth of 10 mm, the vertical force 778.04 N; at a depth of 16 mm, the vertical force is 1064 N, but when the depth reaches the value of 15 mm, the vertical force starts to decrease significantly.

Obtained results—curve and magnitude of puncture force—showed good correlation with modeling results, presented in Fig. 2.91b; thus, theoretical model is verified.

In order to validate the simulated sheet thickness reduction adequacy for the experimentally obtained SPIF profile, the wall thickness of the formed aluminum specimen was measured at several points of the cross-sectional cut (Fig. 2.105) using Topex 31C629 μm with 0.01 mm accuracy.

Measurement results—wall thickness of incrementally formed aluminum sheet at measurement points—are presented in Table 2.16.

The results of measurement and modeling (Fig. 2.98a) correlate with each other.

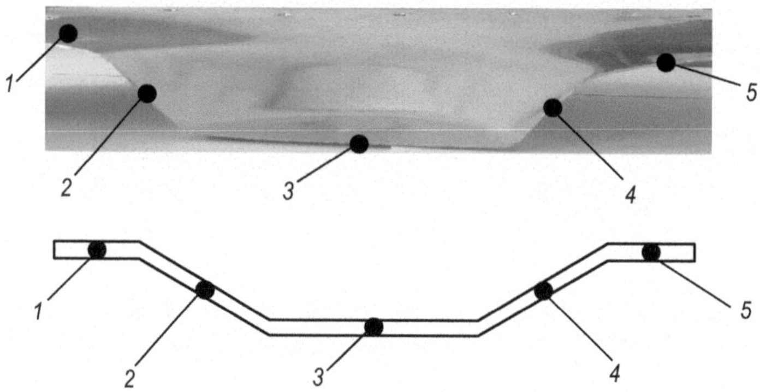

Fig. 2.105 Section of the incrementally formed aluminum specimen with the schematical representation of measurement points

Table 2.16 Thickness of incrementally formed aluminum sheet at measurement points

Measurement point	Wall thickness (mm)	Δ from initial thickness (mm)
1	0.51	0.01
2	0.36	0.14
3	0.49	0.01
4	0.37	0.13
5	0.51	0.01

2.6.2.3 Physical Twin Experimental Data Exploration

The data should be analyzed to determine whether the data collected are free of data quality problems that could adversely affect the intended prediction models [34]. Common issues such as missing values and outliers should be calculated because it is impossible to train error-based models with data that contains missing values. Furthermore, data that contains outliers can provide incorrect predictions. As depicted in Table 2.17, there are two types of features: continuous numeric and categorical binary. Two features, namely "Tool end diameter" and "Wall angle" have constant values; therefore, they must be eliminated.

First of all, data exploration should be performed in order to determine whether or not the collected data suffer from any data quality issues that could negatively affect prediction models that are intended to build. No outliers or missing data have been identified in the data set. After the first iteration of data quality issues analysis, the data set contains five inputs—raw features, that come directly from data sources, and one output—vibro excitation of the sheet (Table 2.18).

The SPIF experiment was performed under dry, lubricated, and vibration-excited friction on the contact surfaces of the tool and the sheet metal, and the vertical force dependences on the process conditions are given in Table 2.19.

Table 2.17 Parameters for data exploration

No	Input parameters	Units	Value (min, max)
1	Forming depth	mm	0…26
2	Tool End diameter	mm	20…20
3	Step depth, (Δz)	mm	0.25…0.5
4	Step width, (Δx)	mm	0.25…0.5
5	Wall angle	°	45…45
6	Sheet thickness, (Θ, °)	mm	0.5…0.8
7	Tool type		Rotating sphere on the end/not rotating
8	Vertical Force Component (VFC) Dry friction	N	4.91… 343.35
9	VFC Oil lubricated	N	1.96…341.39
Output parameter			
1	VFC Vibro excitation	N	4.91… 338.45

Table 2.18 Training parameters

Input parameters				
Forming depth	Step depth, (Δz)	Step width (Δx)	Sheet thickness (Θ)	Tool type
Output parameters				
VFC Vibro excitation				

Table 2.19 Vertical force dependence from different friction when sheet thickness is 0.5 mm

	s = 0.5 mm, steps 0.5 and 0.5 mm, velocity varied to 100%		
Forming depth, mm	Vertical force, N		
	Dry friction	Oil Lubricated	Vibro excitation
0	98.10	80.93	73.58
2	98.10	115.76	79.46
4	112.82	100.55	99.08
6	116.74	117.72	112.82
8	138.32	127.04	137.34
10	144.21	141.07	154.51
12	159.41	146.17	156.96
14	166.28	148.33	161.87
16	166.08	149.60	161.87
18	167.75	151.86	165.79
20	169.22	152.35	166.28
22	169.71	152.84	167.26
24	171.18	153.23	167.75
25	173.15	153.33	168.24

Next, it is reasonable to calculate the correlation coefficients r, which indicate the strength of the linear relationship between the two features. The values of the Spearman correlation coefficients [35] vary between -1 and 1, whereas if $r = 0$, then the variables have no relationship; the closer the coefficient is to $+1$ or -1, the stronger the relationship. The sign indicates whether the relationship is positive or negative, e.g. if $r = 1$, then the two features have an ideal positive relationship. A coefficient close to 0 shows a weak correlation. It has been noted that Step depth and Step width have the same values and a correlation coefficient $r = 1$. It is, thus, reasonable to remove one of these two features. No outliers or missing data were identified in the data set:

$$r = \frac{\frac{1}{n}\sum_{i=1}^{n}\left(\left(R(x_i)-\overline{R(x)}\right)\cdot\overline{R(y_L)-(R(y))}\right)}{\sqrt{\left(\frac{1}{n}\sum_{i=1}^{n}\left(R(x_i)-\overline{R(x_L)}\right)^2\right)\cdot\left(\frac{1}{n}\sum_{i=1}^{n}\left(R(y_i)-R\overline{(y_L)}\right)^2\right)}}, \tag{2.57}$$

where $R(x)$ and $R(y)$ are the ranks of the x and y variables; $\overline{R(x)}$ and $\overline{R(y)}$ are the mean ranks.

The correlation coefficient between "Tool type" and the output "VFC Vibro excitation" is very low, because this input attribute is binary, having only two possible values (rotating sphere on the end; not rotating); therefore, it is eliminated. After the first iteration of data quality issues management and analysis, the data set contains six raw features obtained directly from data sources. These correlation coefficients r are provided in the Spearman correlation matrix (heatmap) and presented in Fig. 2.106.

It has been observed that "VFC Dry friction" has the strongest correlation with the output value, $r = 0.998$. Meanwhile, step depth and sheet thickness, with values

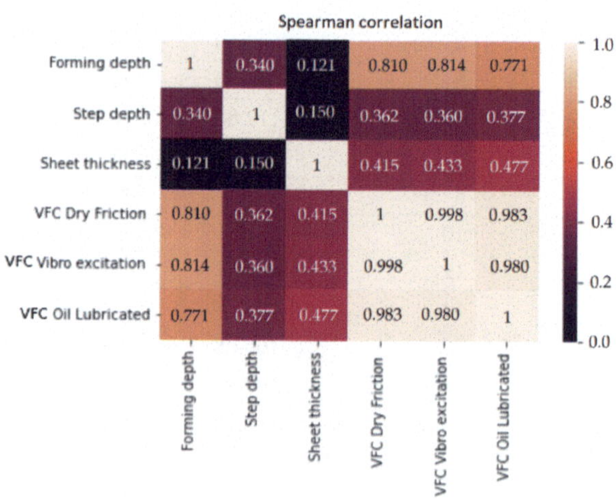

Fig. 2.106 Spearman correlation matrix and r coefficient

$r = 0.360$ and $r = 0.433$, respectively, have shown a moderate correlation with "VFC Vibro excitation". However, it is reasonable to include all six parameters for the further prediction investigation employing Artificial Neural Networks (ANN).

2.6.2.4 Machine Learning-Based Prediction

In the field of mechatronics and bioengineering, the size of experimental data sets is often insufficient; thus, the prediction task requires Machine Learning (ML) algorithms capable of generalizing the data properly. Simple models (such as linear regression and decision tree), feature selection, k-fold for cross-validation [36], ensemble learning, regularization, or, possibly, generation of synthetic data [37, 38] can be used for this purpose. A number of experimental studies have shown that ANN used for correlation analysis and prediction can yield good results even with a small sample of data [39, 40], but other ML algorithms, such as Support Vector Machine (SVM) or Random Forest (RF), are often used as well. In this study, five different supervised machine learning methods were used for the comparative analysis of the prediction results: Gaussian Process Regression (GPR); SVM; Decision Trees (DT); K-Nearest Neighbors algorithm (KNN); ANN. The obtained results confirmed that ANN is the most accurate method (according to Root Mean Square Error (RMSE)) in the framework of this task (Fig. 2.107a), although it is the least efficient in terms of training time. The k-fold cross-validation procedure shows that the RMSE values of all ML algorithms do not differ significantly. Evaluating the accuracy results, it can be seen that ANN and GPR provide similar performance, but the training time for both algorithms differs greatly. ANN has an average training time of 12.68 s which is significantly higher compared to GPR (Fig. 2.107b). This is because ANN has more parameters than GPR. ML algorithm may give different prediction accuracy and training duration each time, even when trained on the same data set. It's possible to reduce the variance of the ML algorithm by optimizing its hyperparameters.

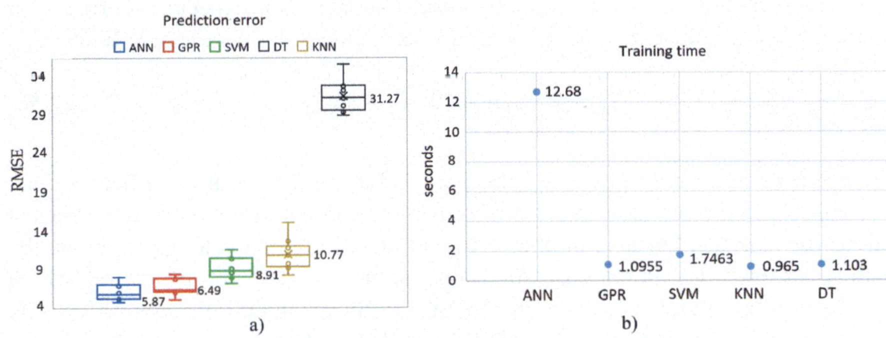

Fig. 2.107 Prediction results using different machine learning algorithms: ANN, GPR, SVM, KNN, and DT: **a** prediction error using k-folds cross-validation, where $k = 10$; **b** training time results

Hyperparameters, in contrast to ML model parameters, are set manually before the model starts training. Hyperparameters cannot be learned within the estimator directly; however, model parameters are properties of the training data that are estimated automatically. For example, the minimum leaf size in a decision tree, or kernel scale and function of SVM are hyperparameters while the weights in an ANN are model parameters learned during training. The choice of hyperparameters in the above models can strongly affect its performance; therefore, the optimization process allows to automatically find the optimal combination of hyperparameters for the ML algorithm [41]. As the result, an optimal model is provided, which reduces a predefined error value and, in turn, increases the accuracy of independent data.

Hyperparameter Optimization

Hyperparameter optimization has been performed on GPR, SVM, DT, KNN, and ANN models using Bayesian optimization [42]. Two other popular hyperparameter tuning algorithms are grid search and random search. Grid search is the simplest algorithm for hyperparameter tuning, which divides the domain of the hyperparameters into a discrete grid. Theoretically, this algorithm should find the best point in the domain, but practically is not used very often, because it is an exhaustive and time-consuming search. Random search, unlike grid search, does not search solution for every possible combination of hyperparameter values but tests only a randomly selected subset of these values. Instead of random searching in the hyperparameter domain, Bayesian optimization enables an intelligent manner of hyperparameters selection, because it uses the results from the previous iteration to decide what is the next set of hyperparameters, which will improve the model performance. Prioritizing hyperparameters is very efficient and allows for finding the best values of hyperparameters' sets much faster compared to both grid search and random search.

The Bayesian optimization method for the tuning of hyperparameters employs the acquisition function with the purpose to determine the next set of hyperparameter values. There are many different acquisition functions such as upper confidence bound, entropy search, probability of improvement, and expected improvement, but the last two functions are most commonly used. In general, the expected improvement function evaluates the expected amount of improvement in the objective function:

$$EI(x) = E\left[\max\left(0, f' - f(x)\right)\right], \tag{2.58}$$

where f' is the minimum value of f observed so far; x is the location of that sample.

The performance of such an optimization process depends not only on the chosen acquisition function but also on the surrogate model that helps to approximate the main target functions. In our case, the Gaussian process (GP) has been used, which is the most often preferred choice. In general, the Bayesian optimization follows the sequence of four cycle steps: 1—use the Bayes rule to obtain the posterior; 2—choose a surrogate model; 3—use an acquisition function to decide the next sample point; 4—add new data to the set of observations and go to step 2.

Four hyperparameters, i.e. sigma value, basic function, kernel function, and kernel scale, have been included in the optimization process of the GPR model. The kernel function plays a significant role because the choice of kernel functions determines almost all the generalization properties of the GPR model. The sigma value σ is selected within the range calculated by Eq. (2.59).

$$\sigma = [\min;\ \max] = \left[0.0001;\ \max(10 \times \sqrt{\sum_{i=1}^{n} \frac{(y_i - \overline{y})}{n-1}}\right], \tag{2.59}$$

where $\overline{y}$ is a sample mean (output sample mean), y_i—the value from the output sample, and n—sample size.

The GPR model kernel scale optimization possibility depends on the kernel function. For the no-isotropic kernel function, the number of the kernel scale l is usually equal to the number of inputs. For isotropic kernel functions, the kernel scale l is selected from a range of values calculated according to the following equation:

$$l = [\min;\ \max] = [0.001(\max(X) - \min(X);\ (\max(X) - \min(X))], \tag{2.60}$$

where max (X)—a maximum value from the input variable matrix; min (X)—a minimum value from the input variable matrix.

	Sigma	Basic function	Kernel function	Kernel scale
Value range	(0.0001; 948.39)	Constant; zero; linear	Isotropic and No-isotropic exponential; Isotropic and No-isotropic Matern 3/2 and 5/2; Isotropic and No-isotropic rational quadratic; Isotropic and No-isotropic squared exponential	(0.33943–339.43)

Different accuracy measures have been calculated from the experiments: Mean Squared Error (MSE), RMSE, and Mean Absolute Error (MAE) [43].

MSE is a measure representing the average of the squared difference between the real and predicted values of the data set. RMSE is simply the square root of the MSE, the only difference being that MSE measures the variance of the residuals, while RMSE measures the standard deviation of the residuals.

$$\text{RMSE} = \sqrt{\text{MSE}},\ \text{where MSE} = \frac{1}{n}\sum_{t=1}^{n} \left|y_t - \hat{y}_t\right|^2. \tag{2.61}$$

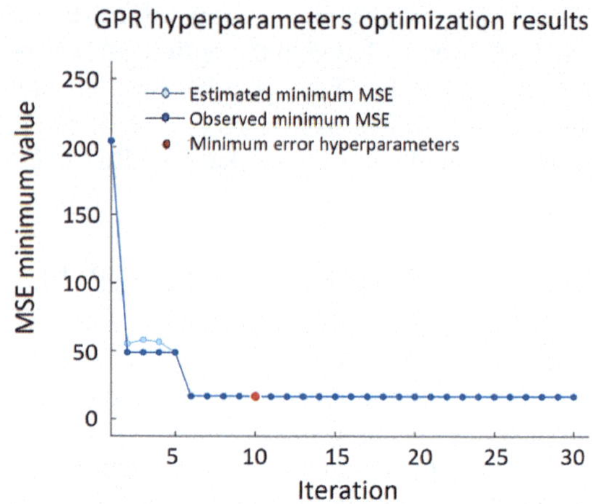

Fig. 2.108 Minimum MSE during GPR hypermeters' optimization process

Here, n—the number of time points, y_t—is the actual value at a given time period t, $\hat{y}_t$—is the predicted value, and t—observation in a data set.

The value of RMSE and MSE penalizes large errors. In contrast, MAE is less biased for higher values and usually does not penalize large errors. MAE is calculated according to the following equation:

$$\text{MAE} = \frac{1}{n}\sum_{t=1}^{n}\left|y_t - \hat{y}_t\right|^2, \tag{2.62}$$

where n—the number of time points, y_t—is the actual value at a given time period t, and $\hat{y}_t$—is the predicted value.

Figure 2.108 shows the minimum MSE of the GPR algorithm, where the red dot indicates the iteration with the minimum MSE, and the light blue dot represents the computed MSE value during the optimization process by varying the GPR hyperparameters. Dark blue dots indicate the observed minimum error minMSE detected up to the current (including current as well) observation:

$$\text{minMSE} = \min\left(\text{MSE}^i\right),\ i = \overline{1, n}, \tag{2.63}$$

where n is the number of iterations.

Figure 2.109 shows the cross-validation results depicting the predicted value of VFC Vibro excitation against the real (true) values. The errors are represented by vertical red-dashed lines, but due to very small error values, the majority of the true and predicted value points overlap.

The best results achieving RMSE = 5.6891, MSE = 31.238, and MAE = 3.872 have been achieved using a linear basic function, no-isotropic rational quadratic kernel function (Eq. (2.63)) with sigma 0.0002.

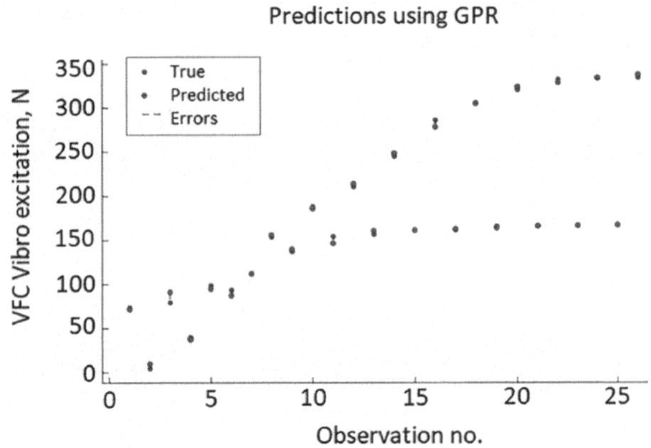

Fig. 2.109 Testing data results of the GPR model

$$k(x_a, x_b) = \sigma^2\left(1 + \frac{\|x_a - x_b\|^2}{2\alpha\ell^2}\right)^{-u}, \tag{2.64}$$

where σ^2—the overall variance, ℓ—the length scale parameter, and α—the scale-mixture (α > 2.0).

Four hyperparameters have been included in the SVM model optimization process: kernel function, kernel scale, box constraint, and epsilon (Table 2.20). The ranges of values for the latter three hyperparameters were selected on the basis of preliminary experiments. Seven different kernel functions have been analyzed: three Gaussian (fine, medium, and coarse), Linear, Quadratic, and Cubic. It has been observed that the Gaussian functions gave the poorest results compared to other functions.

The best results for the validation set—RMSE = 9.124, MSE = 83.253, and MAE = 6.403—were obtained using a linear kernel function, $\varepsilon = 0.105$, with box constrains = 111.25 (Fig. 2.110).

It can be noted that the minimum MSE varies over a wide range depending on the combination of the SVM hyperparameters, and the error can reach almost 3000. Prediction errors are displayed in a response plot in Fig. 2.111.

For the ANN model experimental set-up, we have used a simple feedforward network, Multilayer Perceptron (MLP), presented in Fig. 2.112.

As in the two previous ANN models, four hyperparameters were used in the optimization process: the number of hidden layers, the size of the hidden layer, the activation function of the hidden layers, and regularization strength (Table 2.21). The three most common activation functions were analyzed: Sigmoid, Hyperbolic tangent

Table 2.20 Hyperparameters of the SVM model

	Epsilon	Box constraint	Kernel function	Kernel scale
Value range	(0.01; 1000)	(0.001; 500)	Gaussian, linear, quadratic, and cubic	(0.001; 100)

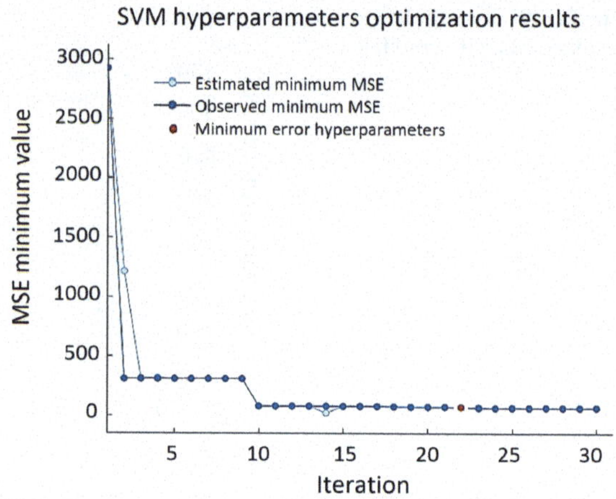

Fig. 2.110 Minimum MSE during SVM hypermeters' optimization process

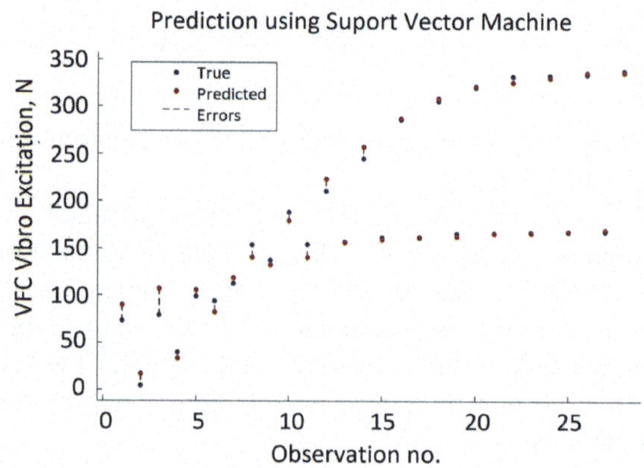

Fig. 2.111 Testing data results of the SVM model

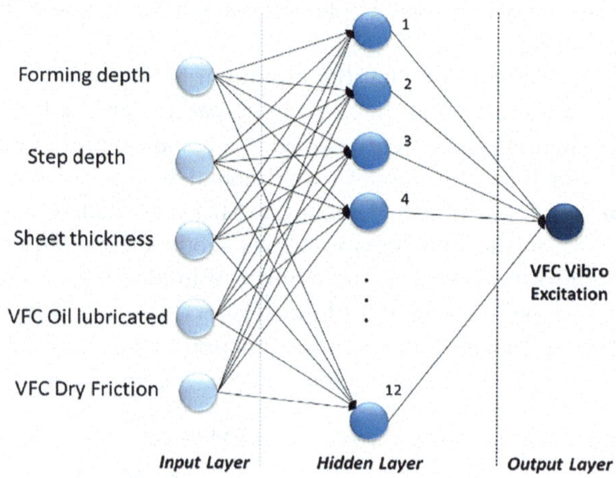

Fig. 2.112 The architecture of used MLP

Table 2.21 Hyperparameters of the ANN model

	Number of hidden layers	Hidden layer size	Activation function	Regularization strength
Value range	(1; 3)	(10; 100)	Sigmoid, Tanh, ReLU	(0; 0.001)

(Tanh), and Rectified Linear Unit (ReLU). The range of regularization strength was chosen based on primary cross-validation results. Value ranges of hidden number layers and hidden layer size were selected according to the size of the data set.

The ANN approach provides very good prediction accuracy, and the best results with RMSE = 4.5337, MSE = 20.573, and MAE = 3.528 were obtained using a single hidden layer neural network with ReLU activation function, 12 neurons, and a regularization strength of zero. The variation of the minimum MSE values during the ANN hypermeter optimization process is shown in Fig. 2.113. The testing data results of the ANN model are presented in Fig. 2.114 providing actual and predicted values of VFC Vibro excitation.

Thus, the minimum leaf size is the only hyperparameter that was included in the optimization. It denotes the minimum number of data points that are required to be present in the leaf node. The search range for this hyperparameter is from 1 to 15 which is chosen according to the size of the data set. The best result of the DT approach: RMSE = 29.567, MSE = 874.19, and MAE = 21.507 (Fig. 2.115) were obtained using a decision tree with a minimum leaf size equal to four. The decision tree approach provides the lowest accuracy compared to GPR, SVM, and ANN. The minimum MSE value varies from 2195.31 to 874.19. As depicted in Fig. 2.116, in half of the observations, the distance between the predicted VFC vibration excitation value and the actual values is more significant than GPR, ANN, or SVM.

Fig. 2.113 Minimum MSE during ANN hypermeters' optimization process

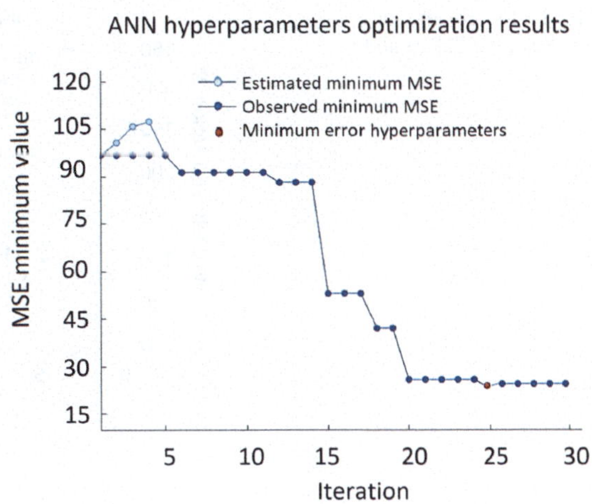

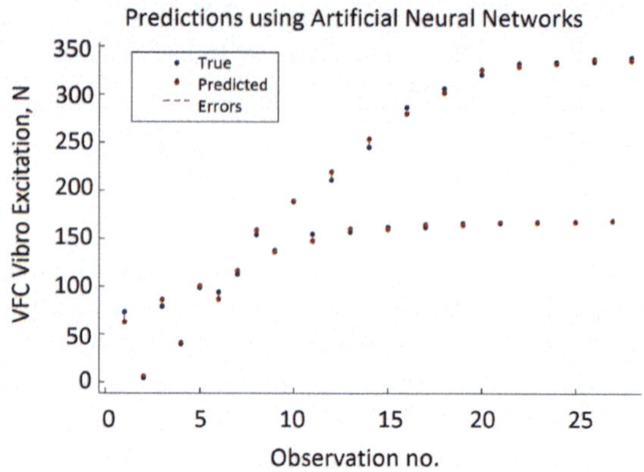

Fig. 2.114 Testing data results of the ANN model

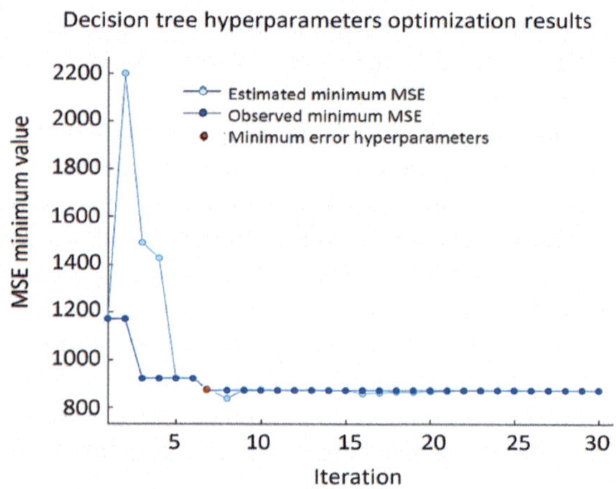

Fig. 2.115 Minimum MSE during DT hypermeters' optimization process

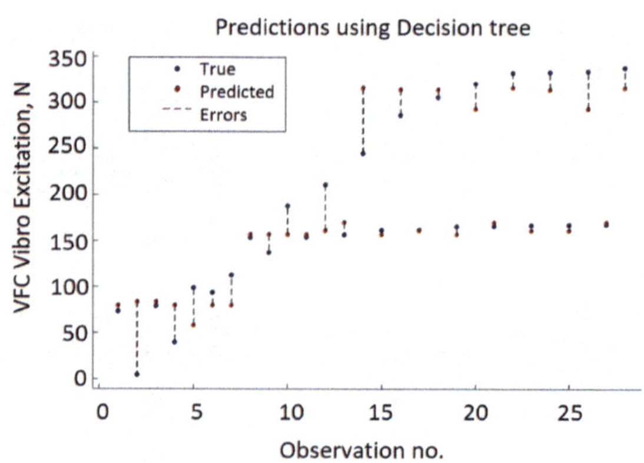

Fig. 2.116 Testing data results of the DT model

Table 2.22 represents the hyperparameters of another, i.e. the KNN approach. Two KNN hyperparameters were included in the optimization process. The first one is K (the number of neighbors to consider) and the second is the employed distance function (most commonly used Euclidean, Manhattan, Minkowski).

For n-dimensional space, the Euclidean distance between the two points x with coordinates ($x_1, x_2, \ldots, x_n$) and y with coordinates ($y_1, y_2, \ldots, y_n$) is determined using the following equation:

$$d_{\mathrm{Eucl}} = \sqrt{(x_1 - y_1)^2 + (x_2 - y_2)^2 + \cdots + (x_n - y_n)^2} = \sqrt{\sum_{i=1}^{n} (x_i - y_i)^2}, \quad (2.65)$$

where ($y_1, y_2, \ldots, y_n$) are attribute values of y data instance and ($x_1, x_2, \ldots, x_n$) are attribute values of x data instance.

The Manhattan distance is also known as city block distance, or taxicab geometry, as well as several other names, because it allows calculating the distance between two data points on a uniform grid, for example, a city block; there may be more than one path between the two points that have the same Manhattan distance. The Manhattan distance between two points x and y is calculated using the following formula:

$$d_{\mathrm{Manh}}(x, y) = \sum_{i=1}^{n} |x_i - y_i|. \quad (2.66)$$

The Minkowski distance is a generalized distance metric. The above formula Eq. (2.65) can be manipulated by substituting "p" to calculate the distance between two data points in different ways. Thus, the Minkowski distance is also known as Lp norm distance:

$$d_{\mathrm{Mink}}(x, y) = \left(\sum_{i=1}^{n} |x_i - y_i|^p \right)^{1/p}, \quad (2.67)$$

where p is the order of the Minkowski metric. With different values of p, the distance between two data points can be calculated in different ways: $p = 1$—Manhattan distance; $p = 2$—Euclidean distance, $p = \infty$-Chebyshev's distance. A value such as $p = 1.5$ provides a balance between the two measures.

Table 2.22 Hyperparameters of the KNN model

	K	Distance metric
Value range	(1; 10)	Euclidean, Manhattan, Minkowski ($p = 1, p = 1.5, p = 2, p = \mathit{Infinity}$)

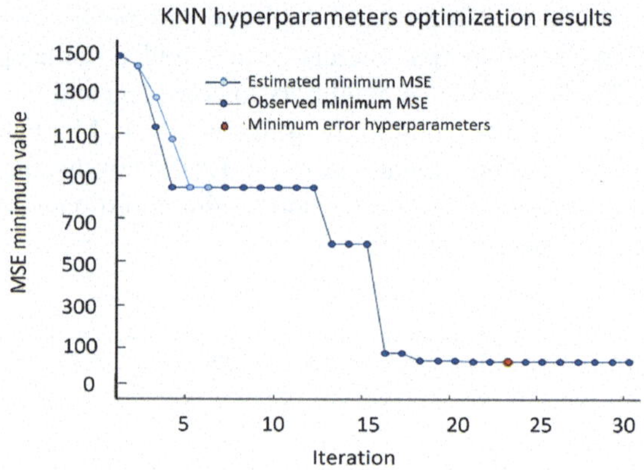

Fig. 2.117 Minimum MSE during KNN hypermeters' optimization process

The best KNN results, including RMSE = 6.0757, MSE = 36.915, and MAE = 3.528, were obtained using the Manhattan distance for two neighbors, $K = 2$ (Fig. 2.117).

The most important step in KNN is to determine the optimal value of *K*. The optimal value of *K* reduces the effect of noise on the classification. A technique called the "elbow method" helps to do this, selecting the optimal *K* value. Different values of *K* are applied to the same data set and the change in *K* is initially observed. In the data set characterizing the SPIF process, the error rate (RMSE) curve obtained by applying the KNN with respect to the *K* value is shown in Fig. 2.118.

The graph presented in Fig. 2.118 denotes that initially the error rate decreases to 2, and then it starts to increase. Thus, the value of *K* should be 2, i.e. it is the optimal *K* value for this model. This curve is called an elbow curve because it has a shape of an elbow and is commonly used to adjust the *K* value.

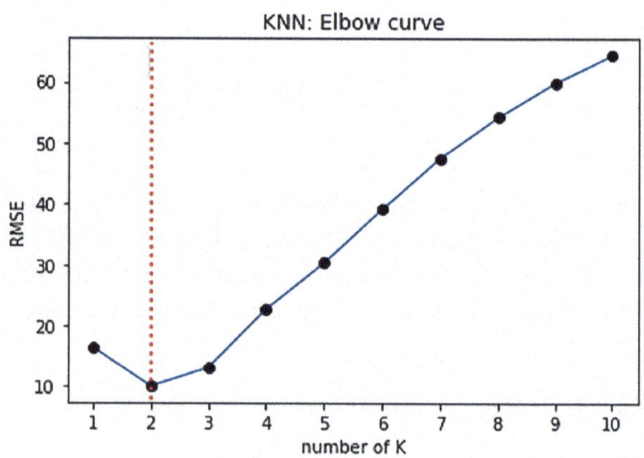

Fig. 2.118 A visual curve with an explicit elbow point in the K range (1; 10)

R^2 (coefficient of determination) is a regression score, which is a statistical measure indicating how close the data are to the fitted regression line. In regression, it is a measure showing how well the regression predictions approximate the real data. An R^2 of 1 indicates that the regression predictions perfectly fit the data:

$$R^2 = \frac{\text{SSR}}{\text{SST}} = \frac{\sum_{i=1}^{m} \left(y_i - \hat{y}_l\right)^2}{\sum_{i=1}^{m} \left(y_i - \overline{y}\right)^2}, \tag{2.68}$$

where SSR is the sum of squares of residuals, SST is the total sum of squares, y_i is the actual value, $\hat{y}_l$ is the predicted value, and $\overline{y}$ is the mean value.

R^2 is always between 0 and 100% (or 0 and 1.0). The higher the R^2, the better the model. The goal is not to maximize R^2 because model stability and adaptability are equally important. When checking the adjusted R^2 value, it is preferred to have the values of the R^2 and adjusted R^2 close to each other. From the graphical representation of R^2 values of the five prediction models (Fig. 2.119), it can be seen that the DT algorithm gave the worst result ($R^2 = 0.878$) compared to others.

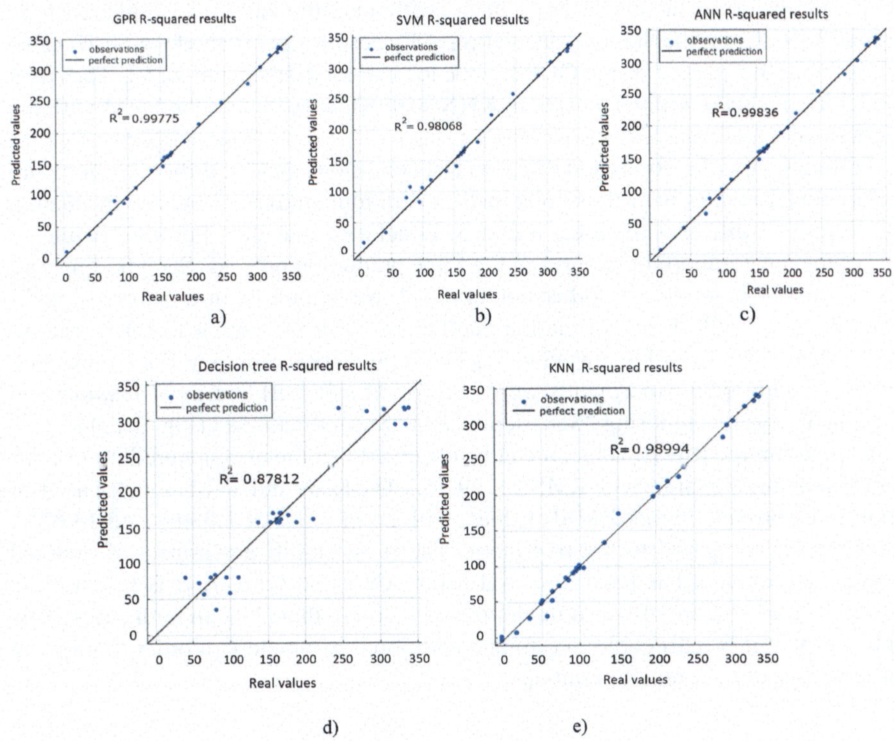

Fig. 2.119 R-squared representation for all five ML algorithms used for the prediction task

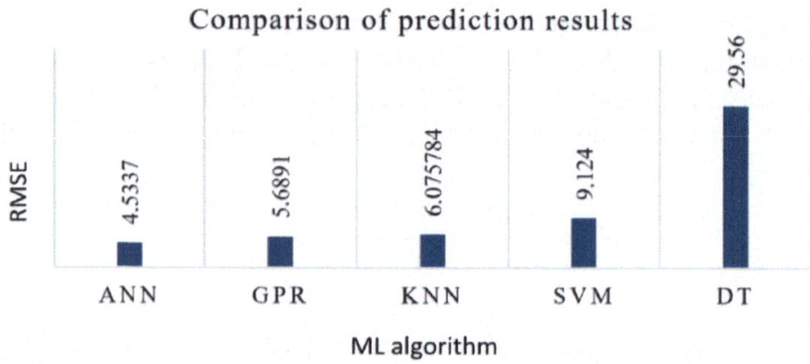

Fig. 2.120 Comparison of different ML algorithm prediction error results

A series of five experimental runs have been carried out. Summarizing the experimental results, ANN and GPR were identified as the most efficient methods for developing VFC Vibro excitation prediction models, giving the lowest prediction error (RMSE) of 4.5337 and 5.6891, respectively (Fig. 2.120). It should be noted that the DT algorithm is inappropriate for this task and for the available data set, as the prediction errors in both cases (with and without optimization) are high, reaching around 30%. As for the Standard Deviation (ST), despite the DT model with a 0 value of ST for all five iterations, the GDR has the lowest standard deviation, ST. The ANN model has resulted with ST = 0.616, KNN with ST = 0.78, and the SVM with the highest value, ST = 2.531.

Different input functions and different test conditions make it difficult to compare experimental results with those obtained in different studies. However, evaluating the impact of different features, it can be concluded that the "Forming depth" is one of the most important features in the force prediction process. Such findings are also observed in studies by other authors [44]. Nevertheless, in most cases, additional features with those correlation coefficients with the output features that are higher than $r = \pm 0.3$ allow for achieving better prediction accuracy. For comparison purposes, additional experiments were performed using all five input features and only three features, ignoring "Step depth" and "Sheet thickness" (Fig. 2.121).

Comparing the RMSE value, it can be observed that the average prediction error for DT remains the same, with RMSE = 29.562. The KNN and SVM models based on input three provide higher RMSE values, with an increase of 1.939% and 13.67%, respectively, compared to the five-input. The most significant impact of denoted features elimination has been observed using ANN and GPR models because the RMSE of the three-input-based model increases more than 25% in both cases. This indicates that it is reasonable to estimate additional parameters in order to improve the accuracy of the prediction model.

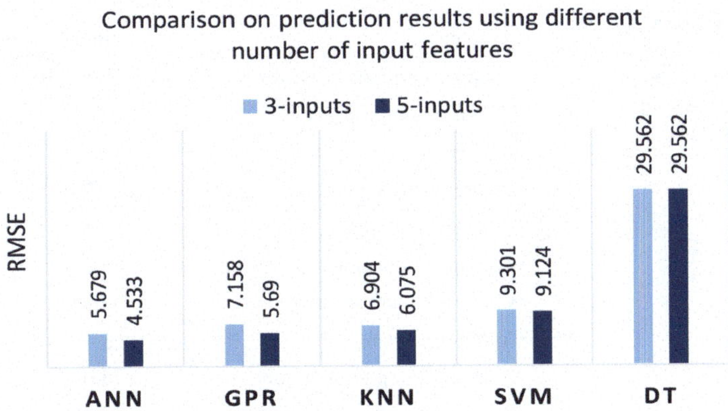

Fig. 2.121 Comparison of prediction results using a different number of input features: five input features (forming depth, step depth, sheet thickness, VFC dry friction, and VFC Vibro excitation) and three inputs (forming depth, VFC dry friction, and VFC vibro excitation)

2.6.3 Evaluation Methodology of Polymer Sheet Forming Process

The numerical and experimental investigation of polymer sheet SPIF parameters have shown that the focus should be on polymer heating techniques that ensure a fast and high-quality SPIF process. An innovative method of thermal deformation of a polymer sheet is proposed using a tool with a ball mounted in a circular magnet, the friction of which with the tool holder is reduced by ultrasound, and the heating performed by a laser.

2.6.3.1 Virtual Twin for Polymer Sheet Parameters Investigation

The numerical research results of the polymer sheet SPIF process parameters—heating temperature dependency from time and displacement of polymer sheet from gravity—are presented. By using Ansys Transient Thermal together with Transient Structural analysis, a computational FE model of the Polyvinyl Chloride (PVC) Trovidur ESA-D sheet, which geometric dimensions and properties of the elements used for the numerical calculations are collected in Table 2.23, was created [45].

Numerical analysis was conducted in two stages: in the first stage, transient thermal analysis was carried out and thus temperature dissipation on the sheet polymer was obtained. During the second stage, this dissipation of temperature was transferred to the transient structural analysis as the input parameter of the thermal load and after the analysis displacement of the polymer sheet from earth gravity was obtained. During the first stage of numerical analysis, PVC sheet, which parameters are presented in Table 2.23, was excited with convection on the surface area of 25 mm offset from all edges of one surface, which corresponds to area affected by the

Table 2.23 Properties of the PVC Trovidur ESA-D material and geometric dimensions of the sheet used in calculation [46]

Parameter	Value	Unit
Length *x* width of the sheet	300 × 300	mm
Thickness of the sheet	3	mm
Density	1.41	g/cm^3
Tensile stress at yield	47.75	N/mm^2
Elongation at break	30.3	%
Modulus of elasticity in tension	2643	N/mm^2
Notched Impact strength	9.09	mJ/mm^2
Compressive strength	65.4	MPa
Vicat-softening temperature	75.0	°C
Coefficient of linear thermal expansion	70	10^{-6}/K

thermal gun during the experimental research. Ambient temperature and conditions were evaluated by adding another convection with different parameters on the rest of the surface of the sheet. A scheme of this stage of numerical analysis is presented in Fig. 2.122a and a computational model with boundary conditions is presented in Fig. 2.122b. The FE model numerical simulation data of this stage are collected and presented in Table 2.24.

During both stages of calculations, analysis time was divided into steps. The total time of simulation was 780 s. The time of the first step was 5 s and the stage was divided into 50 sub-steps. The second stage was 780 s and this stage was divided into 100 sub-steps.

The simulation results are the temperature dissipation of the upper surface of the sheet, on which the ambient convection was added, so this surface is opposite to the heating surface, 2 s after the start of heating, shown in Fig. 2.123a. Temperature dissipation of the same surface of the sheet after 780 s from the start of heating is presented in Fig. 2.123b.

The simulation results allowed us to determine the average and maximum dependence of the polymer sheet surface temperature on time (Fig. 2.124), from which it

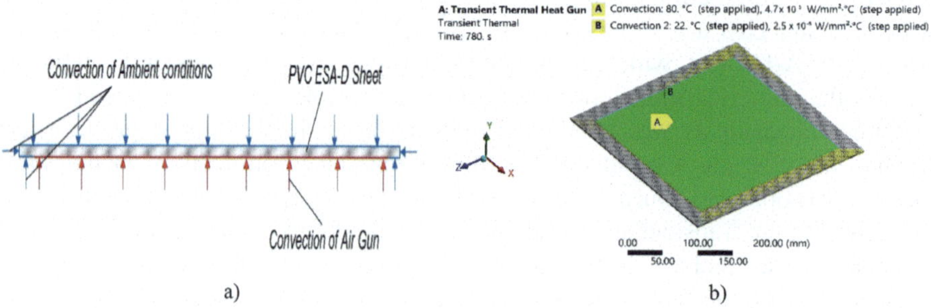

Fig. 2.122 Transient thermal numerical analysis: scheme (**a**) and computational model with boundary conditions (**b**)

Table 2.24 FE numerical simulation data of the first stage of analysis

Parameter	Value	Unit
Mesh elements method	Hex Dominant	–
Number of finite elements	3600	–
Number of nodal points	25,803	–
Convection film coefficient of the heat gun surface area	47	W/m^2
Convection temperature of the heat gun surface area	80	°C
Convection film coefficient of the rest ambient surface area	25	W/m^2
Convection temperature of the rest ambient surface area	22	°C
Total time of calculation	780	s

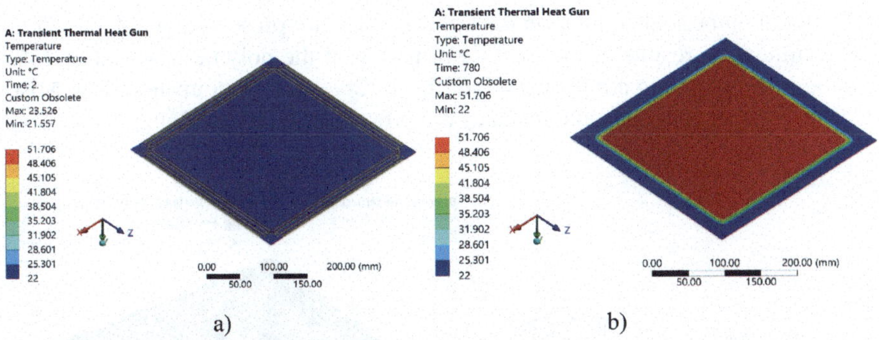

Fig. 2.123 Temperature dissipation on the opposite to the heating surface of the sheet: after 2 s after the start of heating (**a**); 780 **s** after the start of heating (**b**)

Fig. 2.124 Average and maximum temperatures of the opposite to the heating polymer sheet surface

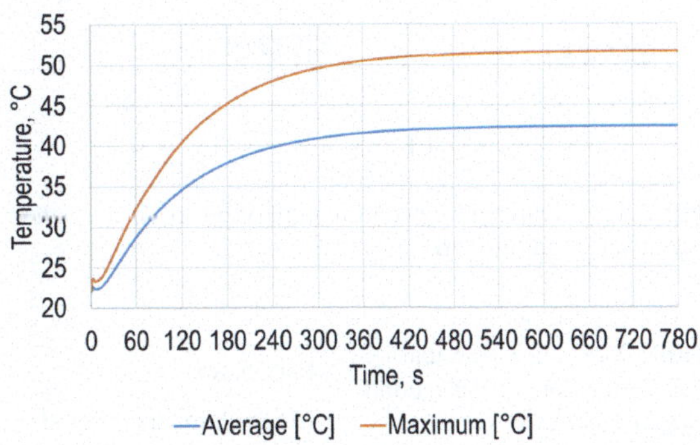

can be seen that the sheet surface temperature changes nonlinearly over time until it settles.

Results of this simulation stage revealed that temperature distribution on the opposite to the heating surface is similar to the heating area, which means that in the polymer sheet SPIF process it is very important to carefully select the geometry of the heating gun flow as well as of the surface of the polymer sheet. Simulation results presented in Fig. 2.124 show that the average temperature of the opposite to the heating surface is equal to 42.4 °C after 780 s from the start of heating and the maximum is equal to 51.7 °C at the same instance. By using these results, it is possible to determine the required temperature and, thus, set the required time from the start of heating to begin the SPIF process.

The next stage of the simulation was carried out using these results as input loads for transient structural analysis. During this stage, polymer sheet was fixed on the 25 mm offset from the sheet edge on the same side as the heating gun and a standard earth gravity vector was added perpendicularly to this surface. The calculation scheme of this stage of numerical analysis is presented in Fig. 2.125a and the computational model with boundary conditions in Fig. 2.125b.

FE model numerical simulation data of this stage are presented in Table 2.25.

The simulation results of the total deformation of the polymer sheet after 780 s from the start of heating are presented in Fig. 2.126a, and the equivalent Von Mises stress dependencies at the same instance are presented in Fig. 2.126b.

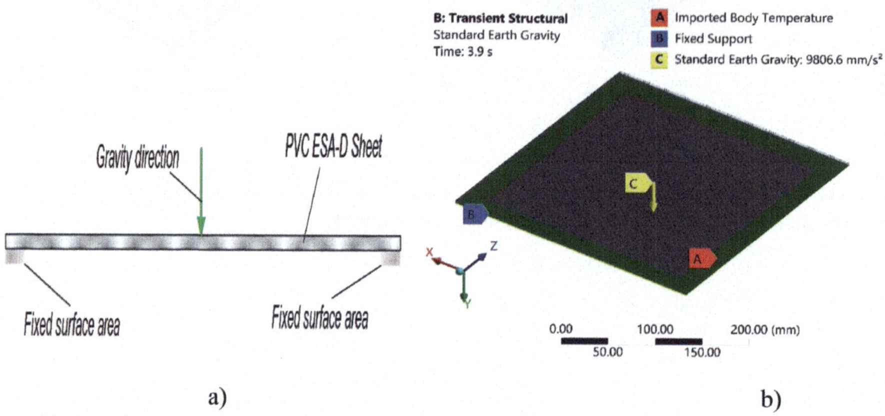

Fig. 2.125 Transient structural numerical analysis: scheme (**a**) and computational model with boundary conditions (**b**)

Table 2.25 FE numerical simulation data of the second stage of analysis

Parameter	Value	Unit
Mesh elements method	Hex Dominant	–
Number of finite elements	3600	–
Number of nodal points	25,803	–
Input load	Temperature	°C
Acceleration of gravity	9806.6	mm/s^2
Total time of calculation	780	s

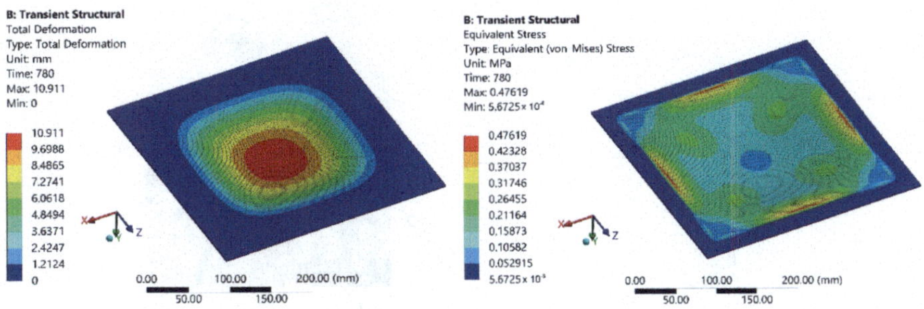

Fig. 2.126 Simulation results of the second stage after 780 s from the start of heating: total deformation (**a**) and equivalent Von Mises stress (**b**)

Fig. 2.127 Average and maximum deformation under the earth gravity versus time of the polymer sheet

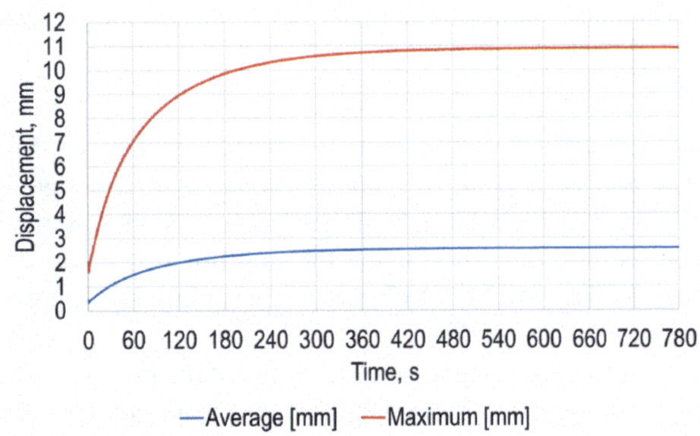

The simulation results of the average and maximum deformation of the sheet under the earth gravity versus time are presented in Fig. 2.127.

The simulation results revealed that the maximum deformation of the polymer sheet from standard earth gravity exceeds 10.9 mm, while the average deformation of the polymer sheet was 2.6 mm. These results show the need to carefully select SPIF parameters such as the initial and forming depth of the tool as well as required forming forces.

2.6.3.2 Physical Twin for the Validation of Simulation Results

The scheme of the experimental set-up is presented in Fig. 2.128a and the experimental set-up view is presented in Fig. 2.128b. During the experimental research, PVC ESA-D polymer sheet 1 was fixed on the frame 2 by 25 mm offset area from its edge, like it was in the theoretical research. The polymer sheet 1 was heated by blowing hot air with blowing system CT-850D (Acifica, Inc, San Jose, USA) 6 from the lower surface through the hole formed in frame 2 and opposite surface displacement was measured with laser displacement meter Kyence LK-G82/3001

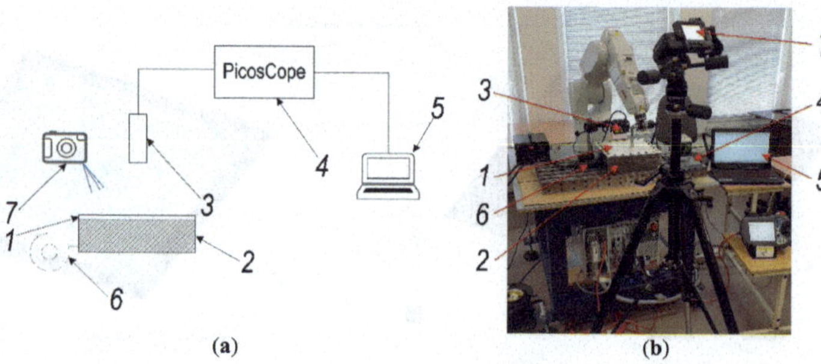

Fig. 2.128 Experimental set-up for polymer sheet forming parameters validation: **a** scheme; **b** set-up view: 1—PVC ESA-D polymer sheet, 2—holding frame, 3—laser displacement meter Kyence LK-G82/3001 (Keyence Corporation, Neu-Isenburg, Germany), 4—digital oscilloscope PicoScope-6403 (Pico Technology Ltd., Cambridgeshire, UK), 5—PC, 6—hot air blowing system CT-850D (Acifica, Inc., San Jose, CA, USA), and 7—thermal imaging camera FLIR T450sc (FLIR Systems Inc., Wilsonville, OR, USA)

(Keyence Corporation, Neu-Isenburg, Germany) 3. Signal obtained from this meter was analyzed using digital oscilloscope PicoScope-6403 (Pico Technology Ltd., Cambridgeshire, UK) 5. At the same instance, temperature of the polymer sheet surface was measured using thermal imaging camera FLIR T450sc (FLIR Systems Inc., Wilsonville, USA) 7. This experimental set-up was applied for patenting by the authors [47].

The experimental research results of the polymer sheet surface temperature versus time and polymer sheet displacement from gravity versus temperature are presented in Fig. 2.129a, b, respectively. As it is seen from the results presented in Fig. 2.129a, the maximum temperature of the polymer sheet after 2 s from the start of heating is equal to 25.8 °C and after 780 s is equal to 45.5 °C. Figure 2.129b shows that displacement of the sheet surface increases significantly after the surface reaches temperature 31 °C and is maximum when the temperature of the polymer sheet is 45.5 °C.

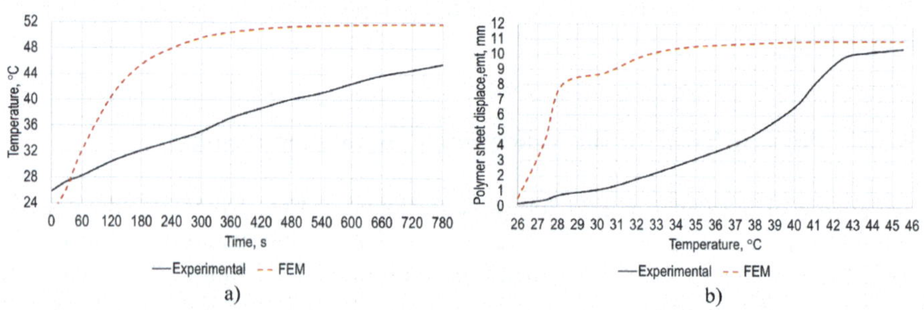

Fig. 2.129 Experimental validation results: **a** polymer sheet temperature versus time and **b** polymer sheet displacement versus temperature

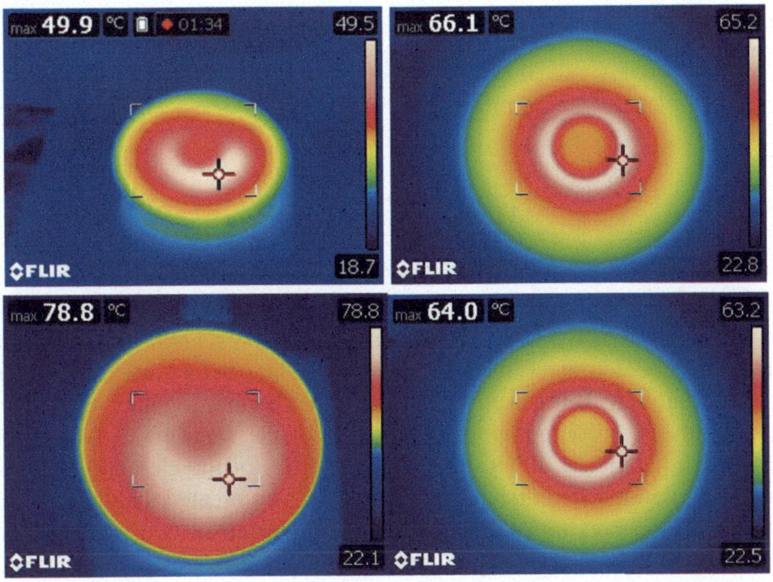

Fig. 2.130 Investigation of the heating efficiency of a robotized polymer sheet SPIF: the temperature distribution is recorded with a thermal imager Flir T450sc

The obtained results in experimental research are close to the results obtained with numerical simulations presented in Figs. 2.124 and 2.127. The temperature rate at the initial zone is lower due to a smaller heating area of the hot air blowing system compared to heating conditions used in FE analysis. This revealed the possibility to use numerical research in the future to determine required parameters for SPIF avoiding experimental determination and thus avoiding expenses related to the manufacturing of the specimens. The experimental investigation results are presented in Fig. 2.130.

2.6.3.3 Investigation of the Polymer Sheet Advanced Heating Device

In Figs. 2.127 and 2.129b, in order to eliminate numerically obtained and experimentally validated self-deformations of the heated polymer sheet, resulting from earth gravity, it is necessary to use backing plates to constrain these deformations. However, this way not only increases production costs but also slows down processes, as backing plates have to be adapted to the forming of specific products, which is undesirable for unit production. A much more efficient method is associated with the use of a tool that, in contact with the polymer sheet, also heats it to a controlled temperature. In addition, the heating of the polymer sheet is much faster. To confirm this idea, an additional set of numerical investigations, which are generally similar to those presented in the previous sub-chapter, were carried out.

Table 2.26 Excitation heat flow parameters

Time (s)	Heat flow power (W)
0	0
0.18	4.5
3	2.2
50	0.05
120	0

During this numerical simulation, the heating temperature dependency from time and displacement of polymer sheet from gravity are presented. In this simulation, the heating is modeled as point heating source, while before was modeled as air flow from a heat gun. By using Ansys Transient Thermal together with Transient Structural analysis, a computational FE model of the same PVC Trovidur ESA-D sheet, which geometric dimensions and properties of the elements used for the numerical calculations are collected in Table 2.23, was created. Numerical analysis as described in the previous sub-chapter was conducted in two stages. The PVC sheet was exposed to a heat flow source defined by heat power varying in time, which corresponds to the point heating source. The properties of excitation heat flow parameters are presented in Table 2.26 and the corresponding curve in Fig. 2.131.

The ambient temperature and conditions were evaluated the same as in previous sub-chapter by adding another convection with different parameters on the rest of the surfaces of the sheet. A scheme of this stage of numerical analysis is presented in Fig. 2.132a, and the computational model and boundary conditions are presented in Fig. 2.132b. The FE numerical simulation data together with the main exposed heat flow parameters of this stage are presented in Table 2.27.

During both stages of the simulation, the analysis time was divided into steps. The total time of simulation was 120 s. The time of the first stage was 3 s and this stage was divided into 50 sub-steps. The second stage was 120 s and this stage was divided into 100 sub-steps.

The simulation results are the temperature dissipation of the lower surface of the sheet, i.e. opposite to the heat flow surface, 1.92 s after the start of heating, shown

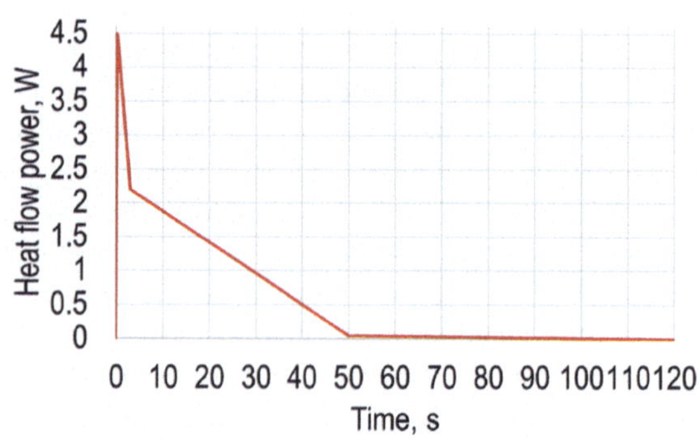

Fig. 2.131 Exposed heat flow curve

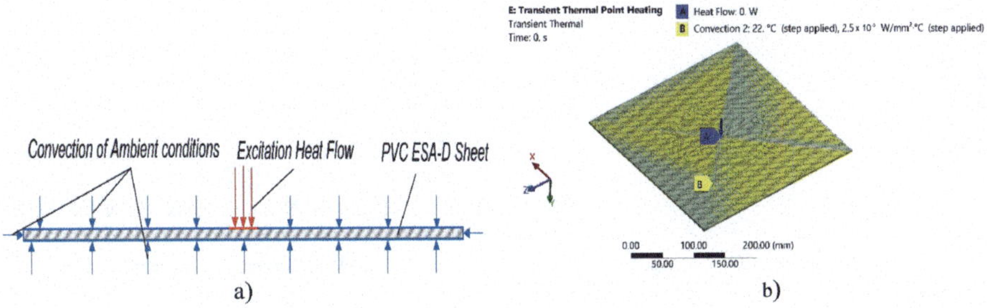

Fig. 2.132 Transient thermal numerical analysis: **a** scheme and **b** computational model with boundary conditions

Table 2.27 FE numerical simulation data of the first stage of analysis

Parameter	Value	Unit
Mesh elements method	Hex dominant	–
Number of finite elements	15,124	–
Number of nodal points	103,231	–
Heat flow application geometry	Ø10 mm circle	–
Heat flow magnitude	Tabular (see Table 2.4)	W
Convection film coefficient of the rest ambient surface area	25	W/m^2
Convection temperature of the rest ambient surface area	22	°C
Total time of calculation	120	s

in Fig. 2.133a. Temperature dissipation of the same surface of the sheet after 120 s from the start of heating is presented in Fig. 2.133b.

The simulation results allowed the determination of the average and maximum dependence of the polymer sheet surface temperature on time (Fig. 2.134a), from

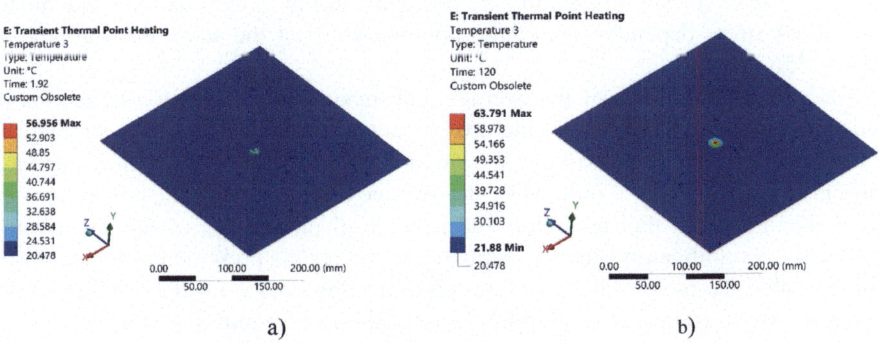

Fig. 2.133 Temperature dissipation on the opposite to the heating surface: **a** after 1.92 **s** and **b** after 120 s from the start of heating

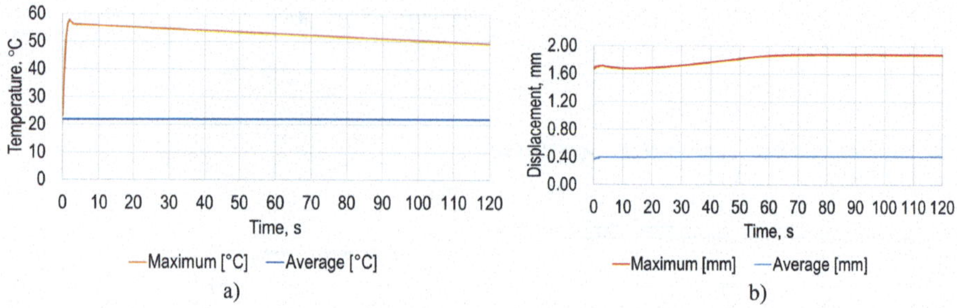

Fig. 2.134 Simulation results: **a** average and maximum temperatures of the opposite to the heating polymer sheet surface and **b** average and maximum deformation under the earth gravity versus time of the polymer sheet

which it can be seen that the sheet surface temperature changes nonlinearly over time until it settles.

The results of this simulation stage revealed that the difference between the maximum and the average temperature of the PVC sheet is significantly higher compared to results presented in Fig. 2.124, which means that the heating occurs in quite small zone of the surface. The obtained results showed that the average temperature of the measuring surface is equal to 22.1 °C after 120 s from the start of heating, and the maximum is equal to 49.3 °C at the same instance. The maximum temperature in the 120 s time period is 57.9 °C after 1.86 s from the start of heating. The shorter heating time and the smaller heating area are more energetically effective compared to heating by air gun. Additionally, by using these results, it is possible to determine the required temperature and, thus, set the required heating flow power for the SPIF process.

The next stage of the simulation was identical to that described in the previous sub-chapter, and its scheme is the same as in Fig. 2.125, and the main parameters were the same as in Table 2.21, and it only differed in simulation time, which was equal to 120 s. The simulation results of the total deformation of the polymer sheet after 120 s from the start of heating are presented in Fig. 2.135a and the equivalent Von Mises stress dependencies of the polymer sheet at the same instance are in Fig. 2.135b.

The simulation results of the average and maximum deformation of the sheet under the earth gravity versus time are presented in Fig. 2.123b. The simulation results revealed that the maximum deformation of the polymer sheet from standard earth gravity exceeded 1.9 mm, while the average deformation of the polymer sheet was 0.42 mm. Both the maximum and average displacements of the sheet from gravity were significantly smaller compared to the results presented in Fig. 2.118b. This revealed the possibility of using a point heating device for the SPIF process; hence, a novel forming tool with point heating ability is required.

An image of the proposed advanced heating device, for which the description was applied for patenting by the authors [48], is presented in Fig. 2.136.

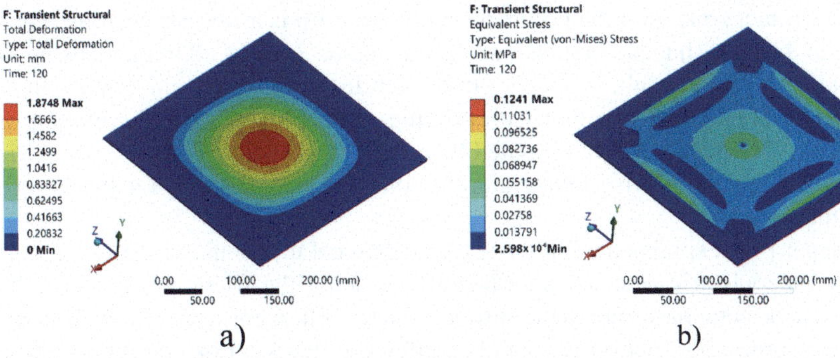

Fig. 2.135 Simulation results of the second stage after 120 s from the start of heating: **a** total deformation and **b** equivalent Von Mises stress

Fig. 2.136 Polymer sheet SPIF tool: 1—metal sphere, 2—ring-shaped magnet, 3—waveguide, 4—ultrasonic vibration transducer—piezoceramic discs, 5—fiber optics, and 6—temperature sensor

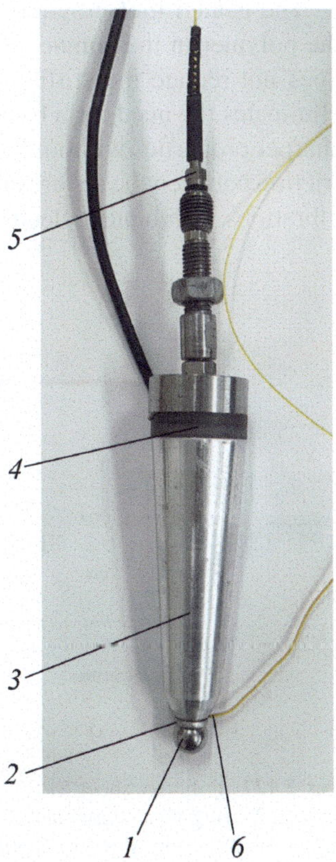

At the contact zone with the polymer sheet, a tool-tip is a free 3D rotating metal sphere 1 in the ring-shaped magnet 2, and a waveguide 3 is placed in the housing of the tool and is excited by ultrasonic vibration transducer–piezoceramic discs 4, thus, reducing the metal sphere friction with the ring-shaped magnet and facilitating its free rotation. The metal sphere is heated by a laser beam which is directed at it via fiber optics 5, and the heating temperature is controlled by means of feedback via a temperature sensor 6.

The paper [49] explains the influence of mechanical ultrasonic vibrations generated by a piezoelectric actuator, significantly reducing the frictional forces in the contact area of metal parts due to the superposition of ultrasonic vibrations. A modification of Coulomb's friction law can be applied to this kind of vibrating friction contact.

A schematics of the polymer sheet SPIF using a local heating tool and a general view of the equipment are shown in Fig. 2.137.

The heated tool moving on the sheet surface locally increases the plasticity of the polymer in the contact zone with less deforming force. This method of heating does not reduce the stiffness of the polymer around the tool contact area, which eliminates the need for a backing plate. This allows the fabrication of products with sharper properties without the need for a supporting structure. In order to reduce the friction between the sphere and ring magnet, the tool housing is excited by ultrasonic vibrations with a piezoelectric transducer.

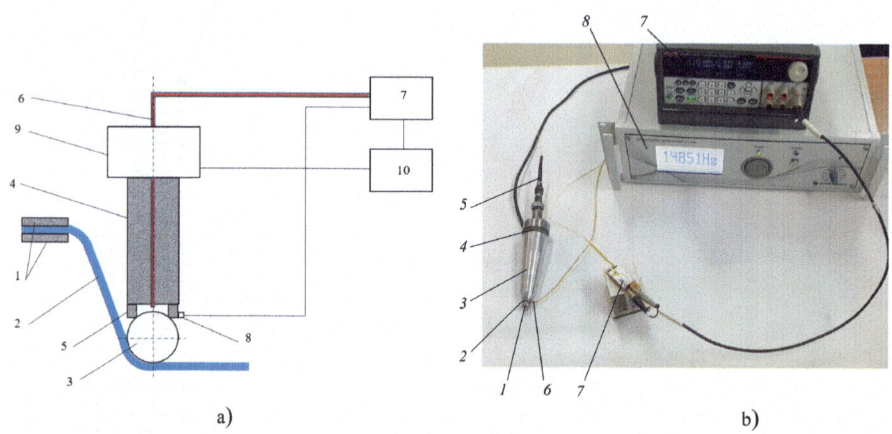

Fig. 2.137 Polymer sheet SPIF with a laser heating: **a** schematics and **b** equipment: 1—metal sphere, 2—ring-shaped magnet, 3—waveguide, 4—ultrasonic vibration transducer—piezoceramic discs, 5—fiber optics, 6—temperature sensor, 7—laser diode module BMU25-940-01-R (Oclaro Inc., San Jose, CA, USA) with laser beam intensity controller Keithley 2230-30-1 (Keithley Instruments Inc., Cleveland, USA), and 8—ultrasonic vibration controller Sensotronica VT-400 (KTU, Kaunas, Lithuania)

2.6.3.4 Physical Twin of the Polymer Sheet Incremental Forming

To verify numerical investigation results, experimental research based on simulation parameters and process presented in the previous sub-chapter were carried out. During this phase of research, experimental investigations of forming four different geometric shapes figure from a polymer sheet, which material properties are presented in Table 2.23, were carried out. Scheme of experimental set-up which description was applied for patenting by the authors [49] and stand view are presented in Fig. 2.138a, b, respectively. The main data of experimental research are collected and presented in Table 2.28.

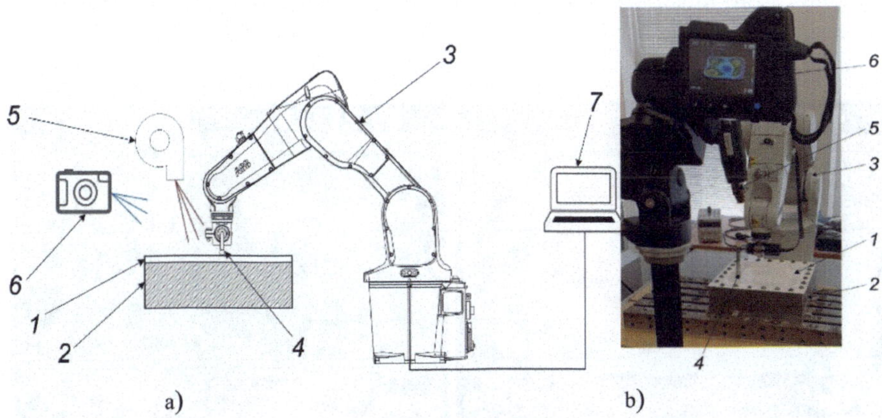

Fig. 2.138 Experimental investigation of robotized polymer sheet SPIF: **a** scheme and **b** stand view: 1—PVC ESA-D polymer sheet, 2—holding frame, 3—robot ABB IRB1200 (ABB Robotics & Discrete Automation, Västerås, Sweden), 4—forming tool, 5—heat gun Toolland PHG2 (Tooland Inc., San Carlos, USA), 6—thermal imaging camera FLIR T450sc (FLIR Systems Inc., Wilsonville, USA), and 7—PC

Table 2.28 Experimental research data for incremental polymer sheet forming

Parameter	Value	Unit
Radius of the forming tool sphere	8.5	mm
Step down	0.5	mm
Radial step	0.5	mm
Total forming depth	30	mm
Feed rate	100	mm/s
Major diameter of the geometric figure	150	mm
Minor diameter of the geometric figure	90	mm
Minimum temperature of the surface	40	°C
Maximum temperature of the surface	60	°C

During the experimental research, the forming tool 4 with a freely rotating sphere on its end was used. Software with four different modifications of circle, square, star, and flower spatial geometric shapes for robot IRB1200 3 was created with a PC 7. Polymer sheet 1 was fixtured on the frame 2 and heated with a Toolland PHG2 hot air gun 5. The temperature within the required range based on the simulation results was controlled using a FLIR T450sc thermal imaging camera 6.

Figure 2.139 shows the transient state of the temperature distribution on the surface of the forming sheet when it is heated using a hot air gun and advanced heating device. The lowest temperature values (40–60 °C) were measured near the tool–sheet interface and in the center of the workpiece. Note that, during the forming process, the temperature must not exceed the softening temperature of 75 °C.

Figure 2.140 illustrates SPIF polymer sheets of different spatial geometry with processing parameters used during the experiment listed in Table 2.23.

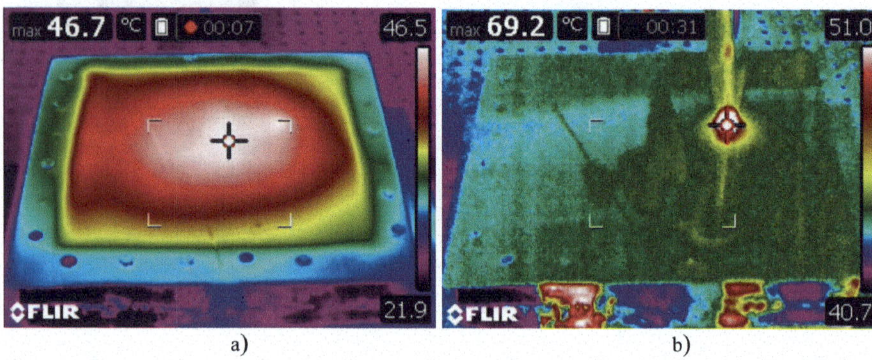

Fig. 2.139 Thermal images of polymer sheet: **a** heated with an air gun and **b** heated with an advanced heating device

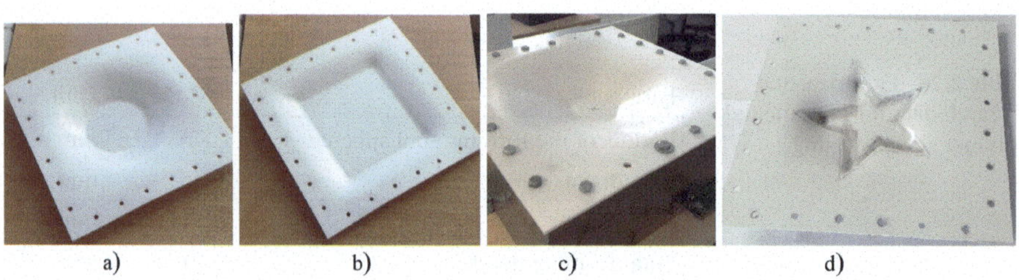

Fig. 2.140 Photos of the incrementally formed polymer sheets of PVC ESA-D material were achieved by elevated temperature forming conditions: **a** spatial circular geometry, **b** spatial square geometry; **c** spatial flower geometry, and **d** spatial star geometry

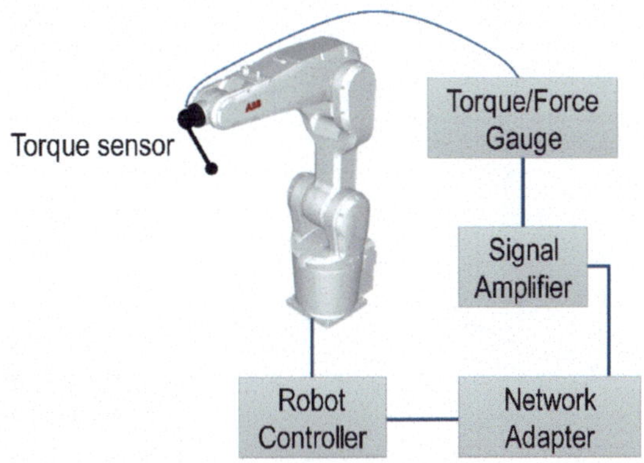

Fig. 2.141 Schematics of a robotized polymer sheet SPIF step-by-step feedback system

2.6.3.5 Physical Twin and Experimentation with Different Forming Tools

A standard version of ABB controller IRC5 M2004 was used to control the industrial robot IRB1200 M2004. Unfortunately, it does not have analog inputs. In order to extend the capabilities of the controller IRCS M2001, a DeviceNet network adapter CREVIS NA-9111 with analog input channels extension PicoScope-3424 (Pico Technology Ltd., Cambridgeshire, UK) was used. The Mark-10 STJ100 (Mark-10 Corp., Copiague, NY, USA) torque sensor that was used in the experiment can provide an analog output of ±1 V at full scale and extended analog input, with a dynamic range of 0–10 V (PicoScope-3424). Therefore, in order to take full advantage of the dynamic range of 12 bits analog input, the signal was amplified five times at the same time providing a positive offset voltage. The schematics of a robotized step-by-step feedback system are presented in Fig. 2.141.

Experimental thermoplastic forming tests were performed with a 2 mm thick PVC Trovidur ESA-D polymer sheet. The geometry of the formed sample is cone-shaped, with large diameter 140 mm, small diameter 80 mm, and embossing depth 30 mm. The vertical and horizontal steps of the pressure are equal to 0.5 mm. The experimental tests were performed with three different forming tools:

1. Forming tool with Ø17 mm freely rotating sphere.
2. Forming tool with Ø10 mm rotating sphere, supported by a ring-shaped magnetic holder.
3. Forming tool with Ø10 mm rotating sphere, supported by a ring-shaped magnetic holder and heated to 46 °C.

The results of the experimental research of the dependence of force on the forming tool and the depth of embossing are presented in Fig. 2.142.

From the presented graphs, it can be seen that in the case of a point polymer heating by forming tool, the vertical component of the force acting on the polymer sheet is the smallest, which allows us to expect the lowest probability of inducing forming defects.

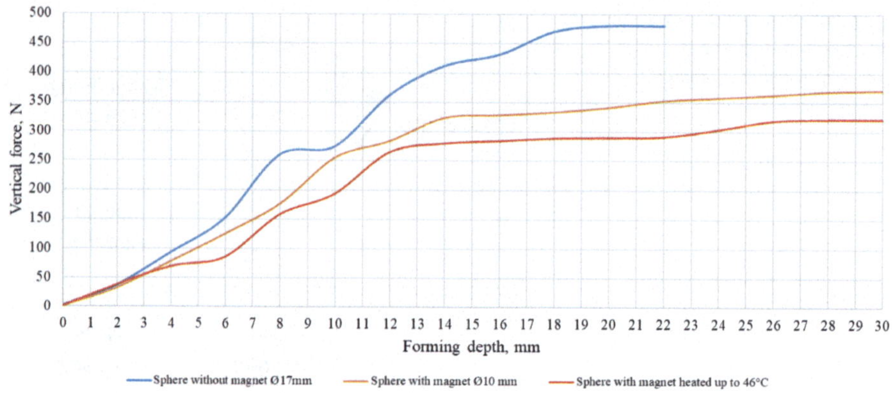

Fig. 2.142 Dependence of the forming force on forming depth

In all polymer forming processes, the first step is ensuring the quality of the polymer materials, by means of controlling the material temperature. Upon heating, most polymers undergo thermal transitions that provide insight into their morphology. The strength of the tested polymer samples decreases with the increasing of heating temperature, and accordingly, the material becomes softer. With the SPIF, a polymer sheet is heated to a target temperature, then formed to a specific shape. A polymer sheet is locally deformed by a forming tool that moves on the surface of the polymer sheet. Since the moving tool is following a defined path in 3D space, this forming process is flexible to be applied to arbitrary shapes. A reasonable localized temperature should be produced below the glass transition temperature, which was ideal to soften the thermoplastic sheet and keep the stiffness at a proper level. A SPIF device is modified through the development of a specialized tool holder and nozzle which heats the polymer sheet to temperatures higher than the room temperature but below the glass transition temperature of the polymer and applies the forming loads. The present process provides localized and controlled heating through the contact of the sheet with a tool and then constitutes a viable way to preserve the flexible nature of this technology. A preliminary experimental campaign consisted of dynamic mechanical analyses on the polymer sheets to get an idea of the influence of the tool shape on the sheet heating, in order to reach temperatures that soften the polymeric sheets and do not compromise their surface quality. The heated tool moving on the sheet surface locally increases the plasticity of the polymer in the contact zone with less deforming force does not reducing the stiffness of the polymer around the tool contact area and eliminating the need for a backing plate.

References

1. Ostasevicius V, Gaidys R, Gylienė V, Jūrėnas V, Daniulaitis V, Liepinaitis A (2011) Passive optimal tool structures for vibration cutting. J Vibroeng 13(4):769–777
2. Haug EJ, Arora JS (1979) Applied optimal design: mechanical and structural systems. Wiley, NY, p 506
3. Ostasevicius V et al, Twist drill with variable cross-section. SU invention No. 1357151
4. Ostasevicius V, Jurenas V, Juskevicius A (2014) Modified tool structures for effective cutting. Mechanika 20(2):171–176
5. Bathe KJ, Wilson EL (1976) Numerical methods in finite element analysis. Prent-Hall, Englewood Cliffs, NJ USA, p 528
6. Ostasevicius V et al, Internal turning tool holder, SU invention Nr.1590206
7. Standards Committee of the IEEE Ultrasonics (1988) Ferroelectrics and frequency control society. In: IEEE standard on piezoelectricity. American National Standards Institute, NY
8. Ostasevicius V, Gaidys R, Rimkeviciene J, Dauksevicius R (2010) An approach based on tool mode control for surface roughness reduction in high-frequency vibration cutting. J Sound Vibr 329:4866–4879
9. Rimkeviciene J, Gaidys R, Jurenas V, Ostasevicius V (2009) Research of ultrasonic assisted turning tool. J Vibroeng 11(1):34–40
10. Rimkeviciene J, Ostasevicius V, Jurenas V, Gaidys R (2009) Experiments and simulations of ultrasonically assisted turning tool. Mechanika 75(1):42–46
11. Jurenas V, Bubulis A, Skiedraite I, Ostasevicius V, Grazevicute J (2010) Tool jaw with ultrasonic transducer. Patent of Rep Lithuania B23C 9/00 No. LT 5586 B: 5
12. Ostasevicius V, Jurenas V, Grazevicute J, Bubulis A, Skiedraite I (2010) Quality control system of processed surface, Patent of Rep Lithuania B23B 1/00 No. LT 5534 B: 6
13. Vaicekauskis M, Gaidys R, Ostasevicius V (2013) Influence of boundary conditions on the vibration modes of the smart turning tool. Mechanika 3:296–300
14. Ostasevicius V, Ubartas M, Gaidys R, Jurenas V, Samper S, Dauksevicius R (2012) Numerical—Experimental identification of the most effective dynamic operation mode of a vibration drilling tool for improved cutting performance. J Sound and Vibr 331:5175–5190
15. Ubartas M, Ostasevicius V, Jurenas V, Gaidys R (2010) Experimental nnvestigation of vibrational drilling. In: Mechanika 2010: proceedings of the 15th international conference, pp 445–449
16. Ubartas M, Ostasevicius V, Samper S, Jurenas V, Dauksevicius R (2011) Experimental investigation of vibrational drilling. Mechanika 17(4):368–373
17. Ubartas M, Ostasevicius V, Jurenas V, Dauksevicius R (2012) Experimental investigation of vibrational tool dynamics. In: Mechanika 2012: proceedings of the 17th international conference, 322–326
18. Bayly P, Metzler S, Schaut A, Young K (2001) Theory of torsional chatter in twist drills: model, stability analysis and composition to test. J Man Sci and Eng 123(4):552–561
19. Gylienė V, Ostasevicius V, Ubartas M (2013) Drilling process modelling using SPH. FEA Inf Eng J 2(6):1–6
20. Gylienė V, Ostasevicius V (2011) Cowper-Symonds material deformation law application in material cutting process using LS-Dyna FE code: turning and milling. In: 8th European LS-DYNA users conference, p 11
21. Ostasevicius V, Jurenas V, Vilkauskas A, Balevicius G, Senkus A, Jotautienė E (2017) A novel excitation approach to ultrasonically-assisted cylindrical grinding. Stroj Vest J Mech Eng 63(12):696–704
22. Ostasevicius V, Jurenas V, Balevicius G, Cesnavicius R (2020) Development of actuators for ultrasonically assisted grinding of hard brittle materials. Int J Adv Man Techn 106:289–301
23. Ostasevicius V, Jurenas V, Eidukynas D, Grigaliunas V, Gudauskis M, Paleviciute I, Ambrasas V (2020) Pecularities of the robotized incremental metal and polimer sheets forming. In: 15th international conference on mechnical system and material MSM'2020 proceedings, IEEE Xplore 19996245, pp 1–3

24. Ostasevicius V, Paulauskaite-Taraseviciene A, Paleviciute I, Jurenas V, Griskevicius P, Eidukynas D, Kizauskiene L (2022) Investigation of the robotized incremental metal-sheet forming process with ultrasonic excitation. Materials 15(3):1024
25. Kumar VC, Hutchings IM (2004) Reduction of the sliding friction of metals by the application of longitudinal or transverse ultrasonic vibration. Tribol Int 37:833–840
26. Ostasevicius V, Jurenas V, Grigaliūnas V, Eidukynas D, Bubulis A, I.Paleviciute I (2020) Incremental forming machine for sheet metal parts. Patent Appl LT2020 516:6
27. Bhattacharya A, Maneesh K, Venkata Reddy N, Cao J (2011) Formability and surface finish studies in single point incremental forming. J Manu Sci Eng 133(6):8
28. Stoughton TB, Yoon JW (2011) A new approach for failure criterion for sheet metals. Int J Plast 27(3):440–459
29. Jawale K, Duarte JF, Reis A, Silva MB (2018) Microstructural investigation and lubrication study for single point incremental forming of copper. Int J Sol Str 151:145–151
30. Gorji M, Berisha B, Manopulo N, Hora P (2016) Effect of through thickness strain distribution on shear fracture hazard and its mitigation by using multilayer aluminum sheets. J Mat Proc Techn 232:19–33
31. Bouffioux C, Lequesne C, Vanhove H, Duflou JR, Pouteau P, Duchêne L, Habraken AM (2011) Experimental and numerical study of an AlMgSc sheet formed by an incremental process. J Mat Proc Techn 211(11):1684–1693
32. Silva MB, Skjødt M, Atkins AG, Bay N, Martins PAF (2008) Single-point incremental forming and formability-failure diagrams. J Str Anal for Eng Des 43(1):15–35
33. Allwood JM, Shouler DR, Tekkaya AE (2007) The increased forming limits of incremental sheet forming processes. Key Eng Mat 344:621–628
34. Emmens WC, Van den Boogaard AH (2007) Strain in shear, and material behaviour in incremental forming. Key Eng Mat 344:519–526
35. Ostasevicius V, Paleviciute I, Paulauskaite-Taraseviciene A, Jurena V, Eidukynas D, Kizauskiene L (2022) Comparative analysis of machine learning methods for predicting robotized incremental metal sheet forming force. Sensors 22(18):22
36. Hauke J, Kossowski TM (2011) Comprison of values of Pearson's and Spearman's correlation coefficients on the same sets of data. Quaest Geogr 30(2):87–93
37. Browne MW (2000) Cross-validation methods. J Math Psych 44(1):108–132
38. Pietersma A, Lacroixa R, Lefebvre D, Wade K (2003) Performance analysis for machine-learning experiments using small data sets. Comp Electr in Agric 38(1):1–17
39. Vabalas A, Gowen E, Poliakoff E, Casson AJ (2019) Machine learning algorithm validation with a limited sample size. PlosOne 14(11):20
40. Torgyn S, Lowe D, Daga S, Briggs D, Higgins R, Khovanova NA (2015) Machine learning for predictive modelling based on small data. Biomed Eng 48(20):469–474
41. Shaikhina T, Khovanova N, Mallick K (2014) Artificial neural networks in hard tissue engineering: another look at age-dependence of trabecular bone properties in osteoarthritis. IEEE EMBS Int Conf Biomed Health Inform, 484–487
42. Li Y, Shami A (2020) On hyperparameter optimization of machine learning algorithms. Theor Pract Neurocomp 415:295–31642
43. Wu J, Chen XY, Zhang H, Xiong LD, Lei H, Deng SH (2019) Hyperparameter optimization for machine learning models based on bayesian optimization. J Electron Sci Techn 17(1):26–40
44. Chicco D, Warrens MJ, Jurman G (2021) The coefficient of determination R-squared is more informative than SMAPE, MAE, MAPE, MSE and RMSE in regression analysis evaluation. Peer J Comput Sci 7:e623
45. Najm SM, Paniti I (2021) Artificial neural network for modeling and investigating the effects of forming tool characteristics on the accuracy and formability of thin aluminum alloy blanks when using SPIF. Int J Adv Manuf Technol 114:2591–2615
46. Ostasevicius V, Eidukynas D, Jurenas V, Paleviciute I, Gudauskis M, Grigaliunas V (2021) Investigation of advanced robotized polymer sheet incremental forming process. Sensors 21(3137):1–25

47. Rochling Industrial Materials. https://www.roechling.com/industrial/materials/thermoplastics/detail/trovidur-esa-d-261
48. Ostasevicius V, Jurenas V, Eidukynas D, Grigaliunas V, Gudauskis M, Bubulis A, Paleviciute I, Ambrasas V (2020) Incremental forming machine for sheet plastic parts. Patent Appl LT2020 528:6
49. Ostasevicius V, Jurenas V, Bubulis A, Eidukynas D, Paulauskaite-Taraseviciene A, Paleviciute I (2021) Sheet parts incremental forming device. Patent Appl LT2021 549:6
50. Littmann W, Storck H, Walaschek J (2001) Reduction of friction using piezoelectric vibrations. Smart Str Mat: Damp Isol Int Soc Opt Phot 4331: 302–311

Chapter 3
Integration of Digital and Physical Data to Process Difficult-to-Cut Materials

3.1 Digital Twin for Excited Cutting Tool

A digital twin-driven cutting tool can be a modern solution to significant manufacturing advances, among which the digitization of production is a major contributor to ever-increasing productivity. Digital cutting tool twin as a digital copy of a physical tool, its data is collected throughout the production life cycle using the international standard ISO 13399.

3.1.1 Prevention of Chemical Interactions Between Tool and Workpiece Materials by Contact Time Reduction

To prevent a chemical reaction between the carbon of a diamond turning tool and iron of the hardened steel an ultrasonic tool holder was designed and a prototype of tool holder was elaborated. Based on a longitudinal excited transducer, a longitudinal wave is transfomed through a flexural sonotrode into a transversal wave. The aim of the new development was to reach an operating frequency of more than 100 kHz and minimize disadvantageous features related to weight and geometry of such devices.

With every day diamond turning becomes increasingly more used in various fields of manufacturing. Applying ultrasonic vibrations to diamond cutting it's possible to decrease cutting forces and temperature. If 40 kHz ultrasonic vibration is applied to a single-crystal diamond tool and directed towards cutting, less than 0.03 μm surface roughness can be achieved. Carbon (i.e. the diamond) and iron (i.e. a steel work piece) are affine to each other. The contact with steel under pressure and high temperature leads to graphitization of the diamond. Therefore, mono-crystalline diamonds are unsuitable for conventional ferrous metal applications. Many methods have been applied to minimize this effect, such as the use of turning with cryogenic or carbon saturated atmospheres. However, the efforts to change the atmosphere of the cutting zone do not avoid the chemical tool wear. Scientists from Fraunhofer

V. Ostaševičius, *Digital Twins in Manufacturing*, Springer Series in Advanced Manufacturing, https://doi.org/10.1007/978-3-030-98275-1_3

IPT, Germany, use a different approach for diamond cutting: they claim, that under the usual conditions of ultra-precision diamond turning, the process of removing material is determined by the effect the material of the workpiece and the cutting edge of the diamond (several tens of nanometres) have on each other. Therefore, the micro structure of the workpiece can have a significant influence on the results of the cutting. As optic moulds are made from brittle materials, such as ceramics, profound achievements have been made using high frequency diamond turning. A feasibility study of ductile mode machining employing ultrasonic elliptical vibration cutting technology to process sintered tungsten carbide demonstrated, that ductile mode machining can produce satisfying results, if elliptical vibration cutting is applied at 36 kHz frequency. Here, in comparison to the usual type of cutting techniques the surface is not seriously damaged upon finishing.

The future manufacturing in small, very precise and accurate field belongs to ultrasonic assisted diamond turning tools [1]. To construct precision diamond tools, the modal analysis of the tool is required for the knowledge of the tool excitation frequency. By combining vibration and high-frequency assisted single point diamond turning, an ultrasonic tool holder for diamond turning can be created. The numerical comparison analysis consists of several parts, namely modal analysis and harmonic response analysis, performed individually on each part of the tool, transducer and sonotrode, plus transient analysis of the whole tool. It has also been established, that this method is especially beneficial in the development of micro structures, because it has a smaller cutting force and reduced formation of burrs. As in an ordinary cutting process, the chip is always on the rake surface of the tool tip, which is a region of high temperature and high pressure. Here, it is very difficult for the coolant to reach the cutting area and thus, the coolant can function only around the tool tip and indirectly in the process. During vibration cutting, with frequent separation between the work piece and the tool, the tool-chip contact area will be opened periodically. Ultrasonic excitation of the tool will promote aerodynamic lubrication and thereby reducing friction between the tool and the work piece. Since ultrasonic assisted cutting is an intermittent cutting process, the tool is separated from the work piece and refrigerant enters the cutting area to cool and lubricate the tool tip sufficiently. Currently, the tool will cut the work piece and the liquid will be strongly compressed, which will generate high pressure. This instantaneously forces the liquid on the contact surface between the tool and the chip directly to cool and lubricate the contacting surface sufficiently. This will lead to a significant reduction in cutting forces and cutting temperature. As a result, this will reduce the possibility of crack initiation and propagation.

3.1.1.1 Virtual Twin of Sonotrode for Diamond Cutting

The ultrasonic tooling system was designed supported by using numerical simulation methods, such as modal, harmonic and transient analysis [2]. In total, the system consists of two main parts: A longitudinal actuated transducer and a bending sonotrode. First, both parts had been designed and investigated, separately by applying modal FE analysis. Afterwards, an assembly of both components have been analyzed, using modal and harmonic FE analysis. At last, a cutting force was added to the model, to perform a transient analysis of the assembly during machining conditions.

The transducer of the tooling system is used to generate a longitudinal vibration at high frequencies that induces the vibrations into the tooling system. Basically, a transducer consists of a backing mass and a piezo electric ring. The first part of the mathematical simulation involves a numerical modelling of the transducer, using the FE software Comsol Multiphysics. Here, an axial symmetry FE model of the transducer was created. The created 3D model of the transduceris shown in Fig. 3.1.

The inverse piezoelectric effect is used to generate the excitation of the transducer. Here, the independent variables, describing the inverse piezoelectric effect, are the strain *S* and the electric displacement *E*. The electric displacement is chosen, because the electric boundary condition is given for the system under consideration. Thus, the fundamental relations for the chosen variables *S* and *E* could be expressed as:

$$T = c^E S - e^t E \tag{3.1}$$

$$T_{ij} = c^E_{ijkl} S_{kl} - e_{kij} E_k$$

and

$$D = eS + \varepsilon^S E$$

$$D_i = e_{ikl} S_{kl} + \varepsilon^S_{ij} E_k \tag{3.2}$$

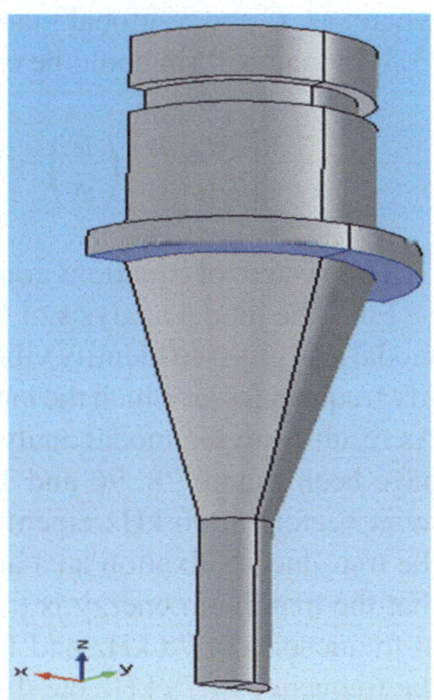

Fig. 3.1 Transducer's 3D model

where E—electric field intensity, D—electric field displacement tensors.

The variables describing the process mechanics are T—stress and S—strain. C^E—describes the elastic stiffnesses at constant electric field, ε^S—dielectric permittivity at constant strain, e—piezoelectric constants matrix. It could be expressed in the matrix form:

$$\begin{bmatrix} T_I \\ D_i \end{bmatrix} = \left[\begin{array}{cccccc|ccc} C_{11} & C_{12} & C_{13} & C_{14} & -C_{25} & C_{16} & -e_{x1} & e_{y2} & -e_{z1} \\ C_{12} & C_{22} & C_{23} & -C_{14} & C_{25} & -C_{16} & e_{x1} & -e_{y2} & -e_{z2} \\ C_{13} & C_{23} & C_{33} & 0 & 0 & 0 & 0 & 0 & -e_{z3} \\ C_{14} & -C_{14} & 0 & C_{44} & 0 & 2C_{25} & -e_{x4} & -e_{y4} & -e_{z4} \\ -C_{25} & C_{25} & 0 & 0 & C_{55} & 2C_{14} & -e_{x5} & -e_{y5} & -e_{z5} \\ C_{16} & -C_{16} & 0 & 2C_{25} & 2C_{14} & C_{66} & 2e_{y2} & 2e_{x1} & -e_{z6} \\ \hline e_{x1} & -e_{x1} & 0 & e_{x4} & e_{x5} & -2e_{y2} & \varepsilon_{11}^S & 0 & 0 \\ -e_{y2} & e_{y2} & 0 & e_{y4} & e_{y5} & -2e_{x1} & 0 & \varepsilon_{22}^S & 0 \\ e_{z1} & e_{z2} & e_{z3} & e_{z4} & e_{z5} & e_{z6} & 0 & 0 & \varepsilon_{33}^S \end{array}\right] \cdot \begin{bmatrix} S_J \\ E_j \end{bmatrix} \tag{3.3}$$

The mathematical formulation of the piezoelectric equations for the finite element method could be expressed by the following system of differential equations:

$$M\ddot{u} + C_{uu}\dot{u} + K_{uu}u + K_{u\phi}\phi = F \tag{3.4}$$

and

$$K_{u\phi}^t u + K_{\phi\phi}\phi = Q \tag{3.5}$$

where: M, C and K-global matrices, and u, F, φ and Q-vectors. Equations (3.4) and (3.5) in matrix form could be expressed:

$$\begin{bmatrix} M_{uu} & 0 \\ 0 & 0 \end{bmatrix}\begin{pmatrix} \ddot{u} \\ \ddot{\phi} \end{pmatrix} + \begin{bmatrix} C_{uu} & C_{u\phi}^t \\ C_{u\phi} & C_{\phi\phi} \end{bmatrix}\begin{pmatrix} \dot{u} \\ \dot{\phi} \end{pmatrix} + \begin{bmatrix} K_{uu} & K_{u\phi}^t \\ K_{u\phi} & K_{\phi\phi} \end{bmatrix}\begin{pmatrix} u \\ \phi \end{pmatrix} \tag{3.6}$$

This system of equations could be solved by direct numerical integration methods.

First, the modal analysis of the transducer was accomplished. The purpose of the modal analysis is to identify vibration modes (in a range of 75 ÷ 115 kHz) and to identify frequencies, at which the largest axial displacement of the transducer's tip occurs. As result from the modal analysis, three eigenfrequencies with such characteristics have been found: 78, 96 and 113 kHz. In Fig. 3.2c, the vibration mode displacement vectors at 96 kHz eigenfrequency are represented. Here, displacements near the transducer's fixation area are the smallest at this vibration mode, which means, that the maximum energy is transported to the transducer's tip. In contrast to this, at frequencies of 78 kHz and 113 kHz (Fig. 3.2a, b) such effect cannot be seen. At the frequency of 78 kHz the displacement passes below the fixation area and at the

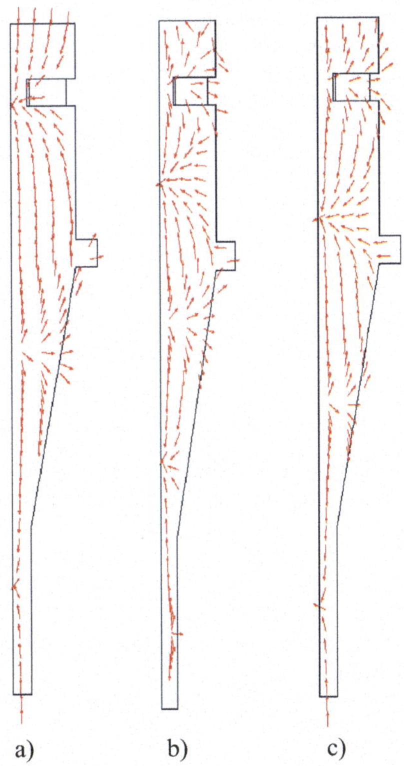

Fig. 3.2 Vibration mode displacement vectors of transducer at: **a** 78 kHz, **b** 113 kHz and **c** 96 kHz

frequency of 113 kHz the displacement appears above the fixation area. In conclusion, it can be stated, that the transducer seems to be operating with advantageous dynamic properties at a frequency of 96 kHz.

The second part of the analysis consists of the transducer's harmonic analysis. Here, the piezo electric rings were excited by a harmonic voltage. Results of the analysis are shown in Fig. 3.3. A frequency response analysis was performed with different voltage values. By applying a voltage of 30 V, a displacement of 0.2 μm at a frequency of 88 kHz occurs at the end of the transducer's tip. Two times higher displacement values can be observed after increasing the voltage up to 60 V.

The sonotrode literally vibrates during the turning process. Thus, it is necessary to find a design and geometry, at which the requirements on the height of the frequency and amplitude are matched. After completing the modal analysis of the sonotrode's shape, modes and nodal points of the sonotrode's transverse vibrations were found. It was determined, that the maximum amplitude of the sonotrode's tip, inside a given spectrum, was obtained at a frequency of 102 kHz. The nodal point of this mode was selected as pinned fixing point of the sonotrode. For a stable, continuous vibration, an advantageous geometrical form must been found. Thus, it is essential to optimize its shape. Based on this, it is necessary, to formulate an optimization procedure

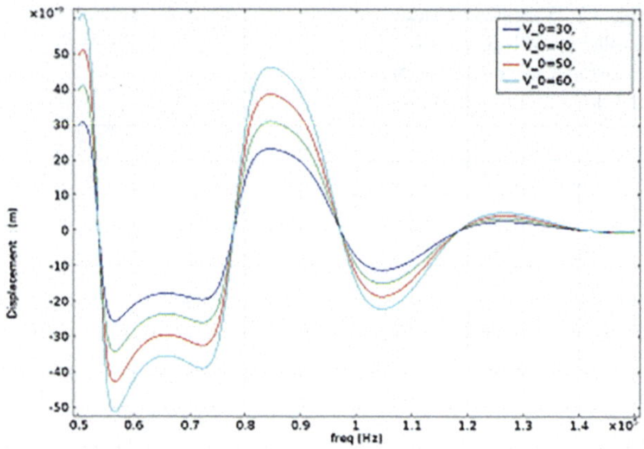

Fig. 3.3 Relation between longitudinal displacement of the transducer's tip and excitation frequency by using different voltages

for the sonotrod's shape: A maximized transversal displacement of the sonotrode's tip—max V_{tip} defined by the state equation:

$$K(x)v = w_i^2 M(x)v \tag{3.7}$$

where w_i is the natural frequency and v is the modal shape of the system, with the constraint

$$x_{\min} < x < x_{\max} \tag{3.8}$$

and eigenfrequency

$$w = w*, \tag{3.9}$$

where x is the vector of the design parameters, w is the eigenfrequency of the cantilever, and $w*$ is the desired frequency.

The designed sonotrode was simulated in the COMSOL environment, by using fixation point areas, as highlighted in (Fig. 3.4). For the ultrasonic tool holder, the sonotrode's function is the transfer of the vibration, generated by the transducer, to the cutting zone.

Analogue to the previous investigation of the tool holder's transducer, a modal and harmonic analysis have been performed for the designed sonotrode. First the modal analysis helped to find the mode for a suitable transversal movement of the sonotrode. Second, the harmonic response analysis provided information about the tool behavior, after harmonic oscillations were applied on it (Fig. 3.5).

Two transversal vibration modes (in a range of 100 kHz) of the sonotrode occur—one at a frequency of 108 kHz (Fig. 3.5a) and another at a frequency of 98 kHz (Fig. 3.5b). As one main result from the modal analysis, it can be seen, that a generated standing wave inside the sonotrode is oscillating around zero displacement point.

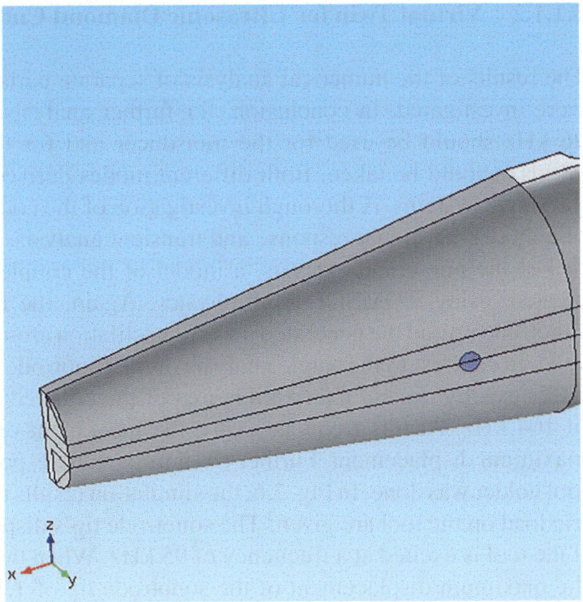

Fig. 3.4 Sonotrode model with highlighted fixation area

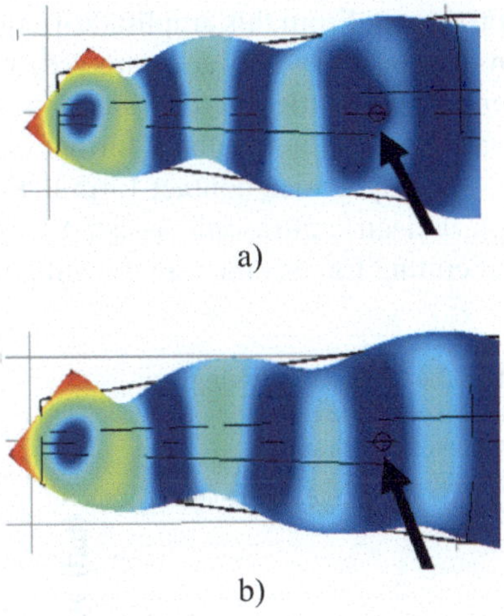

Fig. 3.5 Sonotrode eigen modes at: **a** 108 kHz and **b** 98 kHz (total displacement field, arrow indicates position for fixation area)

Those points lie within the highlighted fixation areas, which are also marked by black arrows (Fig. 3.5a, b). Both displayed modes show a transversal generated movement of the sonotrode tip. While the modal analysis at the mode of 98 kHz shows an oscillation around the fixation areas, the zero displacement points and the fixation area do not coincide with each other at the mode of 108 kHz.

3.1.1.2 Virtual Twin for Ultrasonic Diamond Cutting

The results of the numerical analysis of separate parts of the ultrasonic tool holder were investigated. In conclusion, for further analysis, the mode at a frequency of 96 kHz should be used for the transducer and for the sonotrode a frequency of 98 kHz should be taken. Both different modes derived from a separate analysis of the single systems. A thorough investigation of the coupled system analysis consists of a modal, harmonic response and transient analysis.

For the numerical analysis, a model of the coupled ultrasonic tool holder was created, using COMSOL Multiphysics. Again, the transversal movement of the sonotrode tip and the generated wave of oscillation around the fixation area were analysed. In contrast to the single analysis of the sonotrode, the zero displacement points occur exactly around the fixation area of the coupled system at a modal frequency of 104 kHz. At this frequency, also the movement of the sonotrode tip achieve a maximum displacement. Further on, a frequency response analysis of the ultrasonic tool holder was done. In Fig. 3.6, the simulation results under the harmonic piezoelectric load on the tool are given: The sonotrode tip's displacement is less than 0.1 μm, if the tool is excited at a frequency of 95 kHz. When the frequency reaches 102 kHz, the maximum displacement of the sonotrode tip of 1.1 μm (peak-zero) is reached. From the harmonic response analysis, a gap of low displacement between two peaks is revealed. This zone of operating frequencies should be avoided for later machining processes. From this amplitude-frequency characteristic, we can see that only digital modeling allows to accurately determine the sonotrode excitation frequency that ensure the brittle-ductile transition of the material necessary for processing which is validated by the results of further experiments.

The dynamics contact formulation was implemented into the FE model of the piezoelectric ultrasonic assisted vibration system, in order to simulate the response of cutting forces under a vibro-impact mode of the cutting process. The formulation

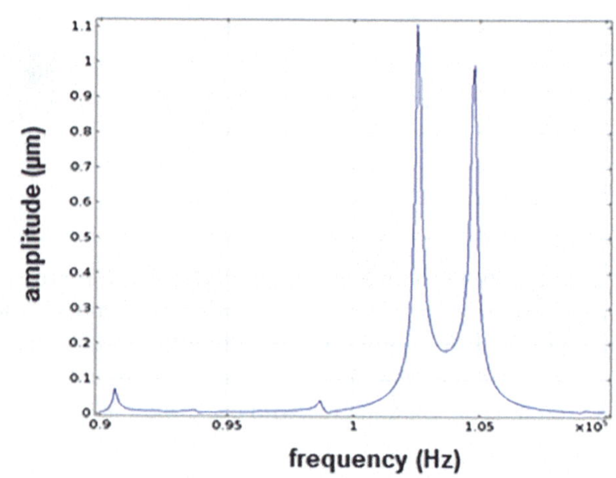

Fig. 3.6 Diamond turning tool (sonotrode tip) transversal vibration amplitude-frequency characteristic

is based on the Kelvin-Voigt rheological model, represented by linear and nonlinear springs, connected in parallel with linear dampers (this coupling element is defined by the stiffness k_c, k_{th} and the damping c_c, c_{th}).

The dynamics of the ultrasonically actuated vibro-cutting are defined by matrix Eq. (3.6), with {F} defining the equivalent cutting force:

$$\{F\} = \begin{cases} \{F(t)\}, \\ \{F(t)\} + \{P_c(\dot{z}), z, t\}, \end{cases} \quad (3.10)$$

$$if\ t > t_1 \text{ and } F_c > 0 \text{ and/or } F_{th} > 0$$

where: $\{F(t)\}$—radial thrust force, $\{P_c\ (z,\ z,\ t)\}$—vector of nonlinear interaction in the contact pair, F_c—cutting force acting in vertical direction, F_{th}—thrust force acting in longitudinal direction, t_1—moment of tool-workpiece contact

$$F_c = if(u_c > 0), k_c * u_c - if(ut_c > 0), c_c * ut_c \quad (3.11)$$

and

$$F_{th} = -if(ut_h \geq 0) * (k_{th} * u_{th})^2 - c_{th} * ut_{th} \quad (3.12)$$

where: u_c and ut_c—vertical displacement and velocity of the tip of the diamond tool, u_{th} and ut_{th}—longitudinal displacement and velocity of the tip of the diamond tool, k_c, k_{th}, c_c, c_{th}—stiffness and damping coefficients of the coupling element, respectively.

The model of ultrasonic cutting contact was introduced into the FE model of the piezoelectric assisted vibration system as a cutting and a thrust force, acting on the selected point, located on the tip edge of the diamond tool. According the results from the harmonic analysis, the system should be excited by the piezoelectric transducer at a frequency of 102 kHz. During the stationary process, the frequencies and phases of the diamond tool's transversal and longitudinal vibrations coincide with each other. When the steady stationary vibration process starts ($t = 0.7 \cdot 10^{-3}$s), the amplitude of the diamond tool tip's transversal vibrations is in a range of 1 μm, which correlates with the following gathered experimental results (see below). The amplitude of the longitudinal vibrations of approximately 200 nm corresponds to this experimental results as well.

When the diamond tool touches the workpiece, the cutting process begins. The vibrating diamond tool is pressed against the workpiece by a force of 1 N at the time ($t > 0.7 \cdot 10^{-3}$s), resulting in the beginning of a new transient process (Fig. 3.7a). Due to the ensuing cutting force, the transversal vibration's axis of the diamond tool budges down. This can be explained be the cutting process to start, when the diamond tool moves to the positive or to the up direction.

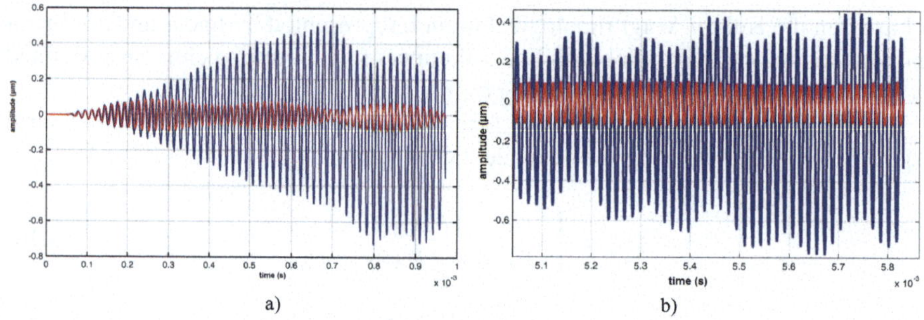

Fig. 3.7 Diamond tool tip transversal (blue) longitudinal (red) vibrations in the initial (**a**) and during the cutting (**b**) stages

During the cutting stage, the amplitudes of the transversal and longitudinal vibrations remain unchanged. But, the phrases of the transversal and longitudinal vibrations become opposite which means, that the difference between them is 180° (Fig. 3.7b). The lower frequencies of the transversal vibration's constituents dominate during this stage. Next, a Fast Fourier Transform (FFT) and spectral analysis of the vibration process was performed, in order to investigate these constituents.

From the diamond tool tip's vertical vibrations power spectral density analysis, the dominant 102 kHz frequency coincides with the piezoelectric transducer's excitation frequency. The lower 5 kHz and 11 kHz frequencies of the diamond tool tip's dynamic process are related to the sonotorde's movement in vertical direction and the first mode of the sonotrode's lateral vertical vibrations. Its influence is not considerable (Fig. 3.8).

The dependency of cutting-thrust forces is presented in Fig. 3.9.

At Fig. 3.9, it can be seen, that the cutting and thrust forces are not synchronized. The period of vibrations is composed of four different intervals if time: Cutting-only interval, thrust-only interval, both acting forces interval and none of them acting interval. When the cutting force is missing, the diamond tool is going down, while the cutting process starts, when the diamond tool tip is going up. Next, only one cutting period is taken into consideration (Fig. 3.10). Inside the initial interval, there

Fig. 3.8 Power spectral density of diamond tool tip steady dynamic process in vertical (cutting direction) vibration

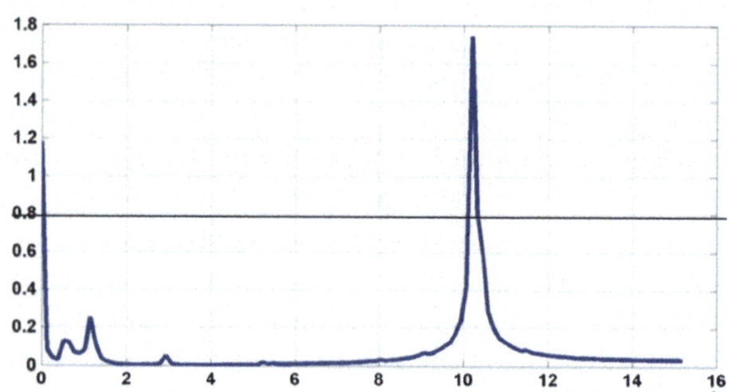

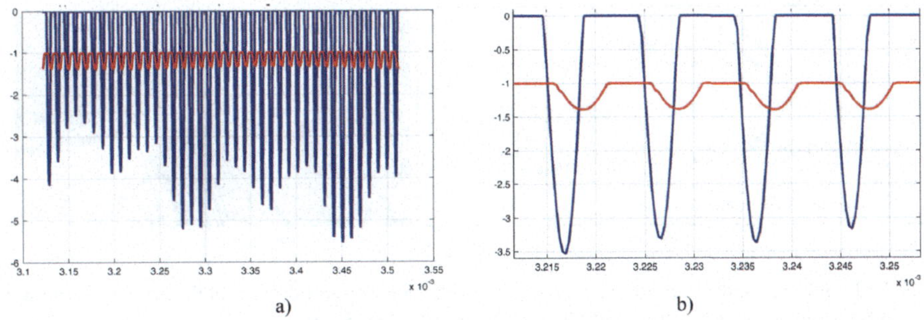

Fig. 3.9 Diamond tool tip and work piece contact forces: **a** in normal and **b** in enlarged scales: blue—cutting force, red—thrust force

Fig. 3.10 Forces acting on the diamond tool during one vibration period: blue—cutting, red—thrust

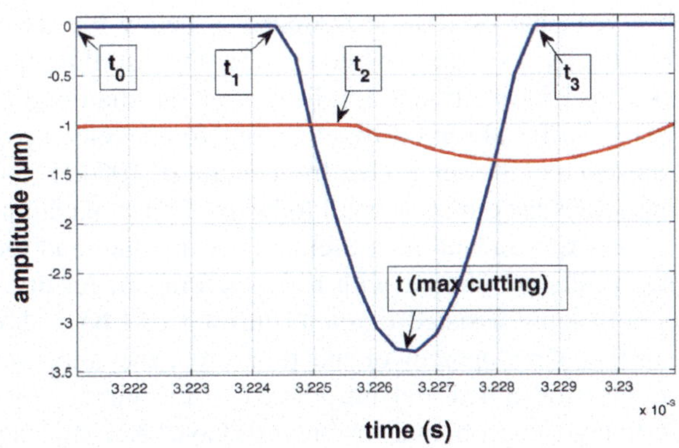

is no cutting or thrust force. This means, that the diamond tool is going down in the vertical direction, but moving opposite to feed direction in horizontal direction. Inside the second interval, the cutting force appears, but the thrust force is equal to zero. Thus, the cutting stage begins, but the dynamical pressing force is missing. Inside the third interval, the maximum cutting and thrust forces are acting, which indicates, that the cutting and thrust forces are present. Inside the fourth interval, the cutting discontinues, but the thrust is remaining.

3.1.1.3 Physical Twin and Experimentation of Diamond Turning

A diamond tool can be mounted on the sonotrode and its vibrational amplitude can be measured, afterwards. This measurement is performed to check whether the device is working properly or not. There might be more than one modal frequency, which can be auto-detected by the used function generator in a range from 95 kHz ÷ 110 kHz. The measurement set-up consists of a laser Doppler vibrometer, a vibrometer controller, an oscilloscope and ultrasonic Generator (Fig. 3.11a).

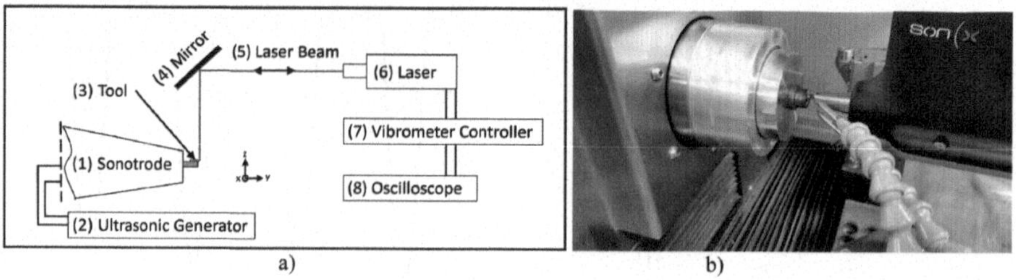

Fig. 3.11 Scheme of the experimental set-up for vibrometer measurements of the amplitude (**a**) and set-up for validation part turning with Moore Nanotech 350 FG (**b**)

Unidirectional peak-to-peak amplitudes of 0.1 ÷ 4.0 μm have been measured using several different oscillation frequencies. During the measurements, it was observed, that oscillation frequencies below 98 kHz generate very small tool tip movements. For theses frequencies, the oscilloscope showed oscillations below 0.4 μm. Unidirectional amplitudes of the diamond tool tip with a magnitude of 3 μm can be observed by changing frequencies up to 103 kHz. By exciting the diamond tool to even higher frequencies of 105 kHz and 110 kHz, it occurs, that the amplitudes become too small to be used for a machining process.

The experimental measurements of the diamond tool tip show, that the FE analysis results are similar with the experimental results. Based on these, a prototype for ultrasonic assisted diamond turning had been developed and tested. Next, as a further development of the prototype, the ultrasonic tooling system 2 (UTS2) has been designed and introduced to the market by the company son-x GmbH, Germany. Experimental turning tests have been carried out with the UTS2, for further investigations of the FEM analysis results and the amplitude measurements.

Two turning experiments of hardened steel standard validation parts (concave/convex) were carried out. All experiments were performed on an ultra-precision lathe, a Moore Nanotech 350FG Ultra-Precision Freeform Generator. The experimental set up is shown in Fig. 3.11b.

One standard procedure for evaluating an ultrasonic tool holder is machining two spherical surfaces (Fig. 3.12a, b). These types of surfaces are commonly used as tools in precision injection molding.

Measurements have been taken for three different diamond tools. The results of the measurements for one tool are summed up in the following Table 3.1.

It can be seen from the Table 3.1 that the amplitudes at a frequency of 102 kHz are significantly higher as for the other two frequencies of 108 kHz and 109 kHz. Comparing to the simulation results the highest amplitudes were received at the frequency of 102 kHz. The next figure (Fig. 3.13) shows the amplitudes of the three different radius diamond tools used at a frequency of 102 kHz.

All three tool's vibrational movements (peak—zero) are above 0.6 μm. By comparing this with the simulation results, the highest amplitude is reached at a voltage of 60 V, which is exactly the same result obtained experimentally. For a stable cutting process applying ultrasonic assited diamond turning with similar

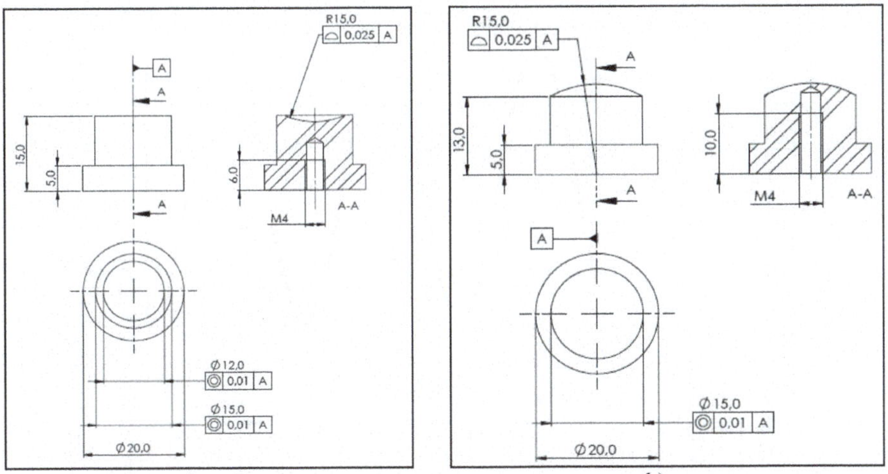

a) b)

Fig. 3.12 Concave (**a**) and convex (**b**) validation part designs in stainless steel with a hardness of 53HRC

Table 3.1 Vibrational amplitude at different frequencies

Tool Radius = 0.208 mm			
Voltage (V)	Amplitude (μm)		
	102 kHz	108 kHz	109 kHz
30	1.36	0.79	0.42
35	1.66	0.84	0.51
40	1.90	0.93	0.63

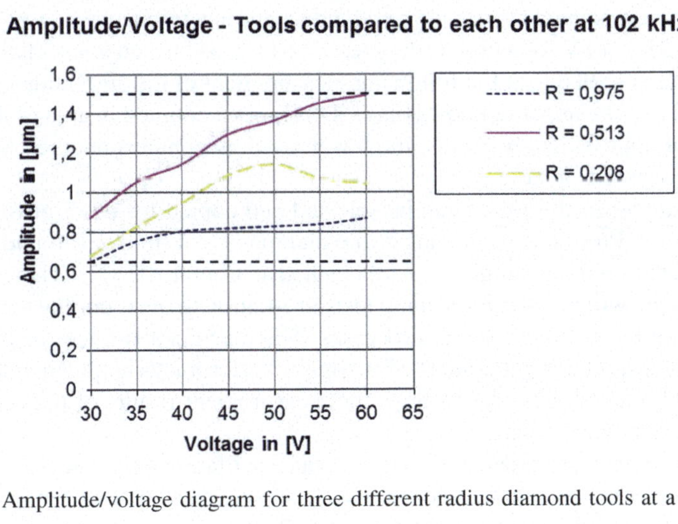

Fig. 3.13 Amplitude/voltage diagram for three different radius diamond tools at a frequency of 102 kHz

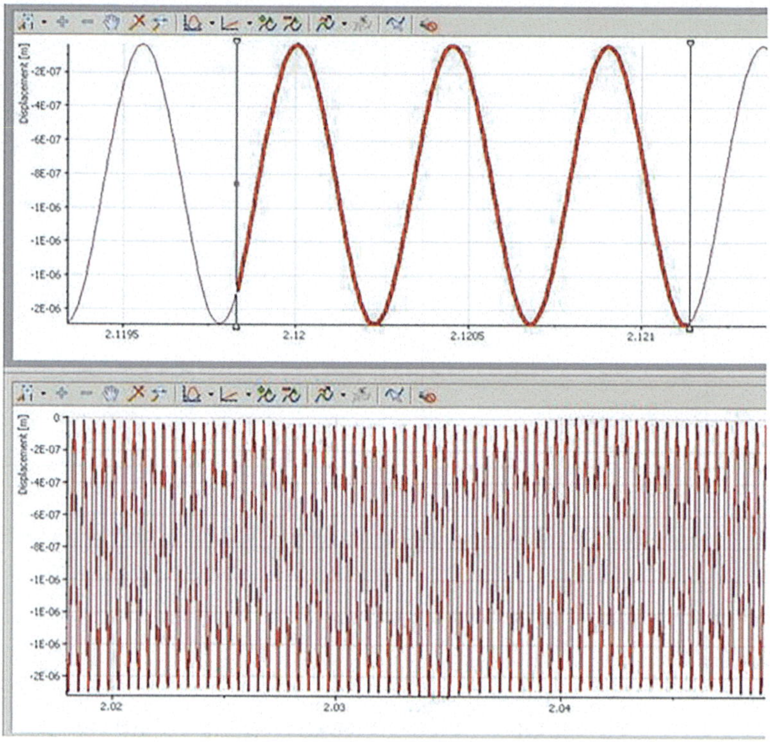

Fig. 3.14 Amplitude over time analyzed by DIA dem 12.0

cutting conditions to 80 kHz process, an amplitude of 0.7 μm is required at 100 kHz. Thus it seemed reasonable to use the designed device at a frequency of 102 kHz with a voltage of 35 ÷ 40 V for a first turning test. Besides, a constant sinusodial shape of the amplitude is seen e.g. at the following measurement of the amplitude over time, recorded for 5 s: the upper window (Fig. 3.14) shows a zoomed in area of the signal in a range of approximately 40 ms, the lower window contains another zoomed in area of the recorded signal.

In there, a part of the signal can be selected and copied for descriptive analysis by the software. From measurements of the diamonds tool tip it was found out, that the highest unidirectional amplitudes are reached at around 102 kHz. Therefore, this frequency value will be used for turning tests in addition to other cutting parameters, corresponding to diamond turning processes. The machined surfaces of both parts have been measured using a Bruker white light interferometer with a magnification of 10 and 50 times. Figure 3.15 shows the measurement results of the concave as well as Fig. 3.16 convex part.

The turning test validated that, while using oscillation frequency of 102 kHz for the UTS2, it is possible to achieve a surface roughness bellow 5 nm, which is considered being a high optical quality surface (Fig. 3.17).

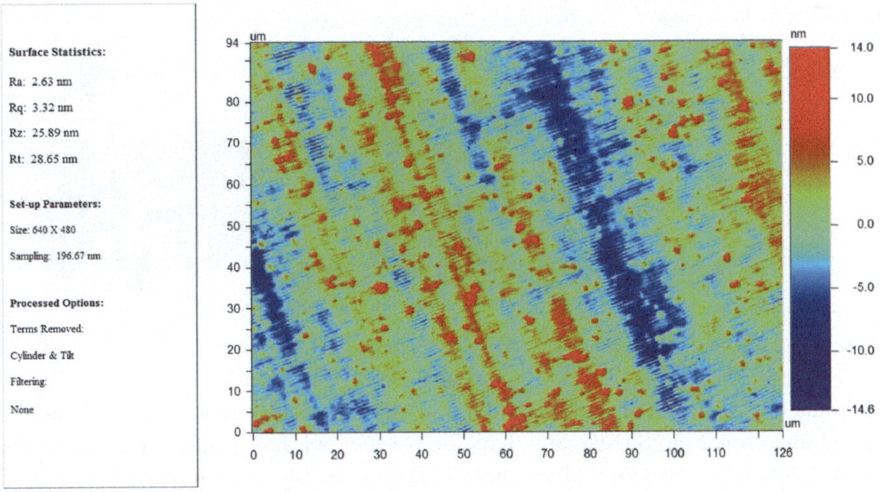

Fig. 3.15 Surface measurements of the concave validation part (concave surface—50 × magnification)

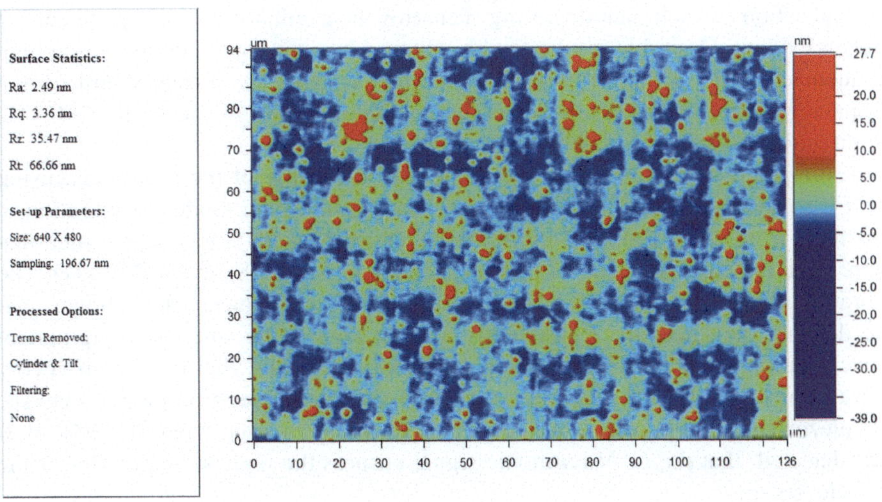

Fig. 3.16 Surface measurements of the convex validation part (convex surface—50 × magnification)

Based on the quality of the surfaces the high frequency seems to have a beneficial influence on the process. This study has provided insights into the design of ultrasonic assisted tool holders for ferrous metals. When comparing conventional with ultrasonic cutting technologies, the interaction between the contacting surfaces of the tool and the workpiece is different. Additional, a reciprocating movement between

a) b)

Fig. 3.17 Finished concave (**a**) and convex (**b**) validation parts with UTS2

the contacting surfaces of the tool and the workpiece results in a periodic change in the frictional force on the rake and flank face of the tool. Ultrasonic excitation of the tool promotes aerodynamic lubrication, thereby reducing friction between the tool and the workpiece. The tool is cutting the workpiece and compressed manufacturing liquid generates high pressure and forces the liquid on the contact surface between the tool and the chip. This leads to a significant reduction of cutting forces and temperature, and reduces the crack initiation and propagation in the workpiece. For the realisation of such manufacturing scenarios the availability of adequate cutting tools is evident. Desirable results can be achieved by exciting diamond tools at high frequencies. State-of-the-art ultrasonic elliptical systems are currently limited to a frequency of up to 60 kHz and thus the challenging task of creating a 100 kHz unidirectional ultrasonic system was proceeded.

The mathematical simulation the numerical modelling of the transducer, using the FE software COMSOL Multiphysics, has identified the modes of vibration in the range from 75 to 115 kHz. From modal analysis results it has been found, that there are three eigenfrequencies with such characteristics: 78, 96 and 113 kHz. The vibration mode displacement vectors at 96 kHz eigenfrequency has shown, that the displacement near the transducer's fixation area is the smallest (longitudinal wave standing point) in this vibration mode, which means, that maximum energy is transported to the transducer's tip. During the transducer's harmonic analysis, the piezoelectric rings were excited with a harmonic voltage from 30 ÷ 60 V. It was detected, that the 60 V harmonic signal ensures the highest amplitudes of the transducers tip.

The Comsol environment was chosen for simulation of the sonotrode. Modal analysis helped to find the mode for the transversal movement of the sonotrode's tip, as well as harmonic response analysis investigated the tool behaviour, while harmonic oscillations were applied. However, the fundamental finding of this research is related to the results of the simulation of the ultrasonic cutting process. As result from the simulation, the optimal operation frequency for the transducer is in a range of 96 kHz and 98 kHz for the sonotrode. In contrast to this result, another optimal frequency for the coupled system was carried out thru simulations. Thus, the results of the harmonic response and transient analysis of the cutting process seem to be more feasible for the evaluation of the vibrational cutting efficiency. From Fig. 3.7a it can be seen, that the sonotrode tip's displacement is less than 0.1 μm while the tool is excited at the frequency of 95 kHz. Such low displacements are barely usable for manufacturing processes. The maximum displacement of the sonotrode's tip is 1.1 μm when reaching a frequency of 102 kHz. From the harmonic response results a gap between two peeks is evident and this low displacement zone should be avoided if high quality machining is required.

To simulate the response of cutting forces under vibro-impacting operation mode of a cutting process, a dynamics contact formulation based on Kelvin-Voigt rheological model, represented by linear and nonlinear springs connected in parallel with linear dampers was implemented into the FE model of the piezoelectric ultrasonic assisted vibration system. When the diamond tool touches the work piece the cutting process begins. Due to the ensuing cutting force, the transverse vibration axis of the diamond tool budges down (Fig. 3.7b). This happens due to the fact, that the cutting process occurs when the diamond tool moves to the positive or to the up direction. During the cutting stage, the amplitudes of the transversal and the longitudinal vibrations remain unchanged. But the phases of the transversal and the longitudinal vibrations become opposite, which means that the difference between them is 180^0 (Fig. 3.8). For the cutting process analysis, four different intervals of time are highlighted: cutting-only interval, thrust-only interval, both cutting and thrust forces interval and none-of-them-acting interval (Fig. 3.11a).

The designed cutting device was manufactured, assembled and experimental cuts were made. Measurements of the amplitude of a mounted diamond tool have been performed, to assure that a required vibrational range was reached. It was found that, when the steady stationary vibration process starts, the amplitude of the diamond tool tip's transversal vibrations are about 1 μm (peak-zero), which correlates with the simulation results. The amplitude of longitudinal vibrations of approximately 200 nm corresponds to the simulation results too. The corellation of the experimental and simulation results testify the adequacy of the mathematical model to the real physical process.

The FE simulation and dynamics analysis for the ultra precision diamond turning tool design was carried out. The ultraprecision diamond turning process was analysed and four different phases of vibration cutting were identified: cutting force, thrust force, both cutting and thrust forces and none of them acting phases. During this process of frequent separation between the work piece and the tool, the tool-chip contact area opened periodically, promoting aerodynamic lubrication and thereby reducing friction between the tool and the work piece. This intermittent cutting process influences the tool—work piece separation and refrigerant is forced into the cutting area, cooling and lubricating the tool tip sufficiently. This leads to a significant reduction in cutting force and cutting temperature, and leads to a brittle-ductile transition phenomena. FE analysis showed, that the most feasible frequencies are in the range of $102 \div 105$ kHz. For the verification of FE simulation results, a diamonds vibration amplitude was measured, along with turning two spherical surfaces, afterwards. The diamond tool tip's amplitude measurements showed, that the best result could be attained while oscillating the system at frequency of 102 kHz with tool tip amplitudes of 1 μm (peak-zero). As result from the cutting experiment, a surface roughness below 5 nm of a ferrous metal workpiece was achieved. Carried out experiments revealed, that numerical calculations fall within the margins of experimental results and could be used in the future for tool modifications and enhancements.

3.1.2 Vibration Milling for Surface Finish Improvement

Treating the milling tool or cutter as an elastic pre-twisted structure and selecting vibration cutting regimes, it is necessary to know the eigen frequencies of the cutters. The qualitative and quantitative characterization of the surface quality of the machined stainless steel and titanium alloys could be performed by excitation of a specific tool mode as a prerequisite for achieving maximal efficiency of the vibration milling process. The statistical analysis of the collected roughness measurement data allows to identify the factors that most significantly contribute to a better surface finish of the workpieces.

3.1.2.1 Virtual Twin of Vibration Milling Tool

The structure of the developed tool prototype (Fig. 3.18a) provides a structural diagram of a prototype of the vibration milling tool [3]. A ring-shaped piezoelectric transducer 8 is embedded in the vibration tool assembly consisting of components $2 \div 9$. The transducer, powered by collector rings 4, is used to excite high-frequency vibrations in the cutting edge of the mill cutter 10. A horn 9 with chuck is fitted onto the end of the assembly in order to augment cuttertip vibration amplitude, which may reach up to 20 μm in this case. The vibration milling tool operates in a resonance mode: its length is equal to the integral number of half-wavelengths (Fig. 3.18b). The tool is mounted in a standard Weldon holder 1 at a nodal point, thereby preventing vibration energy losses by dissipation into the machine tool body.

Laboratory testing of the vibration milling tool has demonstrated that cutter dynamics has a negligible effect on the axial vibrations generated by the tool. This finding justified our model reduction approach, which assumed that the actual excitation provided by the piezoelectric transducer may be represented as an equivalent base excitation, which is imposed on the cutter at the place where it is clamped in the chuck. It should be noted that the vibration tool constitutes a linear dynamic system, therefore its vibrational characteristics may be established by analyzing a numerical model of a single cutter with the boundary conditions that are equivalent to those of the actual vibration tool. Therefore, a structurally complex transducer assembly was discarded in the model resulting in a lower number of degrees of freedom, which, in turn, reduced the computational time of simulations.

SolidWorks software was used for preparation of the 3D models of a mill cutter shaped as a pre-twisted cantilever (Fig. 3.19a). After exporting to FE software Ansys, the models were imposed with the relevant boundary conditions (Fig. 3.19b, c). The

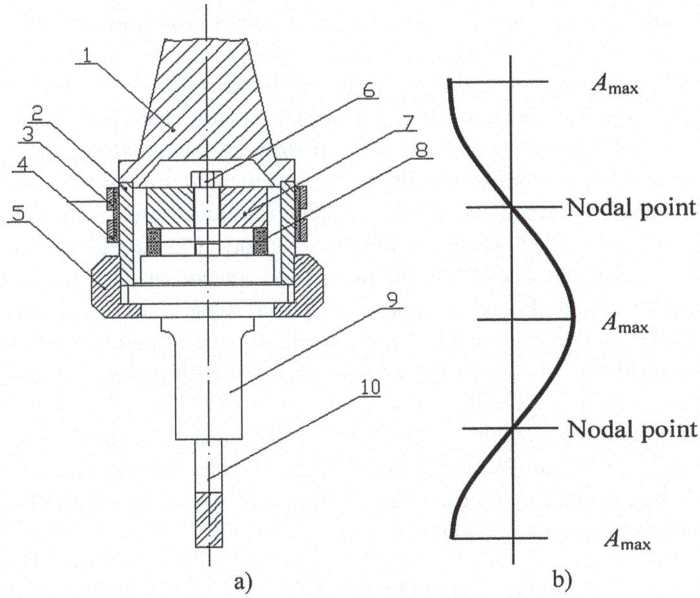

Fig. 3.18 Structural diagram of the vibration milling tool (**a**) and schematics of its operation principle (**b**) 1-standard holder (Weldon) DIN 6359, 2-cylinder, 3-textolite cylinder, 4-collector rings, 5-nut, 6-bolt, 7-collet, 8-piezoceramic rings, 9-horn, 10-cutter

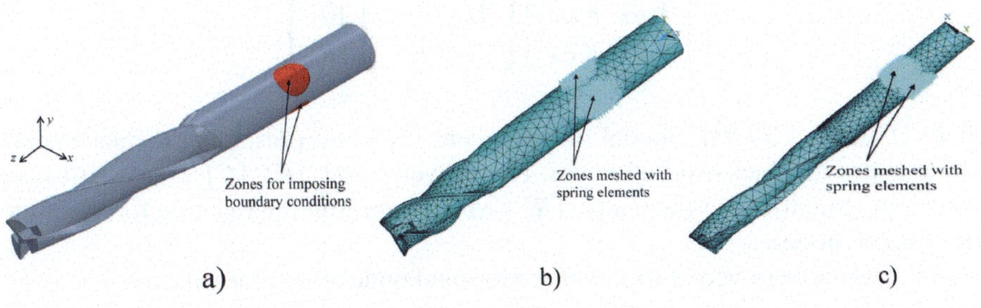

Fig. 3.19 SolidWorks model of a mill cutter with marked zones where boundary conditions are set (**a**) Ansys FE model of (**b**) 74 **mm** and (**c**) 96 mm length end mill with designated zones that are meshed with spring elements Combin14

main properties of FE models: cutter length $l = 96$ mm and $l = 74$ mm, diameter $d = 10$ mm, density $\rho = 8000$ kg/m^3, Young's modulus $E = 207$ GPa, Poisson's ratio $n = 0.3$. The models were meshed with tetrahedral FE SOLID92 characterised by 3 DOFs per node. A Cartesian coordinate system was adopted for modal analysis, while for harmonic and transient simulations a cylindrical system was chosen with the following DOFs: displacements z along the cutter rotational axis (axial direction), displacements r orthogonal to the axis (transverse direction), and rotations φ about the rotational axis (torsional direction).

It was essential to accurately reproduce the actual conditions of elastic cutter clamping in the chuck since they may have a tangible influence on cutter dynamics due to variable clamping force, temperature, etc. To this end, a sub-model consisting of elastic links was inserted into the cutter model. Spring elements Combin14 were selected for modelling the contact interaction in the zones where the cutter is mounted in a chuck (the cutter was secured with three bolts located around the shank at 120° angles). The cutter was fixed elastically with longitudinal spring elements k_r, k_z and torsional elements $k\varphi$ (k_r, k_z stiffness in transverse and axial directions, respectively, $k\varphi$—torsional stiffness). These link elements were placed at each node belonging to the three zones located on the shank (Fig. 3.19a). One node of a spring element was connected to the respective zone node, while the other element node was either constrained (during modal analysis) or was imposed with a base excitation in the axial direction (during harmonic and transient analyses). The developed FE model consists of about 6000 elements Solid92 and 250 elements Combin14 with a total 30,000 dof's. The dynamics of the cutter are described by the equation of motion in a block form by considering that the base motion law is known and is defined by the nodal displacement vector $\{U_K(t)\}$:

$$\begin{bmatrix} M_{NN} & M_{NK} \\ M_{KN} & M_{KK} \end{bmatrix} \begin{Bmatrix} \{\ddot{U}_N\} \\ \{\ddot{U}_K\} \end{Bmatrix} + \\ + \begin{bmatrix} C_{NN} & C_{NK} \\ C_{KN} & C_{KK} \end{bmatrix} \begin{Bmatrix} \{\dot{U}_N\} \\ \{\dot{U}_K\} \end{Bmatrix} + \\ + \begin{bmatrix} K_{NN} & K_{NK} \\ K_{KN} & K_{KK} \end{bmatrix} \begin{Bmatrix} \{U_N\} \\ \{U_K\} \end{Bmatrix} = \begin{Bmatrix} \{0\} \\ \{R\} \end{Bmatrix}, \tag{3.13}$$

where $\{U_N(t)\}$, $\{U_K(t)\}$—nodal displacement vectors representing displacements of free nodes and base-excited nodes, respectively; $[M]$, $[K]$, $[C]$—mass, stiffness and damping matrices respectively; $\{R\}$—vector representing reaction forces of the base-excited nodes.

The displacement vector of the unconstrained nodes is expressed as:

$$\{U_N\} = \{U_{Nrel}\} + \{U_{Nk}\} \tag{3.14}$$

where $\{U_{Nrel}\}$ is a component of the relative displacement with respect to the moving base displacement $\{U_{Nk}\}$.

Vectors $\{U_K\}$ and $\{U_N k\}$ correspond to rigid body displacements exerting no internal elastic forces within the structure. A proportional damping approach is adopted in the form $[C] = \alpha[M] + \beta[K]$, where α and β denote Rayleigh damping coefficients. The final equation in a matrix form is as follows:

$$[M_{NN}]\{\ddot{U}_{Nrel}\} + [C_{NN}]\{\dot{U}_{Nrel}\} + [K_{NN}]\{U_{Nrel}\} = \left[\widehat{M}\right], \tag{3.15}$$

where $\left[\widehat{M}\right] = [M_{NM}][K_{NN}]^{-1}[K_{NK}] - [M_{NK}]$.

The left-hand side of the equation contains matrices of the structure constrained in the nodes of applied base excitation, while the right-hand side denotes a vector of inertial forces acting on each node as a result of the base excitation.

3.1.2.2 Validation of Simulation Results

The model was verified experimentally in order to confirm that it is able to accurately predict the dynamic characteristics of the cutter that is excited by the vibration milling tool. The level of coincidence between the measured and simulated frequency responses was used to characterise model accuracy. The vibrational response of the cutter is largely predetermined by its boundary conditions. Therfore the model was adjusted by varying the stiffness of the spring elements until an acceptably close agreement between the numerical and experimental resonant frequencies was obtained. Figure 3.20 illustrates a comparison between the simulated and measured frequency responses of the cutter in the axial direction (the experimental curve represents the cutter response to a harmonic excitation with a constant relative amplitude).

The responses in Fig. 3.20 reveal that the major measured and simulated resonance peaks coincide fairly well. The accuracy of the FE model is evaluated using the relative error, which is calculated using numerical and experimental values of the resonant frequencies. The relative error was determined to be less than 2%, which confirms the accuracy of the developed FE model. Resonance frequencies of the axial vibration mode of mill cutters of two different lengths were calculated using the FE model, yielding 14.05 kHz for the 74 mm cutter and 18.4 kHz for the 96 mm cutter.

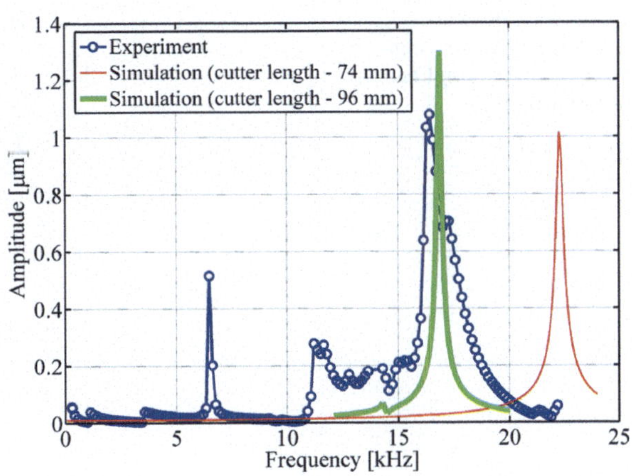

Fig. 3.20 Experimental frequency response of the cutter in the axial direction and the corresponding numerical responses

3.1.2.3 Caracterisation of Workpiece Surface Quality

A series of conventional and vibration-assisted milling experiments were performed in order to determine the effectiveness of the vibration milling tool with respect to improvement in surface quality. Figure 3.21 provides a schematic representation of the measurement setup. Experiments were carried in CNC milling center DMU 35 M with workpieces made of stainless steel (1.4301) and titanium (GOST 22,178–1976) without the use of cooling-lubricating fluids.

Two end mills of different length 74 mm (length of working part is 22 mm) and 96 mm (length of working part is 48 mm) but of the same diameter (10 mm) were used for the vibration milling experiments. The cutters were manufactured by "ASP Arno" from tungsten carbide and are coated with titanium aluminum nitride (TiAlN). This coating is suitable for high temperature and high speed machining of difficult-to-cut metallic alloys with minimal use of lubricant. Therfore this coating was chosen for our milling experiments that were carried out at high temperature without lubricating fluids. The thickness of the TiAlN coating is in the range of 2 ÷ 4 μm and the oxidation temperatures are between 480 ÷ 900 °C. The hardness is typically 2800 HV. Each cutter has its axial resonance at different excitation frequencies due to its difference in length (14.05 and 18.4 kHz). These frequencies were used for tool excitation in the axial direction during vibration milling experiments, which were conducted using the following regimes: milling depth $a_p = 1$ mm, feed rate $v_f = 66.66$ mm/min, milling speed $n_p = 1000$ rev/min. Each material was cut with both mill cutters for

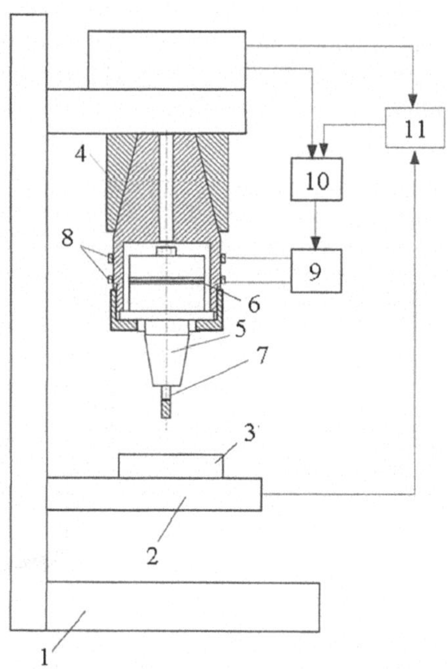

Fig. 3.21 Scheme of experimental set-up for vibration milling: 1-machine construction, 2-desk, 3-workpiece, 4-vibration tool holder DIN 6359, 5-horn, 6-piezoceramic rings, 7-cutter, 8-collector rings, 9-generator, 10-amplifier and 11-multimeter

30 mm with conventional milling and then switched to 30 mm cutting with vibration assistance. The surface of the workpieces was analysed qualitatively by means of a JEOL JSM-IC25S scanning electron microscope. The obtained images shown in Fig. 3.22 provide a visual proof that the surfaces of the workpieces machined with vibration milling are smoother with respect to the conventional conventional process. The image in Fig. 3.22 indicates no microcracks in the case of the vibration machined workpiece. A closer inspection of scanning electron microscope (SEM) images reveals that workpiece surface after conventional milling (Fig. 3.22a) is characterised by a ploughing effect due to cutting process instability. The surface in Fig. 3.22b, on the contrary, shows no signs of ploughing and is characterised by a regular finely meshed structure. Quantitative results of surface quality were obtained using Mitutoyo Surftest SJ-201. The values of the cutoff and evaluation length were 0.8 and 2.4 mm, respectively. Measurement data indicate that the values for surface roughness in the case of vibration milling are approximately one roughness grade number lower with respect to the conventional process (according to DIN EN ISO 1302). Surface roughness data for six machined workpieces was collected during this experimental study. Measurement results demonstrate that difficult-to-cut materials machined using the vibration milling process are characterised by improved surface finish in comparison to the conventional process (Tables 3.2 and 3.3). When the tool was driven with an excitation frequency of 18.4 kHz (96 mm cutter), the best results for the surface finish were obtained in the case of stainless steel milling (Table 3.3). This result is attributed to better machinability of stainless steel 309 in comparison to titanium alloy Ti6-4.

Roughness measurement data was subjected to statistical treatment in order to more thoroughly characterise the effectiveness of the vibration milling process. As a first step, the assumption that the samples come from normal distributions was tested. A normal probability plot provides a quick idea (Figs. 3.23 and 3.24). Both

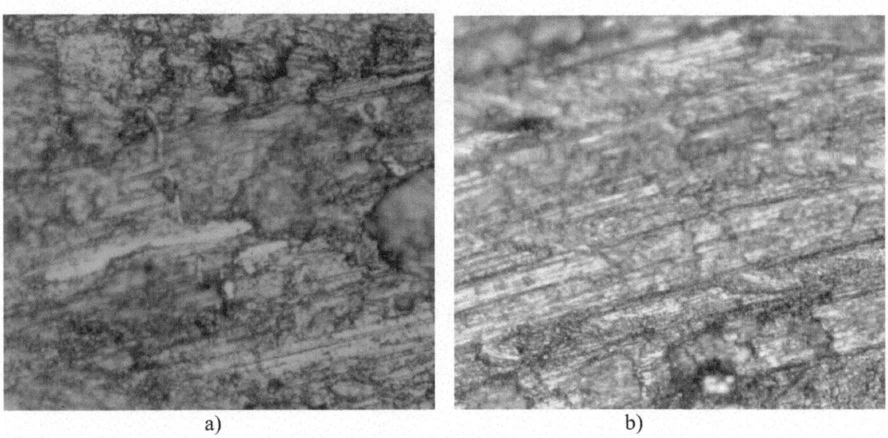

Fig. 3.22 Photos of surfaces of machined workpieces (50 × magnification): after conventional milling (**a**) and after vibration milling (**b**)

Table 3.2 Surface finish results of milled workpieces using a 74 mm length cutter, numbers of measurement (μm)

Materials and milling type	1	2	3	4	5	6	Average (μm)	Difference (%)
Stainless steel (no vibrations)	0.61	0.56	0.75	0.87	0.53	0.58	0.65	21.54
Stainless steel (with vibrations)	0.48	0.39	0.67	0.49	0.43	0.60	0.51	21.54
Titanium (no vibrations)	0.60	0.60	0.56	0.53	0.47	0.56	0.55	6.33
Titanium (with vibrations)	0.49	0.70	0.46	0.50	0.47	0.49	0.52	6.33

Table 3.3 Surface finish results of milled workpieces using a 96 mm length cutter, numbers of measurement (μm)

Materials and milling type	1	2	3	4	5	6	Average (μm)	Difference (%)
Stainless steel (no vibrations)	1.17	1.14	1.15	1.17	1.21	1.11	1.16	22.59
Stainless steel (with vibrations)	1.00	0.86	0.91	0.78	0.95	0.88	0.90	22.59
Titanium (no vibrations)	0.76	1.30	0.88	0.86	1.01	0.85	0.94	12.54
Titanium (with vibrations)	0.83	0.85	0.85	0.80	0.80	0.82	0.83	12.54

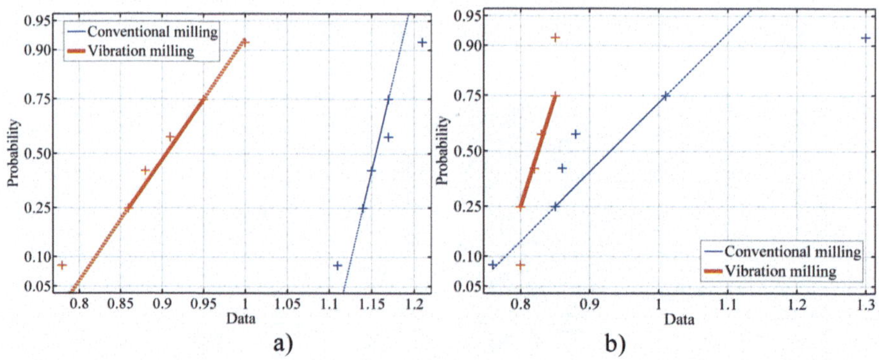

Fig. 3.23 Normal probability plots when 96 mm length cutter is used and the machined material is: stainless steel (**a**) and titanium (**b**)

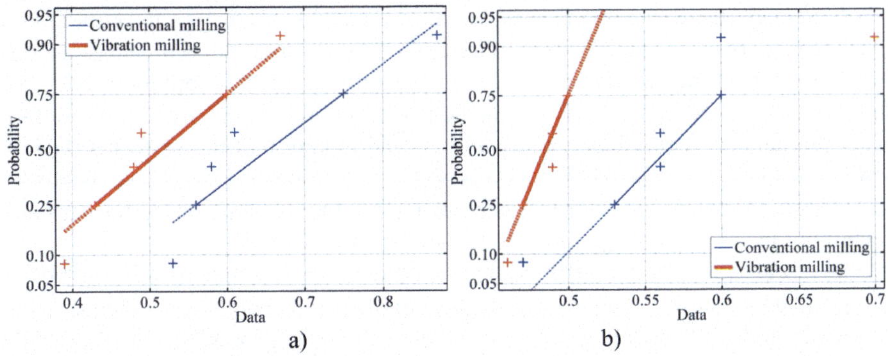

Fig. 3.24 Normal probability plots when 74 mm length cutter is used and the machined material is: stainless steel (**a**) and titanium (**b**)

data scatters approximately follow straight lines through the first and third quartiles of the samples, indicating approximately normal distributions.

A shift in the mean from stainless steel and titanium machined using conventional milling to stainless steel and titanium machined with vibrations is evident. A hypothesis test is used to quantify the test of normality. Since each sample is relatively small, a Lilliefors test is used. The test performs the default null hypothesis that the sample comes from a distribution in the normal family. The results returned by each test indicate a failure to reject the null hypothesis that the samples are normally distributed. This failure may reflect normality in the population or it may reflect a lack of strong evidence against the null hypothesis due to the small sample size. A variance analysis was performed to estimate the relative impact of resonance frequency, applied vibrations, and material type on surface quality (Table 3.4). The *p*-value for the resonance frequency effect is 0.0159, which is highly significant. This indicates that surface roughness is largely dependent on the dynamic characteristics of the mill cutter. In turn, this implies that in order to increase the positive influence of the vibration-assisted cutting process, the tool should be superimposed with vibrations

Table 3.4 Results of variance analysis

Resonance frequency	0.32	1	0.32	1600	0.0159
Vibrations	0.03645	1	0.03645	182.25	0.0471
Material	0.01805	1	0.01805	90.25	0.0668
Resonance frequency × Vibrations	0.005	1	0.005	25	0.1257
Resonance frequency × Material	0.005	1	0.005	25	0.1257
Vibrations × Material	0.00845	1	0.00845	42.25	0.0972
Error	0.0002	1	0.0002	–	–
Total	0.39315	7	–	–	–

Source Sum Sq. d.f. Mean Sq. F, Prob > F

of a frequency that corresponds to the axial resonance frequency of the cutter. As a consequence, there is amplification of axial vibration amplitudes with simultaneous intensification of the twisting motion of the cutting tip due to coupling of the axial and torsional deflections inherent to the helix-shaped mill cutter, which represents a case of parametric vibrations. The *p*-value for the vibration milling effect is 0.0471, which is significant as well. This indicates that the surface quality also depends on the machining means. In other words, statistically, the dynamic characteristics and machining means have the greatest influence on surface quality.

The research results demonstrate that it is crucial to dynamically tailor the excitation frequency of the vibration cutting tool in order to generate the required vibration mode in the mill cutter and thereby achieve the most pronounced improvement in surface finish in difficult-to-cut materials. The proposed approach allows efficient machining of high-strength alloys and could significantly facilitate the treatment of hard and brittle materials such as ceramics, glass, and composite materials. The reported vibration milling experiments were successfully performed under dry machining conditions, which demonstrates that assisting cutting with high-frequency vibrations could benefit the implementation of a minimum quantity lubrication method into industrial manufacturing processes.

3.1.3 Vibration Drilling of Brittle Materials

Vibration assisted machining methods are generally distinguished into low and high frequency applications, those in the low frequency category are several orders of magnitude below the ultrasound threshold, while high frequency applications tend to exceed it. Both categories have found their applications in industry and machining systems can either be designed to be operated at discrete frequencies (resonant system) or on a frequency range (non-resonant system). Vibration assisted machining is increasingly gaining popularity as more rational choice over conventional machining of hard to machine yet desirable materials, as is evident from the multitude of existing and ongoing research in the area. Generally, the investigations can be ranged by different choices of the main process, vibration parameters, and material. A commonly investigated group of methods belong to cutting of metals and other materials. For instance, quite recently, investigated the effect of linear versus elliptical vibration in micro-grooving of 0Cr18Ni9 workpiece, and found that in the linear vibration case, the surface roughness is lower, while in the elliptical case the lower cutting forces were observed.

3.1.3.1 Physical Twin of Vibration Drilling

To perform the drilling experiments "Leadwell" V20 CNC milling machine (Fig. 3.25) was employed [4, 5]. Thin microscope slides were used as drilling samples, while both types of tools were fixed into an ultrasonic tool holder. The operational

Fig. 3.25 "Leadwell" V20 CNC milling machine

frequencies were observed by determining the frequencies producing the highest amplitudes. However, no further attempts into determining the vibration modes were made.

Two types of tool were used for the experiment; the dimensions of choice were Ø25 mm and Ø5 mm for crown drill and spherical abrasive mill respectively. In the case of each tool, a drilling attempt was made under conventional conditions and in presence of vibrational excitation. When subjected to ultrasonic excitation, manual frequency scanning showed the crown drill and spherical abrasive mill to respond strongest at frequencies of 15.9 kHz and 21 kHz respectively. Considering the difference between the tools, amplitudes of the vibration were disregarded in this case and frequencies of 15.9 kHz and 21 kHz were chosen as the operational parameters for the crown drill and spherical abrasive mill respectively. The experimental set-up employed when investigating the effect of vibration drilling application on hard-brittle materials, for measurement of the axial forces and producing the driving signal for the ultrasonic transducer is presented in Fig. 3.26. The axial forces were measured by employing a force-torque sensor Kistler 9365B. The signal from the sensor is amplified by Kistler 5018A charge amplifier 1, and subsequently passed to Picoscope 3424 oscilloscope 2 and subsequently represented by the PC 3. The signal for the tool transducer is generated by the signal generator Agilent 33220A 4 and amplified by the signal amplifier EPA-104 5 before being passed to the transducer, the wide range signal generator allows achieving the desired frequency of the signal, while the amplifier allows adjusting the amplitude to a higher scale.

The experimental procedure itself followed two termination conditions. Drilling was conducted until the fracture occurred or the plate was drilled through. In order to avoid premature fracture of the samples the drilling parameters were limited to 2000 rpm and feed rate of 0.2 mm/min as per industry observations. During the machining process the samples were affixed to a wooden base by a cyanoacrylate glue and were later removed using immersion in an acetone solution.

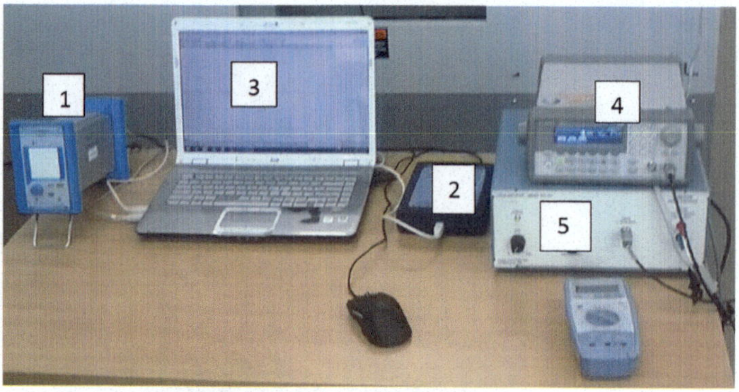

Fig. 3.26 Experimental set-up: 1-charge amplifier Kistler 5018A, 2-oscilloscope Picosope 3424, 3-PC, 4-signal generator Agilent 33220A, 5-amplifier EPA-104

The defining task of the entire experimental set-up is the definition of the metric being observed. The topic of interest to the researchers and medical personnel in the field, is that of the drilling temperature of the bone. It has been shown that elevated temperatures at the bone drill interface may lead to irreversible changes in the bone tissue, a condition called necrosis of the bone, which may be responsible for further complications such as, implant failure and longer healing times. The improvements suggested by the scientific community to the method of machining the bone, namely drilling, incorporate the use of external and internal irrigation systems at the drill site, careful control and selection of favourable cutting parameters, application of automatic or mechatronic systems. Predicament occurs when the need for uniform and consisted samples in the experimental investigation are required considering bone is a non-homogenous material with different properties depending on its type and which particular bone it is. In order to imitate human bone in experimental setups, animal bones (bovine, ovine, porcine), or polymers having similar mechanical properties are generally used. Polymers have the advantage of being homogeneous, easy to shape and form depending on the requirements of the particular experimental setup, as well as allowing to achieve greater repeatability compared that to animal bone.

Therefore, the main objective of this experiment was to determine the drilling temperature using low-frequency vibration-assisted drilling in comparison to conventional drilling. Additionally, bone tissue was substituted by a 4 mm thick Poly(methyl methacrylate)-PMMA sample in order to ensure replicability, as bone properties differ over the investigated domain. The successful effect of the application is judged by how low the measured temperature is when compared to conventional drilling, and the critical temperature value.

Since most bone drilling operations in medical environment are performed by hand, all drilling runs were carried out using Makita 8391DWPE hand drill with 4.2 mm diameter two flute HSS drill bit. The drill was set to 1200 rev/s drilling speed. In order to determine the drilling temperature thermal camera FLIR T450sce was used. Considering the drilling was done by hand, in addition to temperature, forces were measured and monitored during the drilling process as well, using force/torque gauge model BGI-Mark-10, this was done however only to ensure a decent level of replicability, and avoid biases resulting from possible operator error, rather than to draw conclusions related to drilling force changes.

The low frequency electromagnetic vibrator driving set-up was used as follows: sine wave oscillations were generated by signal generator WW5064, they were then amplified with VPA2100MN amplifier to drive electromagnetic vibrator VEB Robotron-Meßelektronik "Otto Schön". Additionally, to obtain consistent vibration amplitudes between samples Schwingungsaufnehmer KD35 accelerometer was utilized in conjunction with PicoScope 4226 digital oscilloscope. As with the previous experimental set-up, the termination conditions remained the same, however, the dependence of vibration frequency on the temperature was of interest. Figure 3.27 presents a graphical representation of the experimental set-up for measuring the temperature during low-frequency vibration-assisted drilling of PMMA plate. Sample was subjected to excitation and drilled using the hand drill, while maintaining the thrust force at the steady level of 30 N, at the same time the thermal camera was used to obtain temperature measurements; the filming was conducted from below the sample *1* (Fig. 3.28), at the predicted position of the exit hole which could be visually determined *2*, taking into account the small thickness and favorable optical properties of PMMA. In order to protect the camera lens from falling chips, it was positioned one meter away and at an angle as shown in Fig. 3.28. At the moment when the drill went through the material, the observed temperature was considered to be the maximum drilling temperature, as at that point all of the frictional energy of the drilling process was expected to be converted to heat.

Fig. 3.27 Scheme of the experiment

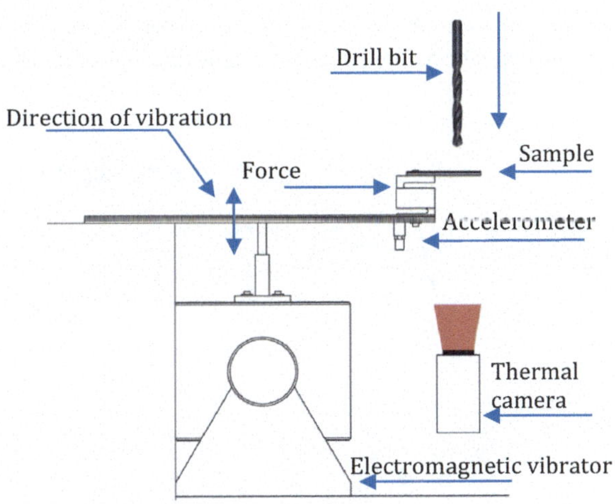

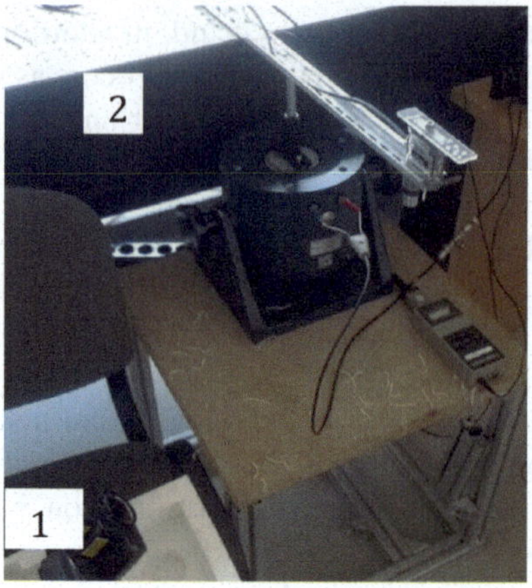

Fig. 3.28 Experimental set-up: 1-thermal camera position, 2-sample position

3.1.3.2 Physical Twin Experimentation Results

The first experiment provided response in the form of axial force measurements. In total four experimental runs were conducted, as in this case resonating modes of the tool were employed. Provided below are smoothed graphs of axial forces. The trends are compared as—excited versus conventional cases for both tools (Figs. 3.29 and 3.30).

The drilling using spherical mill in the conventional drilling was terminated prematurely, due to a fracture of the sample (Fig. 3.31). Under excitation both tools operated without a fracture, however, considering the fact that spherical abrasive ball end mill was considered to represent the harder case scenario, Vibration drilling of

Fig. 3.29 Axial force results for drilling with spherical mill (excited versus conventional)

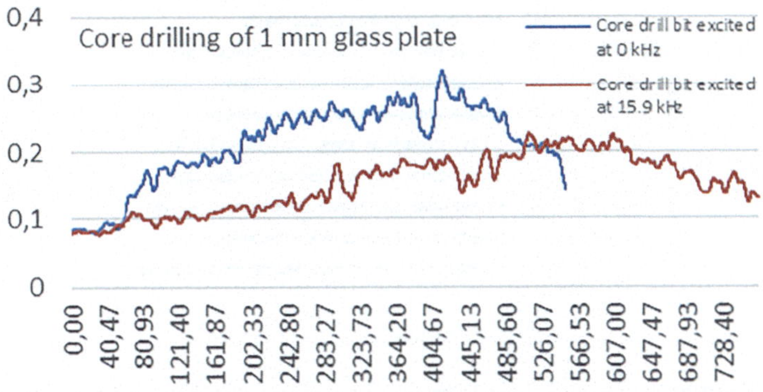

Fig. 3.30 Axial force results for drilling with core drill (excited versus conventional)

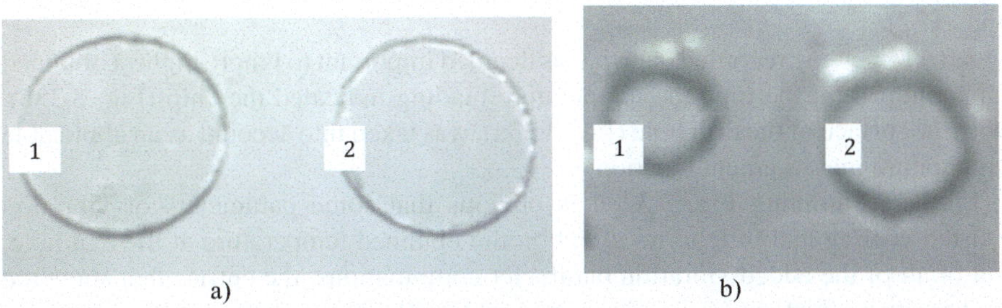

Fig. 3.31 Core drill (**a**) and spherical mill (**b**) drilling samples: 1-conventional case, 2-under excitation

glass plates can be considered to have by default demonstrated superiority over the conventional case.

Core drilling under conventional conditions, however went through with minimal fracture as well, yet this was to be expected, as the tool was specifically designed to handle such materials, however, due to a mismatch between the available sample size and the size of the core drill, the integrity of the samples could not be preserved.

Summing up the results of the first experimental run, vibration drilling has demonstrated superiority over its conventional counterpart in terms of axial drilling force reduction and sample integrity preservation. It is apparent that in both cases, the peak axial force was lower for the case of the excited tool. Additionally, the drop of the axial force was less abrupt in the excited tool case, due to preservation of sample integrity throughout the drilling process.

The response for the second experiment, the bone tissue drilling case, in which PMMA was used as a substitute to ensure consistency of the results, was obtained in the form of temperature measurements at the very end of the drilling run. In grand total the investigation was conducted at 9 different vibration settings, performing 10 drilling runs for each vibration setting. Vibration frequencies for the drilling were

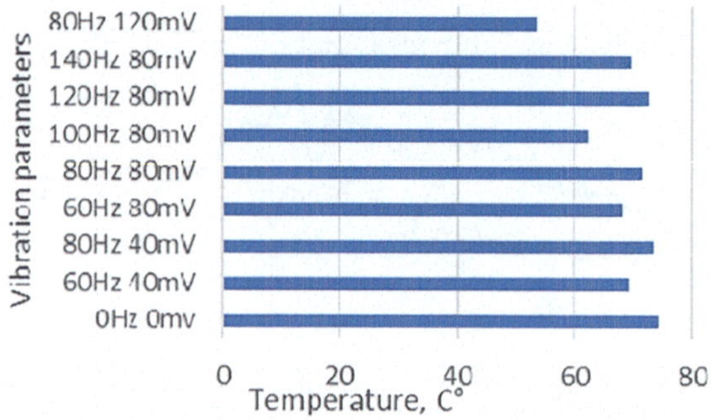

Fig. 3.32 PMMA drilling temperature dependency on vibration assisted drilling parameters

selected every 20 Hz from 60 to 120 Hz disregarding the excitation mode or amplitude (Fig. 3.32).

After reviewing recorded data, it was deemed important to ignore highest obtained temperature if at the time the temperature reading indicated the chip (Fig. 3.33b). Only the surface of the drill hole (Fig. 3.33a), was taken into account as an applicable temperature measurement location.

Further examining Fig. 3.32 it is obvious that some parameters of vibration assisted drilling had little to no effect on the obtained temperature at the exit hole. Out of all of the added vibration parameter combinations, the entries that show the least positive effect are 60 Hz 40 mV and 80 Hz 40 mV (these modes show increase in drilling time as well), in essence one could argue that the amplitude at which these samples were oscillated was not suited to produce positive results. Samples which had their amplitude signal set at 80 mV, more or less showed positive results in comparison. In particular, samples with frequency of 60 Hz and 100 Hz show sizeable decrease in temperature of 6 °C degrees and 11 °C degrees respectively. Vibration parameters of 80 Hz 120 mV showed the greatest decrease, as it not only drastically

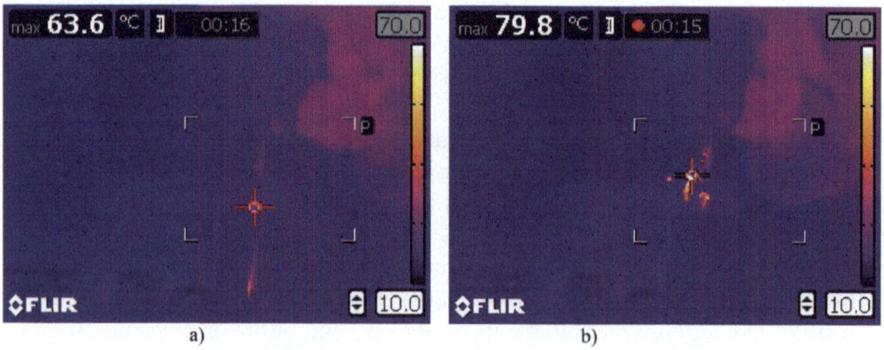

Fig. 3.33 Highest observed temperature: **a**–hole and **b**–chip

reduced drilling temperature by 21 °C, but also had a 1.4 s faster drilling time. Vibration frequencies that show biggest reduction of temperature (80 Hz 120 mV, 100 Hz 80 mV and 60 Hz 80 mV).

From obtained results it is apparent that some vibration frequencies and amplitudes are better suited in lowering drilling temperatures than others. It is also worth noting that some vibration settings produce worse or similar results compared with conventional drilling. Further studies in this topic should investigate in greater depth; the influence the vibration amplitude and excitation mode have on the drilling temperature and time.

3.2 Physical Twin of Vibrationally Excited Workpiece Drilling

Due to the insufficient rigidity in the machining process, the workpiece tends to cause vibrations and deformations, which impair the quality of the machining. However, when performing vibration cutting, it is difficult to increase the frequency of tool oscillations due to the high mass of the cutting tool and the high inertial forces involved. In such cases, vibratory excitation of the workpiece is effective, allowing good workpiece surface quality to be achieved.

3.2.1 Vibration Excitation of a Workpiece for Drilling Force Reduction

Machining of hard and brittle materials, especially glass is still a major problem because of its lower fracture toughness and higher hardness. Glass tends to crack easily during machining under small stress. Many researchers have tried to develop new methods for machining of brittle and hard materials. Vibration cutting was one of the new methods for this purpose. There are two possible cases related to the mechanical properties of the excited workpiece: when it conforms to the mechanical properties of a solid structure and when it conforms to the mechanical properties of a flexible structure.

3.2.1.1 Workpiece Excitation Experimental *Set-Up*

In order to investigate the influence of work piece excitation on the drilling forces in brittle materials, a piezo-ceramic plate of dimensions 30 × 10 mm was used as a sample 1 (Fig. 3.34a). The specific ceramic used was leadzirconate-titanite 4 or PZT 4. It was considered to be suitable for this purpose as it is brittle enough to emulate behavior of most brittle materials, currently relevant in the industry [6]. At

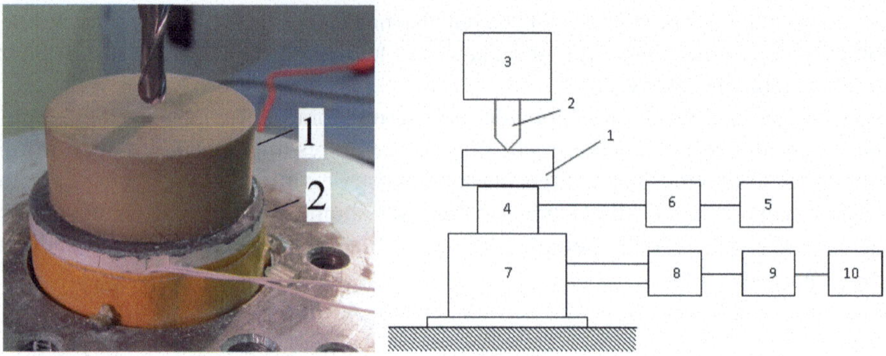

a) b)

Fig. 3.34 Solid structure sample se-tup (**a**) 1–30 × 10 mm workpiece sample, 2-piezo-electric transducer and schematic representation of the drilling force test set-up (**b**) 1-specimen, 2-drill, 3-spindle, 4-piezoelectric transducer, 5-signal generator Agilent 33220A, 6-amplifier EPA-104, 7-force-torque sensor Kistler 9365B, 8-charge amplifier Kistler 5018A, 9-oscilloscope Picosope 3424, 10-PC

such dimensions the sample is expected to behave like a solid structure, therfore the influence of drilling location is negligible, as the distribution of vibrations is uniform.

Every experiment employs a cylindrical piezoelectric transducer of dimensions 35 × 27 × 0 mm as a device for sample excitation *4*. The sample *1* is mounted on top of the cylinder with an adhesive, signal for the transducer is generated by the signal generator Agilent *5* (Fig. 3.35b) and amplified by the signal amplifier *6*. In order to measure the changes in the forces during the drilling process, experiments employ a force-torque sensor Kistler 9365B *7*. The signal from the sensor is amplified by Kistler 9365B charge amplifier *8*, passed to Picoscope 3424 oscilloscope *9* and subsequently observed on a PC *10*.

The drilling is performed with a 4 mm drill bit *2*. The feed rate was kept at 0.05 mm/rev. The drilling was performed at speeds of 145, 290 and 580 rpm. The

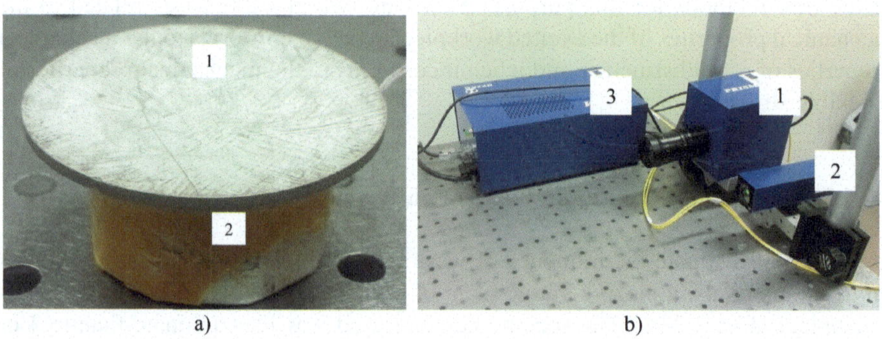

a) b)

Fig. 3.35 Flexible structure sample set-up (**a**): 1–50 × 2 mm sample, 2-piezo-electric transducer and Prism 100 holography equipment (**b**): 1-optic receiver, 2-illumination source, 3-controller chosen speeds are reasonably spaced apart, therefore drilling with and without work piece excitation was performed at each speed to observe different behaviours of the sample. In order to observe the behaviour of a flexible structure when excited at high frequencies (~100 kHz) a piezo-ceramic disc of dimensions 50 × 2 mm was used as a sample 1 (Fig. 3.35a). The sample was clamped to the work table by three bolts equally spaced throughout the perimeter of the workpiece, at a 120° angle from one another. Holographic vibration analysis was employed to observe different vibration modes of the sample [7]. The samples behaviour was recorded by an optic receiver *1* (Fig. 3.35b) while being excited at different frequencies under an illumination source *2*, the process was controlled by the controller *3*.

3.2.1.2 Workpiece Vibrational Simulation and Results Validation

For further investigation of flexible structure behavior under high frequency excitation, a mathematical model adequate to the real one needs to be developed. The simulations would be used to predict the behaviour of the sample without experimental techniques. The dynamics of the circular work piece are described by the equation of motion on a block form by considering that the base motion law is known and defined by nodal displacement vector $\{U_K\}$:

$$\begin{bmatrix} [M_{NN}] & [M_{NK}] \\ [M_{KN}] & [M_{KK}] \end{bmatrix} \begin{Bmatrix} \{\ddot{U}_N\} \\ \{\ddot{U}_K\} \end{Bmatrix} + \begin{bmatrix} [K_{NN}] & [K_{NK}] \\ [K_{KN}] & [K_{KK}] \end{bmatrix} \begin{Bmatrix} \{U_N\} \\ \{U_K\} \end{Bmatrix} = \begin{Bmatrix} \{0\} \\ \{R\} \end{Bmatrix}, \tag{3.16}$$

where $\{U_N\}, \{U_K\}$ are modal displacement vectors. Lower indices N and K represent free nodes and excited (known displacement) nodes respectively, combination pairs of these indices relate the nodes to their elements and their corresponding positions in property matrices $[M]$, $[K]$, which are mass and damping matrices—the column and row positions of elements in the matrices are denoted by these combinations; $\{R\}$ is vector of unknown reaction forces at nodes under kinematic effect.

In modal analysis un-damped resonant frequencies of the work piece are found by solving equation:

$$\left([K] - \omega^2[M]\right)\left\{\hat{U}\right\} = \{0\}, \tag{3.17}$$

here ω is angular frequency, $\left\{\widehat{U}\right\}$ is mode shapes vector.

The simulations were developed using Comsol multiphysics suite, eigenfrequency study. The boundary constraint of the sample is the contour of the piezo-transducer edge, as this part of the sample is rigidly fixed to it. The simulation yields a number of resonant frequencies for further investigation and comparison to the experimentally obtained values. The eigen frequency study of the model is set to return 150 vibration modes in order to cover the expected range of excitation (1 ÷

Table 3.5 Natural frequencies of simulated and experimentally obtained vibration modes

Vibration mode	Hologram frequency (kHz)	Simulation frequency (kHz)
1	4.9	5.2
2	6.3	5.6
3	12.3	12.89
4	43.1	40.1
5	99.8	98.96

100 kHz). The modal analysis of work piece was carried out in range of 0 ÷ 60 kHz by using Block Lanczos mode extraction method. The results of this analysis are shown in Table 3.4. Five modes of workpiece vibrations were found in selected frequency range. The first calculated resonant frequency of the workpiece is 5200 Hz. Due to verification of the FE model adequacy to physical one, the workpiece vibration test was carried out. The workpiece modes of vibration were measured by holography interferometry method.

During the holographic vibration mode analysis, 4 holograms were made at points of most intensive excitation. The number of experimentally obtained frequencies is lesser than the number of analytically obtained ones, therefore, frequency values closest to one another will be chosen for further investigation (Table 3.5). After comparing the vibration mode shapes two of them (4.9 ÷ 5.2 kHz and 88.2 ÷ 98.96 kHz) were dismissed from further investigation, as they clearly did not match.

It can be observed that the flexible structure sample exhibits a resonant vibration mode at a frequency close to 100 kHz (Fig. 3.36). Also, it can be seen that the

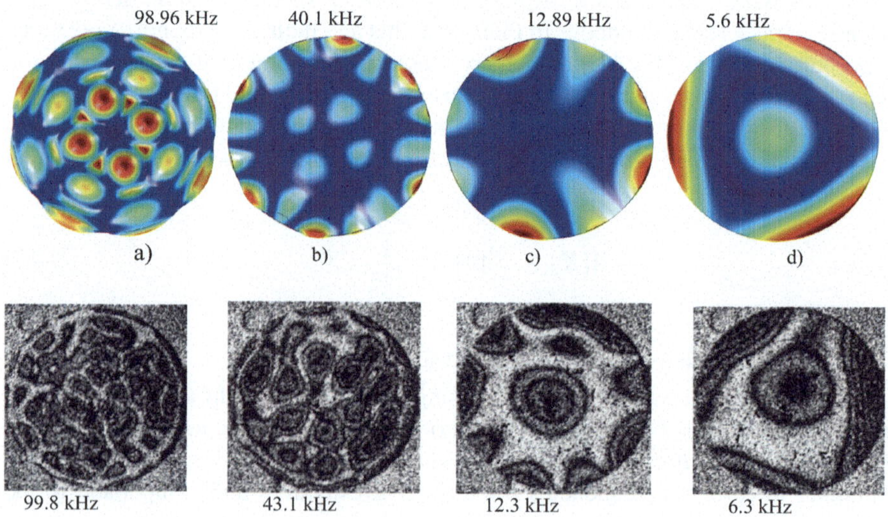

Fig. 3.36 Corresponding vibration modes: **a**—99.8/98.96 kHz, **b**—43.1/40.1 kHz, **c**—12.3/12.89 kHz and **d**—6.3/5.6 kHz

behaviour of the material is relatively adherent to its simulations, as the remaining analytically and experimentally obtained vibration mode shapes (Fig. 3.36a, b, c, d) appear to be similar in overall distribution of excited areas. However, inconsistencies do exist; for instance, the simulated vibration shape in Fig. 3.27c is not excited at the middle of the plate as it can be observed in the hologram. The distribution of excitation areas in the holograms lacks symmetry, which is characteristic of the simulations.

The appearance of the deviations suggests that they are in most part caused by the asymmetrical nature of sample clamping and in part by the unavoidable difference between real and ideal cases. The applicability of the model could be improved by either enhancing the accuracy of the sample clamping setup or tweaking the model to accommodate the existing clamping conditions. However, since the resonant vibration mode at ~100 kHz frequency has been confirmed, further investigation of flexible structure machining while under high frequency excitation is available and can be relatively well predicted through the use of the mathematical model.

The results obtained from the drilling experiment demonstrate that there is a slight difference between torque forces when drilling a still and excited work piece at a low speed of 145 rpm (Fig. 3.37a). However, cutting forces are significantly lower for excited pieces when drilling at higher speeds. From Fig. 3.37a it can be seen that the plots differ only slightly. The averages of cutting force (before maximum peak) differs by 24%. When drilling at a higher speed (290 rpm) the distance between the plots is more consistent (Fig. 3.37b). The difference between the averages of cutting force (before maximum peak) is similar to the previous attempt by 22%. With an increase in drilling speed of up to 540 rpm (Fig. 3.37c) the difference between the plots becomes more obvious. In this case the difference between averages of cutting force is 40%.

According to the obtained results a trend can be observed. The decrease in torque is larger and more consistent at higher speeds. It is important to mention, that it is not only the drilling speed which changes along with each attempt—the feed rate (0.05 mm/rev) also changes its absolute value with each step-up in speed. Therefore, the feed rate should be taken into consideration, as it has a significant effect on the drilling force, increasing the drilling speed the feed rate reduction is obvious due to the difference in cutting forces between vibration and conventional drilling.

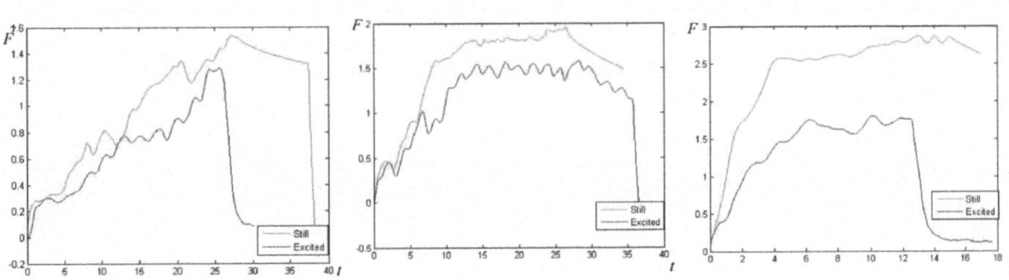

a) b) c)

Fig. 3.37 Dependence of cutting force on time: **a** at 145 rpm, **b** at 290 **rpm** and **c** at 540 rpm

3.2.2 Development of Actuator Enabling a Brittle-Ductile Transition of Warkpiece Material

Brittle-ductile transition is an important phenomenon for particularly precise processing of brittle materials. The theory of such a transition is based on the removal of matter [8]. The basic idea of this theory is that the energy removal rate of brittle mode material is proportional to the second processing scale power while the plastic flow energy is proportional to the third processing scale power, which means that ductile mode removal would be more energy efficient when the processing scale is small enaugh. To this end, the workpiece actuator must be excited in a higher eigen mode, which is possible with the author's patented device [9].

3.2.2.1 Design of Workpiece Actuator

The tool actuator front- and back-masses were designed to be manufactured from an aluminum alloy [10]. Two PZT4, 5 × 35 mm piezoceramic rings served as the piezoceramic stack. The relevant material properties are presented below (Table 3.5). The actuator was designed to exhibit the first longitudinal mode at 40 kHz. This, along with the dimensions of the piezoceramic rings, served as the starting design conditions. Therefore, the front-mass length was determined to be 0.027 m, while the back-mass length was determined to be 0.015 m. A wide-faced actuator, rated for 20 kHz operation, was designed for workpiece excitation using materials described in Table 3.6 and their acoustic properties presented in Table 3.7. Modal shape representations suggest that peripheral zones of the actuator face could be employed at higher vibration modes as well. Because the actuator is attached at the bottom, and not at the nodal point, the impedance condition does not need to be satisfied, as the combined impedance of the back-mass along with the fixture is ultimately always larger than that of the front-mass. Therefore, an aluminum alloy was considered for this case. Numerical analysis was used in order to determine the performance

Table 3.6 Properties of materials used to produce the stationary workpiece actuator

Material	Young's modulus E (GPa)	Density ρ (kg/m^3)
Aluminum alloy	72	2850
PZT4 piezoceramic	78	7600

Table 3.7 Acoustic properties of workpiece actuator materials

Material	Sound velocity c (m/s)	Characteristic impedance Z_0 (kg/m^{2}s)
Aluminum alloy	5026.247	143.248×10^5
PZT4 piezoceramic	3203.616	243.475×10^5

criteria of the actuator and for the validation of design. FE method was used to carry out two types of analyses—modal analysis and frequency response analysis. The first analysis allows determining the modal shapes at different frequencies, while the frequency response analysis helps to identify the dominant modes. This knowledge is crucial to ensuring that the design is correct and that the actuator is capable of operating in modes specified in the design of the experimental study.

If the law of the motion of the excited end of the workpiece is to be considered known and defined by the nodal displacement vector U_K, the block form of the equation of motion of the workpiece is written as:

$$\begin{bmatrix} M_{\mathrm{NN}} & M_{\mathrm{NK}} \\ M_{\mathrm{KN}} & M_{\mathrm{KK}} \end{bmatrix} \begin{Bmatrix} \ddot{U}_N \\ \ddot{U}_K \end{Bmatrix} + \begin{bmatrix} K_{\mathrm{NN}} & K_{\mathrm{NK}} \\ K_{\mathrm{KN}} & K_{\mathrm{KK}} \end{bmatrix} \begin{Bmatrix} U_N \\ U_K \end{Bmatrix} = \begin{Bmatrix} 0 \\ R \end{Bmatrix}, \tag{3.18}$$

where, N and K are indices representing the nodes whose displacements are unknown (free nodes) and known (excited or constrained nodes), respectively. Hence, U_{N} and U_{K} are nodal displacement free and constrained vectors, respectively, while pair combinations of these indices relay the position of these nodes in the respective property matrices M (the mass matrix) and K (the stiffness matrix). Additionally, R is the vector of the unknown reaction forces of nodes under excitation.

The modal analysis of an undamped oscillatory motion determines resonant frequencies by solving the following (Eq. 3.19) where ω is the angular frequency, while $\hat{U}$ is the mode shapes vector:

$$\left(K - \omega^2 M\right)\widehat{U} = \{0\}, \tag{3.19}$$

The modal analysis of the workpieces is performed by employing Ansys workbench and Salome Meca/Code Aster software packages, the modal analysis module. Ansys workbench and Salome Meca/Code Aster are suitable for the use of the Block Lanczos extraction method when seeking to obtain the vibration modes in the prescribed frequency ranges.

The actuator can be driven to reach its second cylindrical mode, which occurs at 85 kHz as apparent by the FE analysis (Fig. 3.38). This allows the workpiece to be vibrated when placed on the periphery of the actuator face, ensuring constant amplitudes at high frequencies. This approach also allows to reliably employ higher modes, since amplitudes in the periphery tend to be dominant as presented in the frequency response diagram (Fig. 3.38a) representations suggest that peripheral zones of the actuator face could be employed at higher vibration modes as well. Because the actuator is attached at the bottom, and not at the nodal point, the impedance condition does not need to be satisfied, as the combined impedance of the back mass along with the fixture is ultimately always larger than that of the front mass. Therefore, an aluminum alloy was considered for this case.

This allows the workpiece to be vibrated when placed on the periphery of the actuator face, ensuring constant amplitudes at high frequencies. This approach also

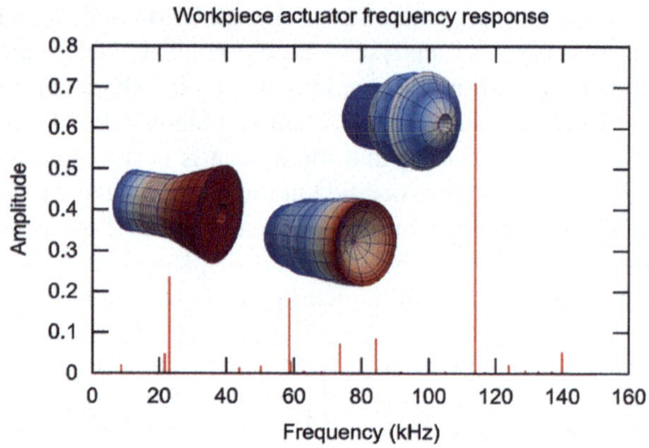

Fig. 3.38 Frequency response and modal analysis of the workpiece actuator

allows to reliably employ higher modes, since amplitudes in the periphery tend to be dominant as presented in the frequency response diagram.

3.2.2.2 Physical Twin of End Grinding

Cutting depth is universally accepted as the limiting factor for surface quality during grinding. Therefore, the experimental studies carried out in this research evaluated the effect of ultrasonic excitation, with respect to increasing grinding depth. The excitation has two contributing factors, whose effect needs to be evaluated—frequency and amplitude. The initial study of the frequency effect on the surface quality was conducted at two different depths (25 and 50 μm), over two frequency values (14.5 and 84.5 kHz). This yielded an experiment with four runs, defined by two controlling factors at two levels (Table 3.8) that can be defined by an *L*4 Taguchi array. The response of the experiment is the average surface roughness, as measured through white light interferometry. The objective of the applied treatment, in this case, is the reduction of the response value, as lower surface roughness is indicative of a surface machined in ductile mode. Therefore, the signal to noise ratio was defined using the "smaller is better" objective function (3.20).

$$S/N = -10\log\frac{1}{n}\sum_{i=1}^{n} y_i^2 \tag{3.20}$$

Table 3.8 Initial experiment outline

Factor	Units	L1	L2
Frequency (f)	kHz	14.5	84.5
Depth (a_p)	μm	25	50

Similarly, upon observing that the higher frequency yields lower surface roughness values, an experiment for observing the effect of depth and amplitude at 84.5 kHz frequency has been constructed. The two controlling factors, in this case, are the depth and excitation amplitude, which are varied over 4 levels (Table 3.8). This resulted in 16 experimental runs defined by an L16 Taguchi array. Further analysis of suitable control variable values was performed using ANOVA. As the actuator system is not contained inside the tool holder, the spindle power is not as important. These experiments were conducted on a Leadwell V-20 CNC milling machine. The sample 1 (Fig. 3.39a) is attached to the surface of the actuator 2, and the actuator is held on a vise 3 attached to a single channel, a Kistler dynamometer 4 that can record either the normal force or the torque and whose signal is transferred through the charge amplifier 1 (Fig. 3.39b) into a digital oscilloscope 2 and recorded on a PC. The actuator's signal is supplied through alligator clips that carry over the signal generated by a signal generator and amplified by a signal amplifier.

The direct excitation of a stationary workpiece was the logical step in the investigation of vibration-assisted machining performance in ductile regime machining of hard-brittle materials. Elimination of the need for the vibrations to pass through the periphery of the machining equipment into the workpiece would allow to achieve higher amplitudes and frequencies of excitation while preventing potential negative effects of vibration on the machine itself. Here, higher excitation frequencies are available as a result of elimination of the design constraints related to electrically driving a system rotating at high speeds as well as the employment of higher order modes that yield peripheral deformation sites at the output surface of the actuator. The workpiece flat surface second mode deformation at 84.5 kHz were measured using 3D scanning laser vibrometer (Fig. 3.40).

The workpiece flat surface second mode deformation at 84.5 kHz presented at Fig. 3.41.

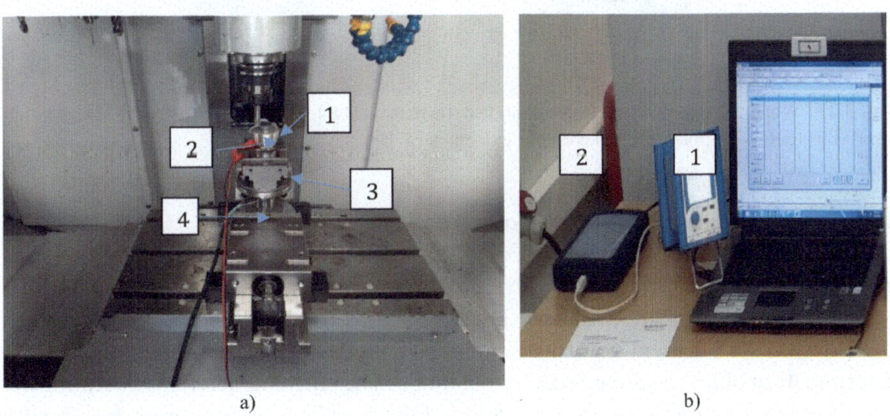

Fig. 3.39 Workpiece excitation study (**a**) and data acquisition hardware (**b**)

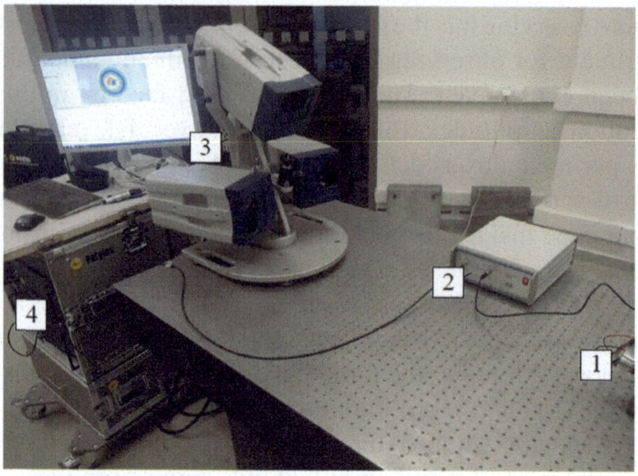

Fig. 3.40 Experimental set-up of Polytec PSV 500 3D scanning laser vibrometer: 1-workpiece on the actuator, 2-signal amplifier, 3-scanning laser head, 4-signal generator/data acquisition system

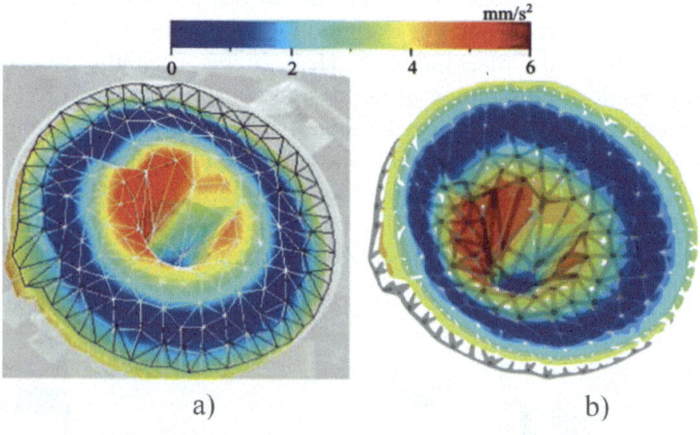

Fig. 3.41 Workpiece flat surface second mode deformation at 84.5 kHz as measured by 3D scanning laser vibrometer at opposite displacement extremes (**a**, **b**)

3.2.2.3 Grinding Chip Morphology Analysis

Where a machined surface is a product of the grinding process, the chips are its byproduct. The chips are produced in discrete quantities, which makes it easier to characterize the different deformation mechanisms taking place while allowing the determination of the dominant machining mode. The chip characteristics defining the ductile machining mode are the continuity of the chip (similar to turning, but smaller in size), whereas the temperature induced ductile mode is often characterized by a spherical chip. Brittle deformation is commonly characterized by fine, powdery residue, and, in the presence of a severe brittle fracture, by a larger-jagged-edged-chip.

While visual analysis is commonly used to characterize the machining mode by chip, it is prone to the same caveats as the visual surface topography analysis. They can be addressed through the use of bulk powder characterization techniques, such as X-Ray Diffraction (XRD), which allows for the determination of the chemical and structural characteristics of the powder, as well as the approximation of the average crystallite size. Based on the previous description on the mode characterizing chip traits, it is assumed that a larger average crystallite size will be indicative of a larger chip, thus suggesting the dominance of the ductile machining mode. In this work, bulk powder dust XRD was performed by using Discovery D8 X-ray diffractometer (Fig. 3.42) with a powder analysis diffractometer head (Fig. 3.43).

The grinding dust was collected after the run (Table 3.10). The ground surfaces were also analyzed via SEM and white light interferometry using a Polytec MSA 500 microsystem analyzer (Fig. 3.44).

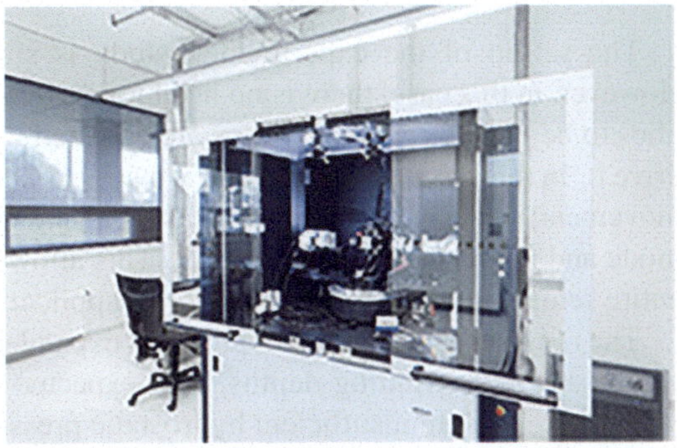

Fig. 3.42 X-ray diffractometer stand Discovery D8

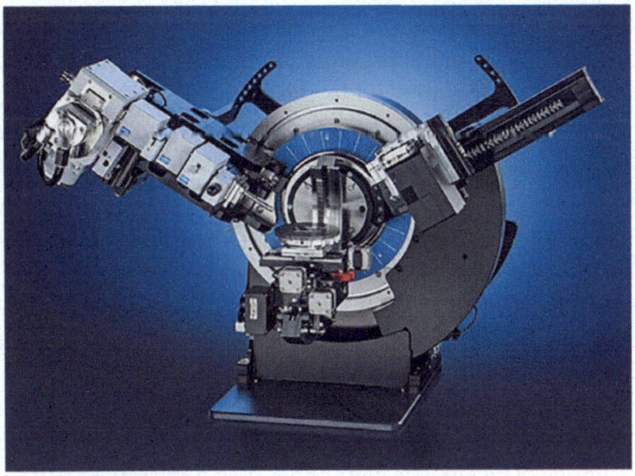

Fig. 3.43 X-ray diffractometer head for XRD analysis of powders

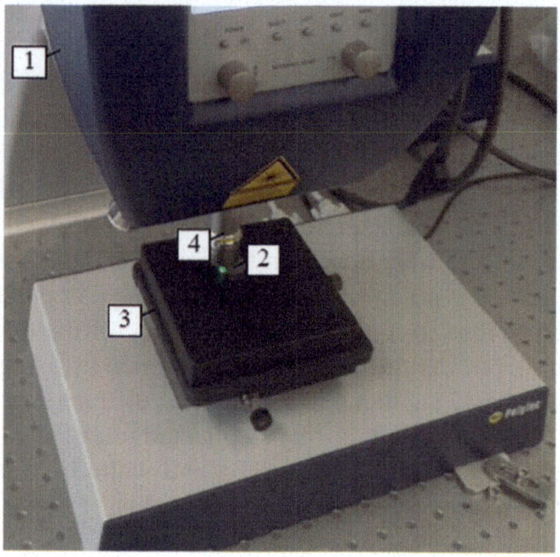

Fig. 3.44 Polytec MSA 500 MICRO microsystem analyzer set-up used for white light interferometry: 1-analyzer head, 2-sample surface, 3-XY positioning table, 4-interferometric lens

The set-up of the experimental study is similar to the tool excitation study. However, in this case, there is no need for carbon brush or commutation ring assemblies to be present. Alligator clips are attached to the actuator electrodes in order to drive it. In this case, the workpiece is stationary, and, in many other applications, the movement, if present, would not be as rapid and would only occur in the translational mode and not in the rotational mode. This allows for a simpler configuration of the entire setup and easier transferability of application from one machine to another.

The range of the grinding depths was expanded to evaluate the machining performance at lower grinding depths. It is expected that the reduction of the grinding depth may result in insufficient hydrostatic pressure, which may cause ductile mode deformation at the grinding site or an improper tool-workpiece contact causing the chipping of the workpiece (i.e., destructive brittle failure) as a result. The achieved frequency of excitation was 85 kHz, which is several times higher than the frequencies employed in the previous studies within this research. In addition to the variation of the depth, the amplitude was varied over 4 values (0, 0.25, 0.5, and 1 μm) as well. The dust was collected from several 50 μm grinding depth runs by using a glass test tube and funnel assembly. The obtained dust samples from ultrasonically assisted and conventional grinding runs were analyzed by using XRD. The ground surfaces were visually evaluated by using SEM and then characterized while using white light interferometry.

3.2.2.4 Grinding Quality Evaluation

The ground surface quality and the mode of deformation were evaluated using several methods. The surface roughness was determined utilizing white light interferometry. White light interferometry utilizes the precise vertical movement of the diffraction lens to identify the surface topology, based on the vertical position of optical interference sites. Further treatment of the surface topology model using computer software allows determining the average surface roughness R_a. However, surface roughness is a poor qualifier for determining the deformation mechanism. While the ductile mode is generally recognized through low surface roughness values, there are no qualifying values established to conclude if the surface was machined in either brittle or ductile mode. Additionally, the brittle mode may occur in some fractions on the machined surface, making it hard to judge the degree of ductile mode achieved. Hence, the obtained topologies were additionally evaluated using an algorithm written in Python programming language by the authors. Its flowchart is outlined below (Fig. 3.45). It was used for identification of brittle deformation mode sites on the surfaces and the evaluation of their impact on the surface quality. Firstly, roughness profile slices, every 1 μm of the obtained typologies, along the feed direction, were produced. Within each profile slice, the peaks and valleys were identified, using existing SciPy functions. The depth and width of each peak were then evaluated. Each peak could be a result of either geometric deviation, ductile deformation at the brittle-ductile transition border, or a brittle deformation. It was arbitrarily decided that, to determine that a peak was a result of brittle deformation, the depth to width ratio of the peak had to be > 0.5. Meaning that the width had to be less than two times greater

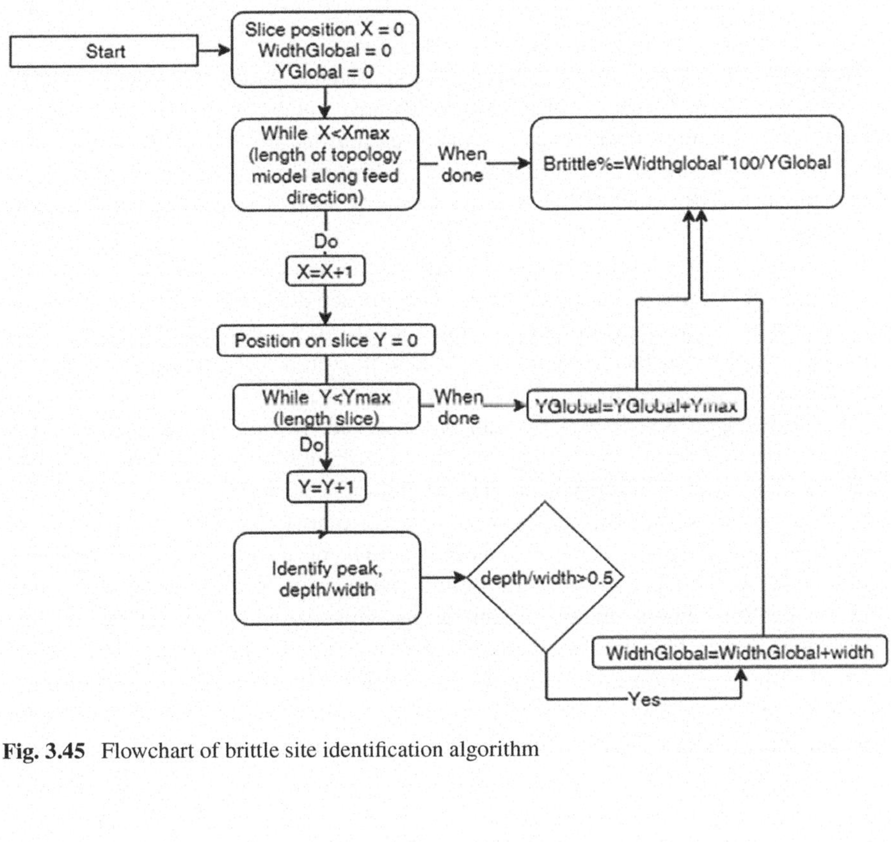

Fig. 3.45 Flowchart of brittle site identification algorithm

than the depth. If a brittle peak was identified, its width was factored in as a brittle deformation zone percentage over the entire length of the profile slices.

The resulting percentage of brittle deformation zones allowed to determine the degree of ductility on the surface. The type of deformation was decided by comparing several results from different factors at the same depth, and by ranking the least brittle to the most brittle result. Also, the mechanism can be identified through chip size. A ductile mode–produced chip should be more continuous and have larger dimensions than the brittle mode–produced chip. XRD analysis can be used for determining various crystalline characteristics of powders. Crystallite size of powders can be estimated using the Scherrer equation:

$$\tau = \frac{K\lambda}{\beta \cos \theta} \tag{3.21}$$

where τ is the mean size of the ordered domains, which may be smaller or equal to the grain size, which may be smaller or equal to the particle size; K is a dimensionless shape factor, with a value close to unity; λ is the X-ray wavelength; β is the line broadening at half the maximum intensity in radians and θ is the Bragg angle.

The tool actuator grinding runs are intended to provide frequency effect evaluation data for the first experiment. Basic practical conditions are used—a depth of 25 and 50 μm, a feed rate of 5 mm/min, and a spindle speed of 5000 rpm. The list of runs conducted with tool actuator is shown in Table 3.9.

The surface of ground samples was characterized using white light interferometry. The surfaces obtained during tool actuator study have undergone brittle deformation, as evidenced by a high occurrence of void irregularities in white light interferometry images. Grinding at 25 μm depth (Fig. 3.46a) yielded average roughness Ra of 0.156 μm and Ra of 0.217 μm at 25 μm and 50 μm grinding s respectively (Fig. 3.46b).

In comparison, workpiece actuator study at the same amplitude and grinding depths yielded improved quality surfaces. The surface at 25 μm exhibited a ductile deformation and at average surface roughness of 0.146 μm produced ductile deformation with similar surface roughness, while at 50 μ m mostly ductile deformation was observed, with a roughness of 0.152 μm (Fig. 3.47).

The influence of grinding depth can be observed in Fig. 3.48 as well. Except for 1 μm grinding depth, the quality of ground surface tended to deteriorate when ultrasonic assistance was introduced. The degree of quality deterioration correlates with the decrease of the amplitude. However, the quality improves when practical 25–50 μm depths are achieved. This is best evidenced by Fig. 3.48a, b; here, the highest and lowest quality surfaces throughout the experiment are presented, through white light interferometry measurements along with scanning electron microscopy micrographs and surface measurements. Results presented in both Fig. 3.48a, b were achieved without ultrasonic excitation. In addition to surface characterization through optical means, the grinding dust from workpiece actuator study runs at 50 μm depth, with and without 85 kHz excitation at 0.5 μm (since it yielded the best surface

Table 3.9 Experimental workpiece excitation study grinding run parameters

Depth	Excitation frequency kHz	Spindle speed rpm	Feed rate mm/min	Excitation amplitude
25	0	5000	5	0
				1
	84.5			0.5
				0.25
50	0			0
				1
	84.5			0.5
				0.25
1	0			0
				1
	84.5			0.5
				0.25
5	0			0
				1
	84.5			0.5
				0.25

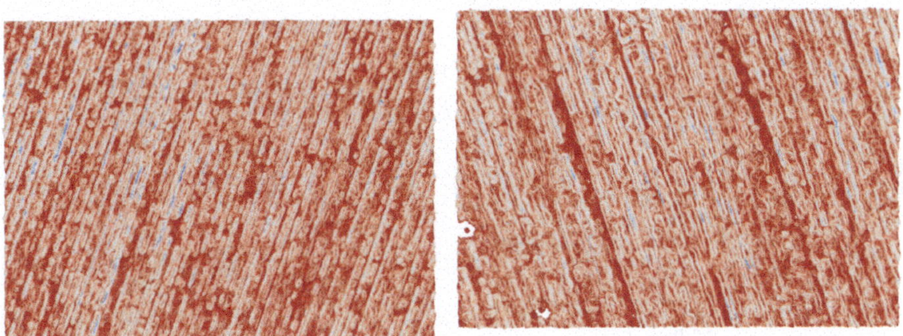

Fig. 3.46 White light interferometry image for tool excitation grinding at 25-μm grinding depth (**a**) and at 50-μm grinding depth (**b**)

quality at 50 μm depth) amplitude was collected for comparison using XRD. It was expected that since in the presence of ductile deformation the chip should be of a higher size, the average crystallite size of dust particles should follow this trend as well. As seen in Fig. 3.48a, b, the average crystallite size is indeed significantly higher with excitation than without it, implying that ductile deformation was induced by ultrasonic excitation.

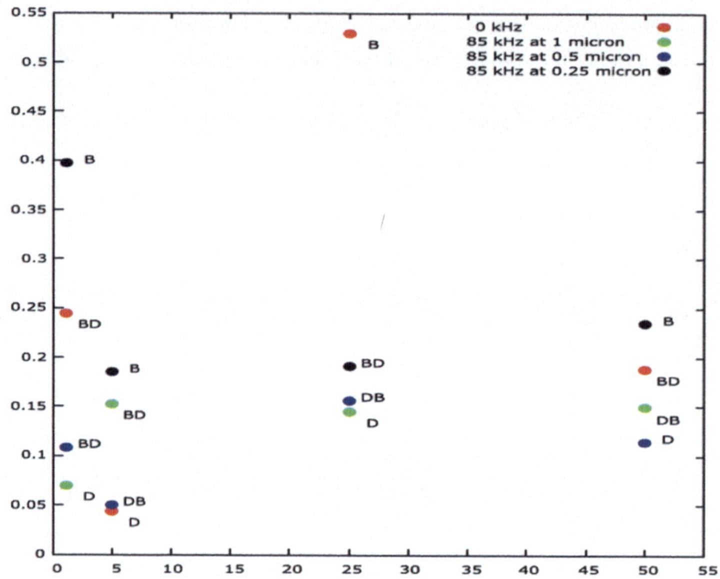

Fig. 3.47 Surface roughness and deformation evaluation for workpiece excitation when grinding at 1-, 5-, 25-, and 50-μm depths. D, ductile deformation; DB, mostly ductile deformation; BD, mostly brittle deformation; B, brittle deformation

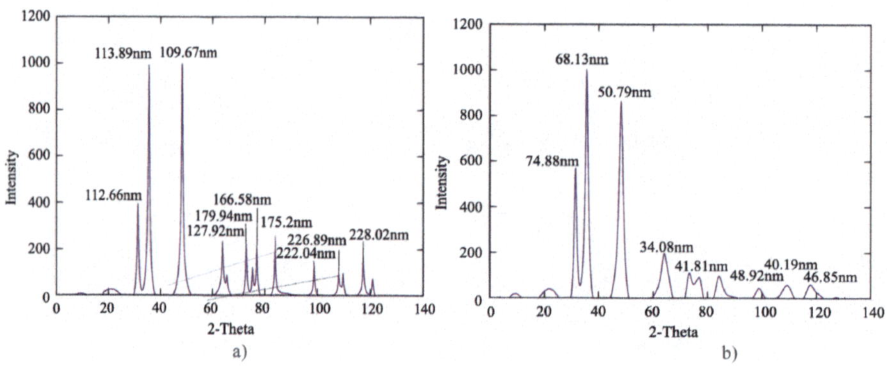

Fig. 3.48 Crystallite size estimation when grinding at 50 µm depth with (**a**) and without (**b**) excitation

Table 3.10 shows the S/N ratios calculated using Eq. 3.20, of all runs from the initial experiment, for evaluating the effect of frequency on surface roughness, whose conditions were summarized previously in Table 3.9. The significance levels of the two factors are outlined in Table 3.11 and it is apparent that the significance levels are similar. However, a clear dependence of surface roughness on the excitation frequency can be seen—as frequency increases, the roughness decreases.

Table 3.10 Test results and S/N ratios based on L4 test design

Exp. no	A	Depth	B	Frequency	Roughness	S/N
		(a_p)		(f)	(R_a)	ratios
1	1	25	1	14.5	0.156	16.138
2	1	25	2	84.5	0.147	16.654
3	2	25	1	14.5	0.217	13.271
4	2	50	2	84.5	0.15	16.479

Table 3.11 Significance levels of control factors for S/N ratios of roughness values

Control	S/N	S/N	Max	Priority
Factors	L1	L2	Min	
Depth	0.152	0.184	0.156	2
Frequency	0.187	0.149	0.147	1

Table 3.12 Results of ANOVA analysis based on *S/N* ratios of roughness values

Control	DOF	Total	Average	F	P	Profit
Factors		Square	Square	–	–	%
Frequency	1	3.466	3.466	1.91	0.398	45.65
Depth	1	2.314	2.314	1.28	0.461	30.48
Error	1	1.811	1.811	–	–	−23.86
Total	3	7.591	–	–	–	100
R^2	76.15%					

Results of ANOVA applied to roughness values are given in Table 3.12. While the most significant control factor on roughness values is the excitation frequency (45.66%), the depth of cut is also significant (30.48%). It is therefore assumed that further analysis of amplitude interactions with cutting depth regarding surface roughness needs to be conducted at the highest available frequency. The proposed workpiece excitation approach is capable of achieving excitation frequencies above 100 kHz; however, reliable indexing of amplitudes is available at 84.5 kHz. Hence, further experiments were conducted at 84.5 kHz frequency, using run conditions outlined previously in Table 3.13.

Table 3.13 Experiment outline

Factor	Units	L1	L2	L3	L4
Amplitude (A)	μm	0	0.25	0.5	1
Depth (a_p)	μm	1	5	25	50

Table 3.14 Test results and S/N ratios based on L16 test design

Exp	A	Depth	B	Amplitude	Roughness	S/N
no		(a_p)		(A)	(R_a)	ratios
1	1	1	1	0	0.245	12.217
2	1	1	2	0.25	0.395	8.068
3	1	1	3	0.5	0.125	18.062
4	1	1	4	1	0.075	22.499
5	2	5	1	0	0.475	26.466
6	2	5	2	0.25	0.185	14.657
7	2	5	3	0.5	0.05	26.021
8	2	5	4	1	0.16	15.918
9	3	25	1	0	0.54	5.352
10	3	25	2	0.25	0.185	14,657
11	3	25	3	0.5	0.16	15.918
12	3	25	4	1	0.15	16.478
13	4	50	1	0	0.175	15.139
14	4	50	2	0.25	0.24	12.396
15	4	50	3	0.5	0.105	19.576
16	4	50	4	1	0.15	16.478

The results of the amplitude effect evaluation experiment along with *S/N* ratios are outlined in Table 3.14. As shown previously in XRD analysis data and as evident by the dependence of the results on grinding depth is observed. While the reduction to lower depths, in the conventional grinding case, should result in better surface quality, it seemed to have a destructive effect, when ultrasonic excitation was involved. At small depths, such as 5 μm, the tool-workpiece contact was unstable, resulting in micro-impacts yielding similar quality deterioration as in the tool excitation case, which was more prevalent with the reduction of amplitude, yet was eliminated, when reducing the amplitude to 0 μm; hence, the best surface quality could be achieved at 5 μm without excitation (Fig. 3.49). An exception could be made, however, for the depth of 1 μm, in this case, 1 μm amplitude yielded ductile deformation and significant improvements to the surface quality. As the grinding depth was increased, the surface quality of no excitation setup deteriorated, while the ultrasonically assisted cases improved (Fig. 3.50).

Due to the destructive effects occurring at lower depths, the signifficance values for amplitude and depth are essentially equal (Table 3.15). ANOVA similarly yields almost equal profit for both factors (Table 3.16).

The results show that the higher excitation frequencies, as well as tool stability inherent to the workpiece excitation approach, have a significant influence on the mode of deformation occurring during the grinding. During excitation, the tool actuator exhibited parasitic low-frequency vibrations, which may have induced tool

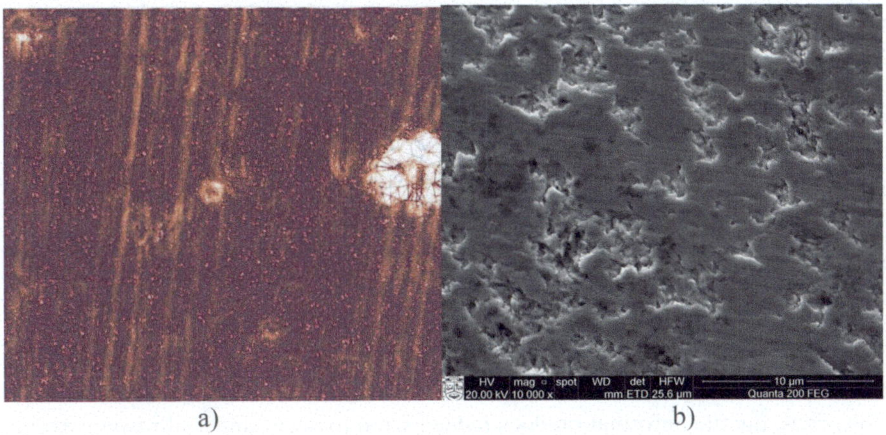

Fig. 3.49 White light interferometry (**a**) and SEM micrograph (**b**) comparisons of tungsten carbide surface after ductile deformation during grinding with workpiece excitation at 0 kHz 0 μm at depth of 5 μm and resultant surface roughness (Ra) = 0.044 μm

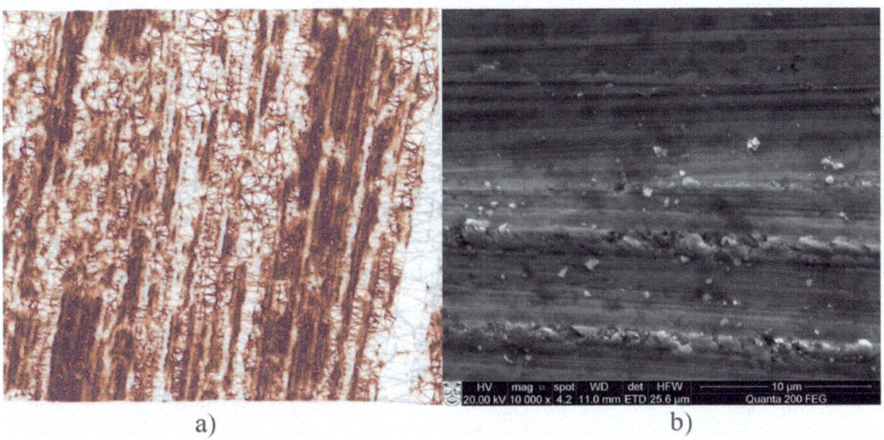

Fig. 3.50 White light interferometry (**a**) and SEM micrograph (**b**) comparisons of tungsten carbide surface after brittle deformation during grinding with workpiece excitation at 0 kHz 0 μm at depth of 25 μm and resultant surface roughness (Ra) = 0.530 μm

Table 3.15 Significance levels of control factors for S/N ratios of surface roughness values during amplitude effect experiment

Control	S/N	S/N	S/N	S/N	Max	Priority
Factors	L1	L2	L3	L4	Min	
Depth	15.21	20.77	13.1	15.9	7.66	1
Amplitude	14.79	12.44	19.89	17.84	7.45	2

Table 3.16 Results of ANOVA analysis based on *S/N* ratios of roughness values in amplitude effect experiment

Control	DOF	Total	Average	F	P	Profit
Factors		Square	Square	–	–	%
Depth	3	126	42.01	1.71	0.234	57.014
Amplitude	3	129.7	43.23	1.76	0.224	58.689
Error	9	221	24.55	–	–	15.701
Total	15	476.7	–	–	–	100
R^2	53.64%					

chatter—resulting in poor surface quality. As evidenced by XRD analysis of crystallite sizes, ductile deformation does indeed seem to occur under ultrasonic excitation. However, the effect is more expressedl at higher grinding depths. Also, there seems to be a variation of results at different amplitudes, which seems to be related to different tool-workpiece contact conditions. Therefore, further research should focus on workpiece excitation at high ultrasonic frequencies (100 kHz), when grinding at high depths, as well as the nature of effects different amplitudes have the surface quality of the workpiece at varying depths.

The research results show, that ductile deformation was achieved during ultrasonically assisted grinding of tungsten carbide, employing workpiece excitation approach under practical conditions, while it was not possible to achieve the same outcome during conventional grinding or ultrasonically assisted grinding employing tool excitation. Ultrasonic workpiece excitation is a more promising approach in ultrasonic machining when using rotary tools, due to its potential to employ higher modes and frequencies, as well as ease of use and no inherent limitations on spindle speed.

References

1. Vaicekauskis M, Gaidys R, Ostasevicius V (2015) Ultra precision high frequency assisted diamond turning device sonotrode dynamics analysis. In: Mechanika 2015 proceedings of the 20th international science conference, pp 272–276
2. Gaidys R, Dambon O, Ostasevicius V, Dicke C, Narijauskaite B (2017) Ultrasonic tooling system design and development for single point diamond turning (SPDT) of ferrous metals. Int J Adv Man Tech 93(5–8):2841–2854
3. Ostasevicius V, Gaidys R, Dauksevicius R, Mikuckyte S (2013) Study of vibration milling for improving surface finish of difficult-to-cut materials. Stroj Vestn-J Mech Eng 59(6):351–357
4. Ostasevicius V, Balevicius G, Zakrasas R, Baskutiene J, Jurenas V (2016) Investigation of vibration assisted drilling prospects for improving machining characteristics of hard to machine materials at high and low frequency ranges. Mechanika 22(2):125–131
5. Ostasevicius V, Balevicius G, Zakrasas R (2016) Development guidelines for models of grinding forces in ultrasonically assisted grinding of binderless tungsten carbide. In: Mechanika 2016: proceedings of the 21st international science conference, pp 193–198
6. Ostasevicius V, Jurenas V, Balevicius G, Zukauskas M, Ubartas M (2014) Vibrational excitation of a workpiece for drilling force reduction in brittle materials. Mechanika 6:596–601

7. Janusas G, Palevicius A, Ostasevicius V, Bansevicius RP, Busilas A (2007) Development and experimental analysis of piezoelectric optical scanner with implemented periodical microstructure. J Vibroeng 9(3):10–14
8. Scattergood RO (1991) Ductile-regime grinding: a new technology for machining brittle materials. J Eng Ind 113(2):184–189
9. Ostasevicius V, Jurenas V, Balevicius G (2019) Workpiece holder with ultrasonic actuator for abrasive finishing. Lithuanian patent LT2019 505
10. Ostasevicius V, Jurenas V, Balevicius G, Cesnavicius R (2020) Development of actuators for ultrasonically assisted grinding of hard brittle materials. Int J Adv Man Techn 106:289–301

Chapter 4
Wireless Connectivity Options for Tool Condition Monitoring IoT Applications

4.1 The New Principles of Energy Harvesting in Macro Level

The ability to avoid replacing depleted batteries is very attractive for wireless sensor networks, where the maintenance costs associated with inspecting and replacing batteries are significant. Vibration energy harvesters based on piezoelectric resonators are promising to power wireless sensor nodes and are attractive for their high power on a small scale. Since most practical vibration sources are frequency-varying or random, how to extend the throughput of vibration energy harvester devices becomes one of the most challenging issues before putting them into practice. The promising energy harvesting possibilities are associated with the excitation of higher vibration modes of elastic links and the segmentation of the piezo layers on them. The results of the segmentation of the piezoelectric layers of the optimally shaped cantilever designed for the second eigenfrequency are described in the [1–3]. The piezoelectric energy harvester links, which are optimal for the given third eigenmode frequency, also efficiently generate energy. A few ways of producing this mode stimulation, namely vibro-impact or forced excitation, as well as its application for energy harvesting devices are proposed.

4.1.1 Virtual Twin of Piezoelectric Energy Harvester

Usually, piezoelectric energy harvesters cannot be cost effectively built and tested without prior modeling of its component, which may be performed with different software packages like Coventorware, Ansys, or Comsol multiphysics. Thus, one of the objectives of this research was to develop a general and universal FE virtual twin, parameters of which could be easily changed. This should help to analyze the dynamic response of various configurations and determine their possible electrical outputs.

V. Ostaševičius, *Digital Twins in Manufacturing*, Springer Series in Advanced Manufacturing, https://doi.org/10.1007/978-3-030-98275-1_4

The ultimate purpose of FE analysis is to mathematically recreate behavior of an actual engineering system [4]. To obtain an accurate mathematical model of a physical prototype, a 2D FE model of vibro-impacting piezoelectric energy harvester was developed with Comsol multiphysics software. Figure 4.1 and Table 4.1 provide principal scheme and geometry data of developed unimorph transducer—cantilever beam of stainless steel covered by piezoelectric layer (of PZT, PVDF or PMN-PT), operating in transverse (d_{31}) mode. It was assumed that ideally conductive electrodes of negligible thickness cover the entire area of top and bottom surfaces of the piezoelectric layer. Modeling was performed with Lagrange-quadratic elements using plane-strain approximation since flexural vibration modes have a greater influence on vibro-impact process in comparison to torsional modes. Boundary conditions were set to represent electrodes enveloping piezoelectric material and structure clamping, meanwhile a simple electric circuit, comprised of a single resistive load,

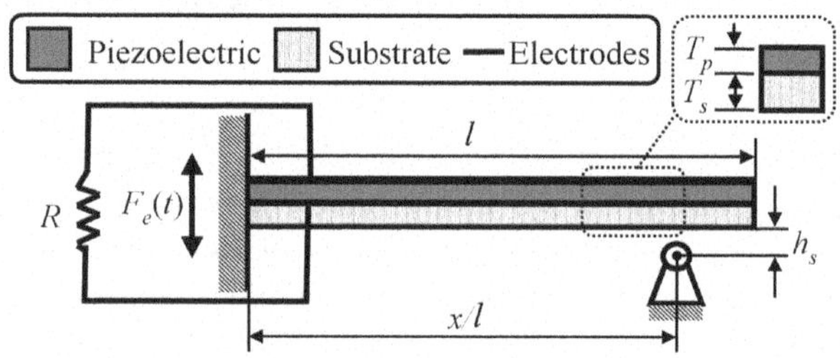

Fig. 4.1 Principal scheme of developed FE model of piezoelectric energy harvester

Table 4.1 Main parameters of the developed model of piezoelectric energy harvester

Model parameters	Value	Description
l_s	100 mm	Substrate length
l_p	100 mm	Piezoelectric layer length
T_s	1 mm	Substrate thickness
T_p	0.2 mm	Piezoelectric layer thickness*
W_s	10 mm	Substrate width
W_p	10 mm	Piezoelectric layer width
A	1.0*g	Applied excitation acceleration
R_{sc}	100 Ω	Connected resistor load, corresponding to open-circuit conditions
R_{oc}	1 MΩ	Connected resistor load, corresponding to open-circuit conditions
F_e	*	Applied excitation frequency (corresponding to the first resonant frequency of the system)
h_s	2 μm	Stopgap size at the support location
C_{vdW}	1×10^{-32}	Adhesion constant

*Numeric expression varied through the course of simulations

Table 4.2 Matlab software

Matlab script	Comments

text = '1_PZT_290.txt';
[time, voltage] = textread (text, '%f%f', 'headerlines',100);
size(voltage)
x = voltage;
t = time;
n = size(x);
y = sqrt (1/n (1)* sum(x.^2))
plot (t, x)
	% determines Comsol multiphysics data file;
	% reads data from file and trims off transient periods;
	% determines the number of rows and columns in the matrix
	%defines variables
	%defines variables
	%determines trimmed voltage data quantity;
	%calculates and outputs RMS;
	%plots time versus voltage graph for trimmed data;

*Numeric expression varied through the course of simulations

was introduced to an electromechanically coupled system via Spice circuit editor enabling closer-to-practice simulations of piezoelectric energy harvester.

Eigenfrequency solver function was used to determine resonant frequencies of harvester prototype, while transient solver function was employed to predict its displacement and voltage output. Simulated voltage output was exported as.txt data file and transferred for processing to Matlab software (Table 4.2). The below script was used to determine voltage generated by operating at certain conditions.

Piezoelectric energy harvester dynamics is described by the following equation of motion presented in a general matrix form:

$$[M]\{\ddot{z}\} + [C]\{\dot{z}\} + [K]\{z\} = \{Q(t,\ z,\ \dot{z},)\} \tag{4.1}$$

where $[M]$, $[C]$, $[K]$—mass, damping, and stiffness matrices of the harvester, respectively, $\{z\}$, $\{\dot{z}\}$, $\{\ddot{z}\}$— displacement, velocity, and acceleration vectors, while $\{Q(t,\ z,\ \dot{z},)\}$ vector represents the sum of external forces.

To define piezoelectric effects, the following considerations and constitutive piezoelectric equations [5] were used in the FE model. As it was already discussed in the introductory section, when a piezoelectric material is subjected to stress T, it produces polarization P, which is the function of stress (direct piezoelectric effect)

$$P = dT. \tag{4.2}$$

In contrast, when electric field E is applied across electrodes of piezoelectric material, it produces strain S which is the function of electric field (inverse piezoelectric effect)

$$S = dE. \tag{4.3}$$

For an elastic material, the relationship between strain S and stress T is given by

$$S = s^E T. \tag{4.4}$$

For a dielectric substance, the relationship of electrical displacement D with electric field strength E is given by

$$D = \varepsilon_0 E + P. \tag{4.5}$$

with ε_0 being dielectric permittivity of vacuum and P being polarization of material due to applied field.

From these relationships, with electric field E and stress T as independent variables, the two constitutive piezoelectric equations can be written as

$$\begin{aligned} S &= s^E T + dE \\ D &= dT + \varepsilon^T E. \end{aligned} \tag{4.6}$$

Since the phenomenon of piezoelectricity is anisotropic, electric field E and electrical displacement D are represented in vector magnitudes, while stress T and stain S are given in symmetrical tensile magnitudes

$$\begin{aligned} S_i &= s_{ij}^E T_j + d_{mi} E_m \\ D_n &= d_{nj} T_j + \varepsilon_{nm}^T E_m, \end{aligned} \tag{4.7}$$

where $m, n = 1, 2, 3$; $i, j = 1, 2{,}6$.

These equations can be well represented in matrix form as:

$$\begin{bmatrix} S_1 \\ S_2 \\ S_3 \\ S_4 \\ S_5 \\ S_6 \end{bmatrix} = \begin{bmatrix} s_{11}^E & s_{12}^E & s_{13}^E & 0 & 0 & 0 \\ s_{12}^E & s_{11}^E & s_{13}^E & 0 & 0 & 0 \\ s_{13}^E & s_{13}^E & s_{33}^E & 0 & 0 & 0 \\ 0 & 0 & 0 & s_{44}^E & 0 & 0 \\ 0 & 0 & 0 & 0 & s_{44}^E & 0 \\ 0 & 0 & 0 & 0 & 0 & 2(s_{11}^E - s_{12}^E) \end{bmatrix} \begin{bmatrix} T_1 \\ T_2 \\ T_3 \\ T_4 \\ T_5 \\ T_6 \end{bmatrix} + \begin{bmatrix} 0 & 0 & d_{31} \\ 0 & 0 & d_{31} \\ 0 & 0 & d_{31} \\ 0 & d_{31} & 0 \\ d_{31} & 0 & 0 \\ 0 & 0 & 0 \end{bmatrix} \begin{bmatrix} E_1 \\ E_2 \\ E_3 \end{bmatrix}$$

$$\begin{bmatrix} D_1 \\ D_2 \\ D_3 \end{bmatrix} = \begin{bmatrix} 0 & 0 & 0 & 0 & d_{15} & 0 \\ 0 & 0 & 0 & d_{15} & 0 & 0 \\ d_{31} & d_{31} & d_{33} & 0 & 0 & 0 \end{bmatrix} \begin{bmatrix} T_1 \\ T_2 \\ T_3 \\ T_4 \\ T_5 \\ T_6 \end{bmatrix} + \begin{bmatrix} \varepsilon_{11} & 0 & 0 \\ 0 & \varepsilon_{11} & 0 \\ 0 & 0 & \varepsilon_{33} \end{bmatrix} \begin{bmatrix} E_1 \\ E_2 \\ E_3 \end{bmatrix} \tag{4.8}$$

In order to simulate harvester operating in vibro-impacting mode, a viscoelastic–adhesive contact formulation was implemented into the FE model. This formulation is based on Kelvin–Voigt rheological model represented by linear spring connected in parallel with linear damper (this coupling element is defined by stiffness k_p and damping c_p). Additionally, adhesion-related parameter c_{vdW} is introduced, which allows taking into account Van der Waals forces acting at the micro-scale before mechanical contact actually occurs. Therefore, the proposed contact model may be used for simulations of both macro- and micro-scale energy harvesting devices. The dynamics of the vibro-impact harvester is defined by the matrix equation below

$$[M]\{\ddot{z}\} + [C]\{\dot{z}\} + [K]\{z\} = \begin{cases} \{F(t)\} + \{F_s(t)\}, & if\ z_{l_s}(t) < h_s - \xi_0 \vee p_{l_s}(\dot{z}_{l_s}, z_{l_s}, t) \geq 0; \\ \{F(t)\} + \{F_s(t)\} + \{P_c(\dot{z}, z, t)\}, & if\ z_{l_s}(t) \geq h_s - \xi_0 \wedge p_{l_s}(\dot{z}_{l_s}, z_{l_s}, t) < 0. \end{cases} \tag{4.9}$$

where

$$\{F_{ls}(t)\} = \begin{cases} \frac{C_{vdW}}{[h_s - z_{l_s}(t)]^3}, & if\quad z_{l_s}(t) < h_s - \xi_0 \vee p_{l_s}(\dot{z}_{l_s}, z_{l_s}, t) \geq 0; \\ \frac{C_{vdW}}{\xi_0^3}, & if\quad z_{l_s}(t) \geq h_s - \xi_0 \wedge p_{l_s}(\dot{z}_{l_s}, z_{l_s}, t) < 0. \end{cases} \tag{4.10}$$

here $\{\dot{z}(0)\} = \{\dot{z}^0\}, \{z(0)\} = \{z^0\}$, $[M]$, $[C]$, $[K]$—mass, damping and stiffness matrices, $\{\dot{z}\}$, $\{\ddot{z}\}$—velocity and acceleration vectors, $\{z^0\}$—displacement at $t = 0$, $\{\dot{z}^0\}$—velocity at time moment t = 0, $\{F(t)\}$—vector of external forces acting on harvester (in this case this is base excitation), $\{F_s(t)\}$—vector representing the influence of Van der Waals forces, $\{P_c(\dot{z}, z, t)\}$—vector of nonlinear interaction in the contact pair. The developed contact model is defined by the following two components:

$$C_{vdW} = \frac{A_H A_c}{6\pi} \tag{4.11}$$

and

$$p_{l_s}(\dot{z}_{l_s}, z_{l_s}, t) = -k_p[z_{l_s}(t) - (h_s - \xi_0)] - c_p \dot{z}_{l_s}(t), \tag{4.12}$$

where $z_{l_s}(t)$, $\dot{z}_{l_s}(t)$—displacement and velocity of harvester surface point, h_s—stopgap, ξ_0—distance between surfaces when it is assumed that mechanical contact has occurred (~1 nm), k_p, c_p—stiffness and damping of the coupling element, respectively, A_H—Hamaker's constant, A_C—contact area, p_{ls}—contact pair's nonlinear interaction force at the contact point (l_s—contact point position along the longitudinal axis, measured from the clamped end, where $l = 0$).

The developed contact model was introduced into the FE model as a transverse force acting on a selected point located on the bottom edge of the harvester cantilever.

The model allows variation of both vertical and horizontal positions of the "virtual" support with respect to the transducer surface.

Thus, the above described developed harvester FE model is universal (i.e. it takes into account a number of different constituents that can be easily adapted) and allows performing complex and integrated simulations, namely: (i) evaluating dynamic and electric response of harvester and the way it is dependent on the variation of geometric parameters of the device; (ii) the influence of the external circuit impedance on the dynamic and electrical response of the harvester; (iii) analyzing nonlinear dynamic effects occurring as harvester is impacting on incorporated rigid support, location of which may be easily adjusted.

4.1.1.1 Evaluation of Harvester Response to Harmonic and Random Excitation

The aim of this sub-chapter simulations is to ascertain the way electrical outputs of harvester change once it is subjected to harmonic and random base excitations. FE model of basic configuration with incorporated support, scheme of which is already depicted in Fig. 4.1 and the main characteristics presented in Table 4.3 are employed to perform subsequent simulations. For the first round of simulations, the harvester was subjected to sinusoidal base excitation, which was defined as vertically acting body load with magnitude controlled by imposed acceleration and excitation frequency. Meanwhile, for the second round of simulations, a random base excitation signal was introduced to the model. In both cases, a resistive load representing open circuit (1 MΩ) was introduced to the electromechanically coupled system via Spice circuit editor, keeping geometric parameters of the system constant. The mechanical response of the harvester was evaluated and resonant frequencies of the system were determined. Determined first eigenfrequency was further introduced into the FE model of the harvester to define harmonic excitation signal. Root Mean Square (RMS) voltage of impacting harvester was determined for a number of relative support positions and compared to the voltage output when the same harvester operates in the non-impacting mode.

For harmonic excitations of harvester, effect of support location on electrical outputs is demonstrated in Fig. 4.2, which provides a plot of impacting/non-impacting

Table 4.3 Geometric characteristics of FE model used to evaluate harvester electric response to harmonic and random base excitations

Parameters	Value
Stainless steel substrate dimensions ($L_S \times W_S \times T_S$) (mm)	100 × 10 × 1
Piezoelectric PZT—5H dimensions ($L_P \times W_P \times T_P$) (mm)	100 × 10 × 0.2
External electric circuit resistance (MΩ)	1
Stopgap size at the support location (μm)	2

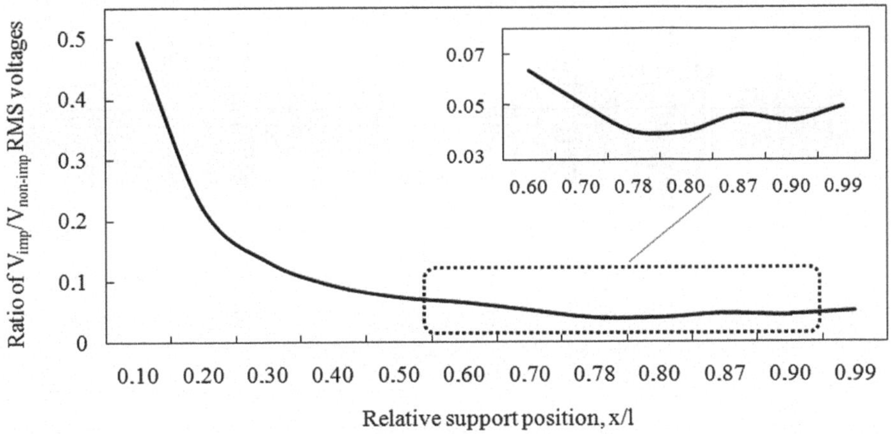

Fig. 4.2 Harvester generated impacting/non-impacting RMS voltage ratio as a function of relative support position (harmonic excitation signal)

RMS voltage ratio as a function of relative support position. One may note that for the harmonic excitation signal case, the presence of support limits harvester displacements leading to the 50–90% reduction in the generated RMS voltages if compared to the RMS voltage when the device is operating in non-impacting mode. However, the magnified view in Fig. 4.2 suggests that if relative support position is in the vicinity of the nodal points of the second and third transverse vibration mode (0.78*l* and 0.87*l*), harvester generated RMS voltages slightly increase. This implies that if support location coincides with the nodal points of the second and third transverse vibration modes, it enables improvement of the overall reliability of the considered system at the smallest expense of the generated voltage; however, the expense is still very high.

For the second round of simulations, random excitation signal was introduced into the FE model of the system, as it would represent a more realistic case of vibro-impacting harvester utilization. In order to define random excitation signal the below described steps were taken: (i) time-acceleration curve of operating industrial heating fan was registered by employing accelerometer, oscilloscope, and PicoLog software set-up (measured real signal is presented in Fig. 4.3a); (ii) registered signal was mathematically processed with Matlab software, which resulted in a few mathematical expressions of the excitation signal, qualitatively defined by the coefficient of determination R^2 (the more value of R^2 approaches 1, the more accurate random excitation signal mathematical approximation is (example of real excitation signal compared to approximated one is given in Fig. 4.3b); (iii) chosen random signal approximation was introduced in the FE model as vertically acting body load (mathematical expression of approximated random excitation signal are presented in Fig. 4.3c).

Generated RMS voltages of impacting harvester, excited by random signal were determined for a number of relative support positions and compared to the RMS voltage output of harvester operating in non-impacting mode (Fig. 4.4). As one may

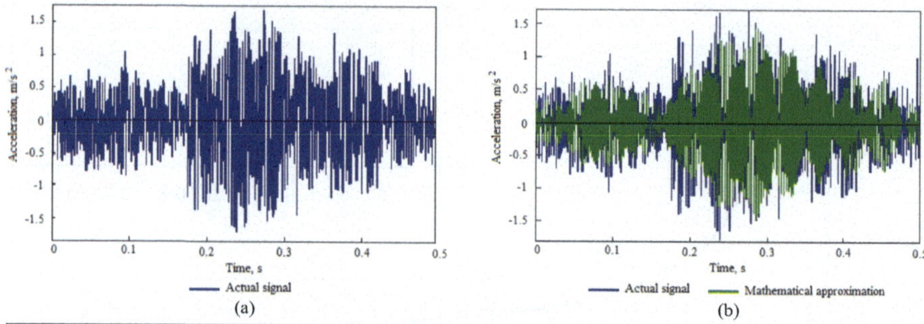

F=0.1322*cos(2*pi*300*t-1.6153)+0.1164*cos(2*pi*302*t+0.235)+0.2796**cos(2*pi*320*t-2856)+0.3476*cos(2*pi*322*t+2.8952)+0.3383*cos(2*pi*324*t--0.9567)+0.344*cos(2*pi*326*t+1.3189)+0.0898*cos(2*pi*384*t-2.2426);

Fig. 4.3 Steps of random excitation signal mathematical approximation: **a** measured real random signal, **b** comparison or real and approximated random excitation signal; $R^2 = 0.7869$ and **c** mathematical expression of approximated random excitation signal ($R^2 = 0.7869$)

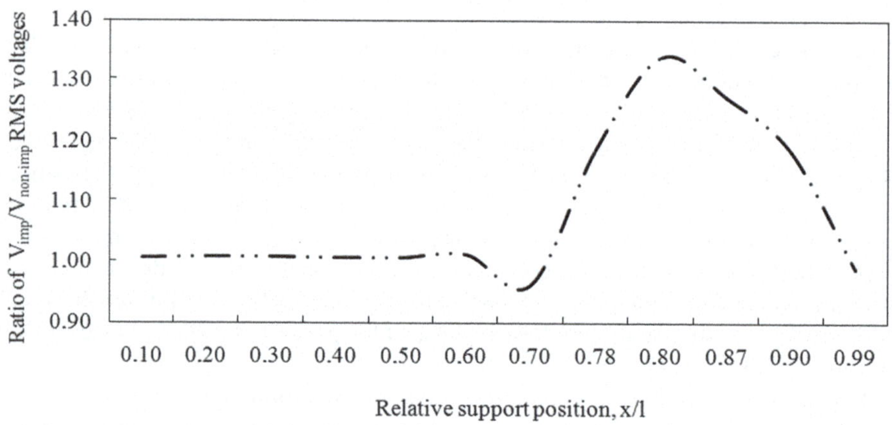

Fig. 4.4 Harvester generated impacting/non-impacting RMS voltage ratio as a function of relative support position (random excitation signal)

note from Fig. 4.4 that the presence of support does not limit the performance of the harvester once it is excited by the random signal. Furthermore, if the support is located in the vicinity of nodes of the second and the third vibration mode, the generated RMS voltages may increase up to 1.3 times for the analyzed harvester configuration.

One should also note that the characteristics of generated voltage curve become quite different and have higher absolute voltage values as support is placed in the vicinity of nodal points, if compared to the generated voltage curve when support is placed elsewhere (example presented in Fig. 4.5a is time-voltage characteristic for

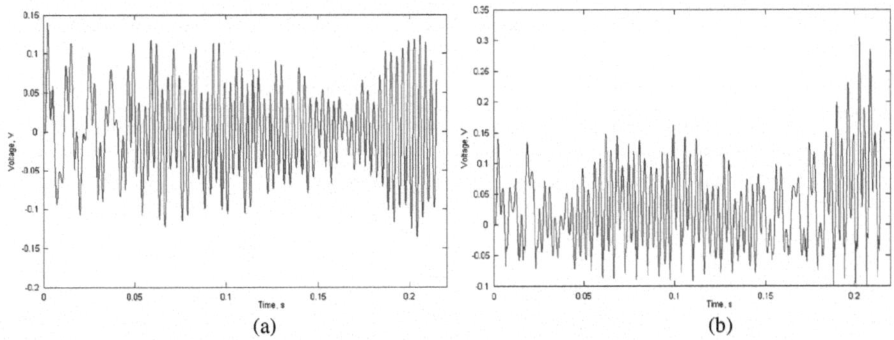

Fig. 4.5 Time-voltage characteristics of generated voltage for different support locations: **a** support is located at $0.1l$ and **b** support is located at $0.87l$

support located at $0.1l$ and presented in Fig. 4.5b is time-voltage characteristic for support located at $0.87l$).

The advantages achieved when the support is positioned in the nodal point ($0.78l$) of the second eigenmode or nodal point ($0.87l$) of the third eigenmode are related to the intensification of transverse vibrations of the respective modes. It should be noted that the amplification of the second and the third transverse eigenmodes (when the support is located at their nodal points) does not terminate the first eigenmode.

If one slightly increases the stopgap size between the support and harvester surface, the effect of support on generated RMS voltage output becomes less pronounced (as may be seen in Fig. 4.6, where stopgap size between the support and harvester surface is increased from 1 to 3 μm, yet increases in harvester generated RMS voltages may still be noted in the vicinity of nodal points.

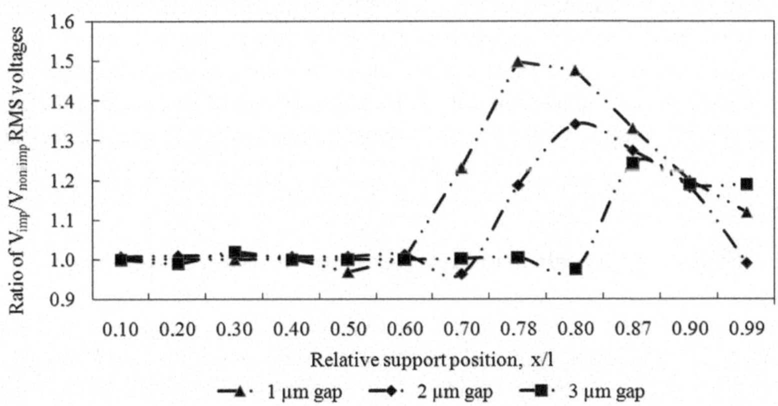

Fig. 4.6 Harvester generated impacting/non-impacting RMS voltage ratio as a function of relative support position for different stop gap sizes (1, 2, and 3 μm) in case of random excitation signal

Table 4.4 Comparison of harvester generated RMS voltages excited by the harmonic and random excitation signal

Relative support position, x/l	Harmonic excitation signal		Random excitation signal	
	Generated RMS voltage, V	$V_{imp}/V_{non\text{-}imp}$ ratio	Generated RMS voltage, V	$V_{imp}/V_{non\text{-}imp}$ ratio
0.10	4.5589	0.4944	0.0522	1.0077
0.20	2.0375	0.2209	0.0523	1.0096
0.30	1.2273	0.1331	0.0523	1.0096
0.40	0.8557	0.0928	0.0522	1.0077
0.50	0.6687	0.0725	0.0522	1.0077
0.60	0.5844	0.0633	0.0525	1.0135
0.70	0.4695	0.0509	0.0499	0.9633
0.78	0.3650	0.0395	0.0615	1.1872
0.80	0.3662	0.0397	0.0694	1.3397
0.87	0.4249	0.0460	0.0660	1.2741
0.90	0.4063	0.0440	0.0613	1.1833
1.00	0.4564	0.0495	0.0513	0.9903
Generated RMS voltage in non-impacting mode, V				
–	9.22		0.0518	

To summarize this stage of simulations, one may compare harvester generated RMS voltages as harvester is excited by harmonic or random signals (Table 4.4). One will be able to note that an unsupported harvester, exited harmonically at its first resonance, generates significantly greater RMS voltages; generated RMS voltages drastically drop if support is introduced in such a system.

Meanwhile, when the harvester is excited by a random signal, support positively influences its' performance, especially if it is located in the vicinity of the nodal points of the second and the third eigenmodes. This implies that if support location coincides with these nodal points, it enables improvement of overall reliability as well as the performance of harvester as long as the system is excited by random signal.

4.1.1.2 Effects of Piezoelectric Material Type on Generated Voltage

The objectives of this research were to analyze the effects of piezoelectric, elastic, and dielectric piezoelectric material constants on harvester generated RMS voltages and to explore if only large piezoelectric strain constants (especially d_{31}) account for substantially larger voltage generation. Or, stating in other words, to clarify if d_{31} constant alone is a sufficient parameter to select piezoelectric material. While reviewing piezoelectric material properties, it was noted that large piezoelectric constants are always followed by large elastic stiffness. Thus, it was assumed that

Table 4.5 Characteristics of FE model used to evaluate how different piezoelectric materials influence harvester electric response

Parameters	Value
Stainless steel substrate dimensions ($L_S \times W_S \times T_S$) (mm)	100 × 10 × 1
Piezoelectric dimensions ($L_P \times W_P \times T_P$) (mm)	100 × 10 × 0.2
External electric circuit resistance (MΩ)	1
Stopgap size at the support location (μm)	2
Piezoelectric materials introduced to the FE model	PMN-28%PT, PZT-5H, PVDF

for the latter reason, very large d_{31} constants may not necessarily lead to very large generated voltages as the stiffness of piezoelectric material affects generated voltages via electromechanical coupling.

Three different piezoelectric materials-polymer PVDF (polyvinylidene fluoride), piezoceramic PZT-5H (lead zirconium titanate), single-crystal PMN-28% PT (lead magnesium niobate)—were introduced one by one to the developed FE model of harvester (Fig. 4.1) in order to explore the effect of piezoelectric material type to the magnitudes of generated RMS voltages. The main geometric characteristics of the FE model used for subsequent simulations are presented in Table 4.5.

Main material properties of PVDF, PZT-5H, and PMN-28% PT are listed in Table 4.6. There are quite many sources in literature and the internet describing piezoelectric material properties, most of which are devoted to PZT ceramics. Meanwhile, polymer and single-crystal material properties are more difficult to find, as these are relatively new materials if compared to the PZT, which is most commonly met in energy harvesting applications. As one may note from Table 4.6 that the piezoelectric, elastic, and dielectric properties of these active materials differ from each other considerably. For example, piezoelectric constants increase in the orders of magnitude if comparing PVDF and PMN-28% PT. One may also note that large piezoelectric constants come with large elastic stiffness; meanwhile, elastic compliance constants reduce comparing PVDF to PMN-28% PT. The large piezoelectric material stiffness may overweight the advantages of high piezoelectric strain constants and harvesters with PMN-PT layers may not necessarily generate higher RMS voltages.

However, simulation results reveal that selection of material with the highest piezoelectric constants (as well as highest stiffness), i.e. PMN-28% PT, results in the greatest harvester generated RMS voltage values. FE model with PVDF as piezoelectric material displays the smallest generated RMS voltages, meanwhile harvester performance with PZT-5H as active material is somewhat mediocre. This may be seen as well in Fig. 4.7a, b, which depict generated RMS voltages of harvesters with piezoelectric material layers made of different materials versus relative support position when the harvester is excited by harmonic (Fig. 4.7a) and random (Fig. 4.7b) excitation signals, respectively.

Table 4.6 Material properties of PVDF, PZT-5H, and PMN-28% PT [6]

	PVDF	PZT-5H	PMN-28% PT
Elastic stiffness constants: c^E (10^{10} N/m^2)			
c_{11}^E	0.38	12.72	19.36
c_{12}^E	0.19	8.02	8.48
c_{13}^E	0.10	8.46	2.22
c_{22}^E	0.32	12.72	11.00
c_{23}^E	0.90	8.46	9.55
c_{33}^E	0.12	11.74	13.88
c_{44}^E	0.07	2.29	6.70
c_{55}^E	0.09	2.29	0.73
c_{66}^E	0.09	2.34	4.87
Elastic compliance constants: s^E (10^{-12} m^2/N)			
s_{11}^E	365	16.50	12.38
s_{12}^E	−192	−4.78	−19.44
s_{13}^E	−209	−8.45	11.38
s_{22}^E	424	16.50	53.09
s_{23}^E	−192	−8.45	−33.41
s_{33}^E	472	20.70	28.36
s_{44}^E	–	43.50	14.93
s_{55}^E	–	43.50	136.97
s_{66}^E	–	42,60	20,54
Piezoelectric constants: d (10^{-12} C/N)			
d_{15}	−23	741	2070
d_{24}	−27	741	150
d_{31}	21	−274	420
d_{32}	2.3	−274	−1140
d_{33}	−26	593	850
Dielectric constants: ε (ε_0)			
ε_{11}	12.50	1704.40	696
ε_{22}	11.98	1704.40	1090
ε_{33}	11.98	1433.60	716
Density, ρ (kg/m^3)			
ρ	1780	7500	8095

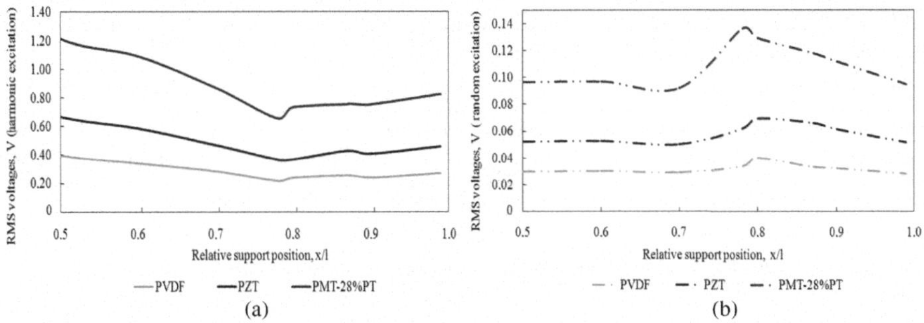

Fig. 4.7 Generated RMS voltage as a function of relative support position **a** harmonic and **b** random excitation signal for harvesters with piezoelectric layers of different material

Summarizing the simulation results, it is important to note that despite the superior performance of harvester with piezoactive PMN-28% PT material layers, it was decided to carry on most of the further investigations with piezoactive layers of PZT-5H, since PMN-28% PT is extremely expensive, brittle, and may not demonstrate longevity in actual PEH application environments, especially when the harvester is operating in vibro-impacting mode. It may also be concluded that despite the type of piezoelectric material, all generated RMS voltage curves display the same trend, which was already noted: if support is placed in the vicinity of nodal points ($0.78l$ and $0.87l$) of harvester eigenmodes, one may note an increase in generated RMS voltages, which is especially prominent in case of the random excitation signal.

4.1.1.3 Evaluation of Effects of Electric Circuits

Harvesters are usually connected to an electric circuit that might consist of capacitors to store energy, rectifiers to convert from AC to DC, diodes, load resistors, etc. All these components should transform harvested energy into a usable form. Nevertheless, most current FE models of harvesters assume that the vibration amplitude of the device is independent of the connected circuit. The voltage and current through the load resistor, and hence the power generated by the harvester, and its mechanical displacement are related. It is therefore necessary to determine the effect of the external electrical circuit and explain the processes/phenomena involved.

In order to assess the effect of the connected load resistor on the mechanical and electrical characteristics of the harvester, a schematic very similar to the one shown in Fig. 4.1 was used. The developed FE model of harvesting possesses geometrical characteristics listed in Table 4.7. The FE model of the cantilever beam consists of PZT-5H piezoelectric layer with electrodes deposited on a stainless steel substrate. Since continuous electrodes covering piezoceramic layers are assumed to be very thin if compared to the overall thicknesses of the harvester, their contribution to the thickness dimension is assumed negligible. These electrodes are thought to be perfectly conductive so that a single electric potential difference can be defined across

Table 4.7 Characteristics of FE model used to evaluate effects of an external electric circuit on the dynamic and electric response of the harvester

Parameter	Value
Stainless steel substrate dimensions ($L_S \times W_S \times T_S$) (μm)	6000 × 10 × 1
Piezoelectric dimensions ($L_P \times W_P \times T_P$) (μm)	6000 × 10 × 0.5
External circuit electric load resistance, representing short- circuit conditions (Ω)	1
External circuit electric load resistance, representing open- circuit conditions (MΩ)	100

them. Therefore, the instantaneous electric field induced in the piezoceramic layer is assumed to be uniform throughout the length of the beam. For this research, the harvester is assumed to undergo bending vibrations due to harmonic base excitation only and no rigid supports are introduced in the system.

External electric load is introduced to an electromechanically coupled system by employing Spice circuit editor. Spice Circuit Import feature is commonly used to add circuit elements as variables to FE models created with Comsol multiphysics software, as this allows variables to be connected to a physical device model in order to perform co-simulations of circuits and multiphysics. For example, to introduce an external electric circuit with a resistor of 50 MΩ resistance to the operating harvester, the following command prompt should be used in Spice:

R1 0 1 50Meg X1 0 1 Piezo.SUBCKT Piezo sens1 sens2 COMSOL: *ENDS

Piezoelectric harvester is subjected to harmonic base excitation, thus continuous electrical outputs can be extracted from the electromechanical system (only the fundamental eigenfrequency and no higher eigenmodes are considered for these simulations). Simulation results presented in Figs. 4.8, 4.9, and4.10 reveal the expected: increasing load resistance of the connected circuit not only influences electrical characteristics of the harvester, but also the dynamic response of the system changes.

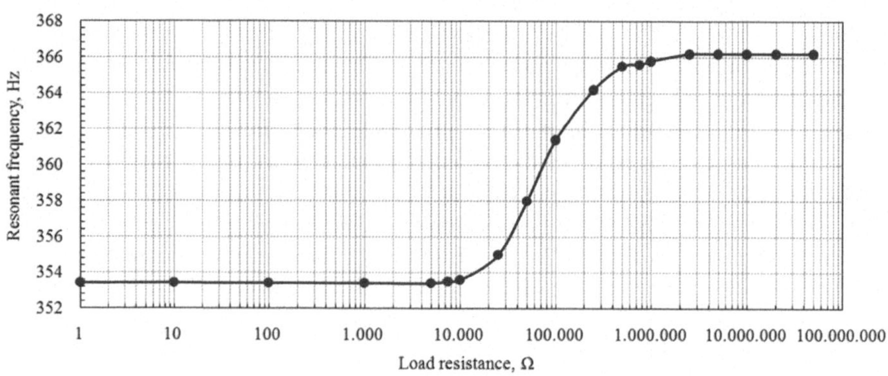

Fig. 4.8 Resonant frequency of the device as a function of connected load resistance

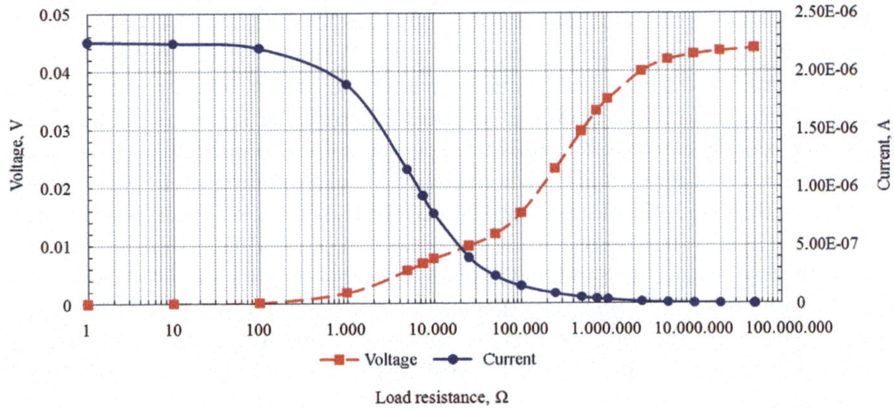

Fig. 4.9 Harvester generated voltage and current values as a function of connected load resistance

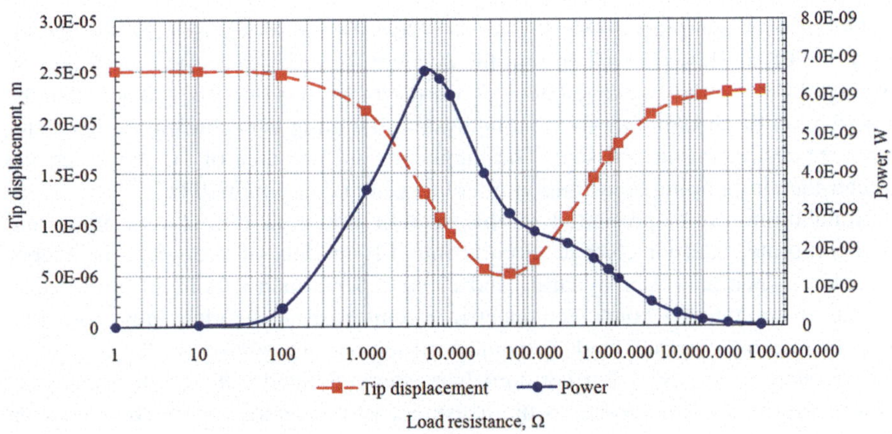

Fig. 4.10 Harvester generated power and tip displacement as a function of load resistance

Figure 4.8 depicts the shift of 15 Hz in resonant frequency as the resistance of connected electric circuit load increases from nearly short to nearly open-circuit conditions. One may note that the structure becomes stiffer as a higher resistance load is introduced to the external circuit. It is assumed that this frequency difference is resulting from backward coupling. The generated voltage values increase incrementing the resistance of the connected load, meanwhile generated current values decrease, as load resistance increases from short to open-circuit conditions (Fig. 4.9). Figure 4.10 combines generated power and tip displacement plots as functions of connected load resistance at constant acceleration levels. Generated power plot is based on power calculations per equations below

$$P = I^2 R_{load} \text{ or } P = V^2 / R_{load} \tag{4.13}$$

Figure 4.10 suggests that harvested energy can be visualized as a decrease of harvester displacement due to electrical damping. When harvester operates at short or open-circuit conditions, energy dissipation in the load is low. At the optimum load resistance, maximum power is transferred from the harvester to the load, meanwhile, displacement at the maximum power point is reduced (load resistance reduces harvester motion amplitude at the short-circuit conditions until the so-called optimum load enlarges the motion amplitude toward open-circuit conditions). Energy harvesting may also be considered as electrical damping of harvester tip displacement. One may observe that both in the open and short-circuit condition cases, electrical damping is minimal and mass displacement is maximal. A vibrating harvester operating at a short-circuit conditions is only mechanically damped as no electric power is consumed. As the connected load resistance approaches optimum load, mechanical energy is partly transferred to electrical energy. This harvested electrical energy is considered electrical damping which adds to the still present mechanical damping. Eventually, the total damping (mechanical and electrical) leads to attenuation in harvester tip displacement.

Mechanical damping caused by air is a loss factor which reduces the output power and thus should be taken into the account. In order to prevent air damping, one could use vacuum packaged devices. Furthermore, it is important to note that the contribution of air damping to the total system damping is dependent on harvester substrate configuration and size. The bigger the surface area of the harvester, the greater the air damping contribution on total system damping will be.

Simulation results presented in this section reveal that it is very important to incorporate external electric circuits into FE models of piezoelectric energy harvesters since harvester dynamic and electric performance is highly influenced by the magnitude of connected electric circuit resistance. One may note that with increasing external load, resistance structures are getting stiffer (i.e. resonant frequency increases). Furthermore, harvester generated voltage values increase, while current values decrease, as the resistance of connected circuit changes from nearly short to nearly open-circuit conditions.

4.1.1.4 Evaluation of Piezoelectric Layer Segmentation on Its Electric Output

The main principle of harvester operation relies on the fact that dynamic strain field induced throughout piezoelectric material layers due to excitation results in alternating voltage output across electrodes covering piezoelectric material. Most of the currently analyzed harvesters do not perform efficiently in real environments, since it is assumed that they are excited only by harmonic signal and at their fundamental frequency. However, if one analyzes real environmental vibration energy source, it usually would not consist of one single harmonic, and higher eigenmodes of harvester may be excited. These higher eigenmodes of cantilever beam have strain nodes—i.e. positions on the device where bending strain distribution curve changes sign (examples of normalized strain distribution curves are presented in Fig. 4.11a).

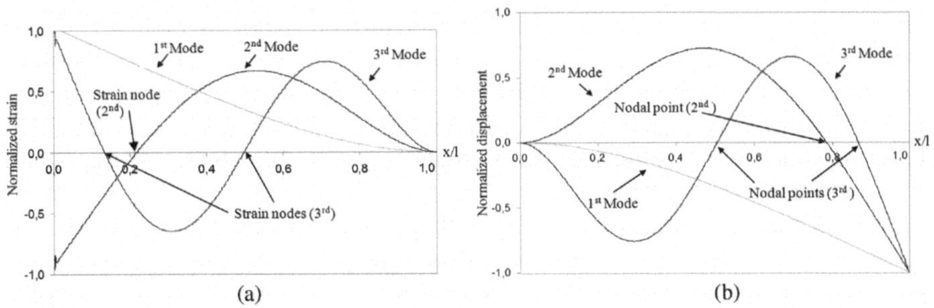

Fig. 4.11 Examples for normalized strain nodes **a** and displacements **b** with highlighted 2nd and 3rd mode nodal points for the first, second, and third eigenmodes

Mathematically, curvature eigenfunction, which is a measure of bending strain, is the second derivative of displacement eigenfunction (examples of normalized displacement curves are presented in Fig. 4.11b). Authors of [7] suggested that if these strain nodes are covered by continuous electrodes, cancelation of electric outputs occurs, resulting in overall harvested energy reduction.

For the purpose of clarifying the effects of piezoelectric layer segmentation on electrical parameters, the FE model was slightly modified: one continuous piezoelectric (PZT) layer on stainless steel substrate was segmented either in two segments (location of segmentation coincided with the strain node of the second eigenmode) or in three segments (location of segmentation coincided with the strain nodes of the third eigenmode). In both cases, perfectly conductive electrodes of negligible thickness covered the entire area of the top and the bottom surfaces of piezoelectric material layers. They were directly connected to the resistive load, introduced to the electromechanically coupled system via Comsol Spice circuit editor. This circuit editor was used, as it would allow including additional and more complex circuit elements (e.g. diode bridges) in future research, aiming to develop a complete energy harvesting system. Locations of strain nodes were determined from bending strain distribution functions, as depicted in Fig. 4.11a. Table 4.8 lists the obtained dimensionless positions of nodal points and strain nodes of modeled harvester for the first three eigenmodes.

FE models were created for two harvester configurations—for the first one, piezoelectric layer was segmented at the location of the second eigenmode strain node,

Table 4.8 Dimensionless positions of nodal points and strain nodes of the harvester for the first three eigenmodes

Mode	Dimensionless positions (x/l) on x-axis of			
	Nodal point		Strain node	
1	–	–	–	–
2	0.783	–	0.217	–
3	0.505	0.868	0.133	0.498

presented in Fig. 4.12a, with the main characteristics presented in Table 4.9, meanwhile for the second one—segmentation occurred at the strain nodes of the third eigenmode (as depicted in Fig. 4.12b) with the main characteristics presented in Table 4.10. It was assumed that these configurations would allow one to avoid undesired charge cancelation effects in the piezoelectric material, which, in turn, would increase the generated voltage

RMS voltages generated by harvesters with segmented piezoelectric layers (calculated by adding RMS voltages generated by each piezoelectric segment) were plotted for relative support positions in the range of $0.5l$ to $1.0l$ in Fig. 4.13a for harmonic and in Fig. 4.13b for random excitation signals. As one may note from these graphs, harvesters with piezoelectric layers segmented at the location of second (Segmented, II mode) and third (Segmented, III mode) strain nodes generate much greater RMS

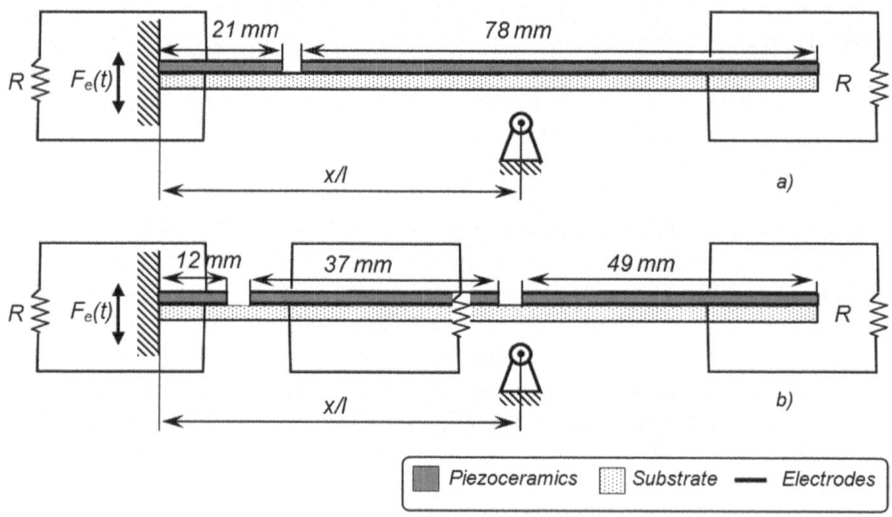

Fig. 4.12 Principle schemes of segmented piezoelectric energy harvesters: **a** with piezoelectric layers segmented at the strain node of the second and **b** with piezoelectric layers segmented at the strain nodes of the third eigenmodes

Table 4.9 Characteristics of harvester model with piezoelectric layer segmented at the strain node of the second eigenmode (0.217 *L*)

Parameter	Value
Stainless steel substrate ($L_S \times W_S \times T_S$) (mm)	100 × 10 × 1
1st piezoelectric segment ($L_p \times W_p \times T_p$) (mm)	21 × 10 × 0.2
2nd piezoelectric segment ($L_p \times W_p \times T_p$) (mm)	78 × 10 × 0.2
External circuit electric load resistance (MΩ)	100

Table 4.10 Characteristics of harvester model with piezoelectric layers segmented at the strain nodes of the third eigenmode (0.133*L* and 0.498*L*)

Parameter	Value
Stainless steel substrate ($L_S \times W_S \times T_S$) (mm)	100 × 10 × 1
1st piezoelectric segment ($L_P \times W_P \times T_P$) (mm)	12 × 10 × 0.2
2nd piezoelectric segment ($L_P \times W_P \times T_P$) (mm)	37 × 10 × 0.2
3rd piezoelectric segment ($L_P \times W_P \times T_P$) (mm)	49 × 10 × 0.2
External circuit electric load resistance (MΩ)	100

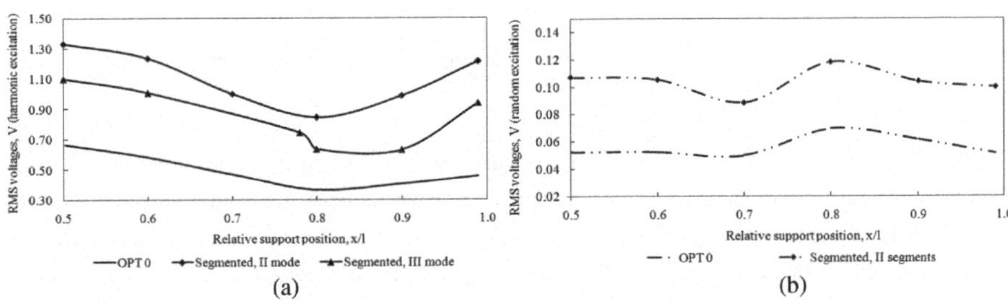

Fig. 4.13 RMS voltages generated by different configuration harvesters as functions of relative support position: **a** harmonic and **b** random excitation signal

voltages if compared to their counterpart with continuous piezoelectric layer (when all are having the same geometric dimensions). All RMS voltage curves display the same trend, which was already revealed in the sections above: the presence of the support limits harvester performance when it is operating under harmonic excitation, yet improves its performance if it is excited by random excitation signal (especially, if support is placed in the vicinity of nodal points). Comparing RMS voltage curves generated by the two-segment (Segmented, II mode) and three-segment (Segmented, III mode) harvesters excited by harmonic signal, one may conclude that the two-segment harvester demonstrates better performance—i.e. the values of generated RMS voltages are higher when the relative support position ranges from 0.5*l* to 1.0*l*.

Although advanced electric circuits are not a part of this research, some considerations below are provided on relatively simple electric circuits that could be connected to harvesters operating in different environments (and thus excited by different frequencies). If the operating environment is dominated by the first natural frequency of the harvester, the energy can be harvested by non-segmented continuous piezoelectric layers as the strain distribution over the length of the device does not change sign. For the latter case, a fairly simple AC-DC conversion electric circuit (scheme of which is depicted in Fig. 4.14a) can be employed, where electrodes of piezoelectric layers are connected to a diode bridge to eliminate electrical output sign alteration

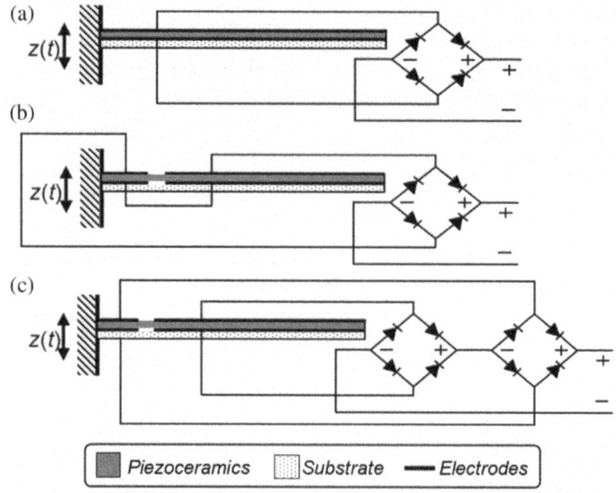

Fig. 4.14 Electrode and electric circuit configurations suitable for energy harvesting once **a** first, **b** second, and **c** both eigenmodes are excited

[8]. For the case when second eigenfrequency is predominant, two separate piezoelectric segments could be used to cover substrate at regions 0-strain node, strain node L and, in such way, cancelation in charge would be avoided, as voltage outputs from these segment pairs would be out of phase with each other. A relatively simple electric circuit could be used to collect harvested energy, where bottom electrodes of piezoelectric layers would be connected to each other and top electrodes of piezoelectric layers would be connected to the diode bridge (as per Fig. 4.14b). Yet the latter circuit configuration would be effective only for energy harvesting from the second eigen mode, as it would result in charge cancelation for the first one. To harvest energy from both the eigenmodes, i.e. first and second eigenmodes, a more complex electric circuit, comprised of two separate diode bridges, connected in series (as per Fig. 4.14c) is suggested to be used.

The other way to avoid charge cancelation using continuous piezoelectric layers is to apply "patterned polling" [8] to the piezoelectric material—i.e. change the direction of polarization of pre-planned piezoelectric regions. This process involves etching electrodes of piezoelectric material at certain regions, corresponding to the strain nodes of vibrating harvester, applying strong electric fields at these desired portions (i.e. polling them), and reconstructing the electrodes back. The sentence is good bearing in mind, that the piezoelectric material segments under the electrode regions are poled oppositely so the charge can be collected with a contiguous electrode. However, it is important to note that practically patterned poling is only effective if the harvester is excited at certain vibration mode for which the patterning is performed (e.g. if the harvester has patterned polling applied bearing in mind the strain nodes of second vibration mode, a strong charge cancelation would occur once it operates at the first resonant frequency), which means that the approach is not flexible.

Thus, by summarizing, one may come to the conclusion that the use of segmented piezoelectric layers is more preferable than patterned polling, as it is practically easier to implement by combining the leads of electrodes accordingly and joining them to different electric circuits. Yet, whichever case is chosen, the main concern remains not to cover strain nodes of vibrating harvester with continuous piezoelectric layers in order to avoid charge cancelation problems. Simulation results reveal that harvesters with segmented piezoelectric layers are able to generate almost twice greater RMS voltages if compared to their counterparts with continuous piezoelectric layer.

4.1.1.5 Experimental Set-Up for Vibro-Impact Investigation

The specially designed and custom-built experimental set-up [9] for vibro-impact investigation of a steel cantilever is featured in Fig. 4.15. The cantilever specimen is formed of a rectangular stainless steel bar ($E = 209$ GPa, $\nu = 0.3$, $\rho = 7917$ kg/m^3) $HxB = (0.5 \times 8.8) \times 10^{-3}$ m and length $l = 80.0 \times 10^{-3}$ m. [10].

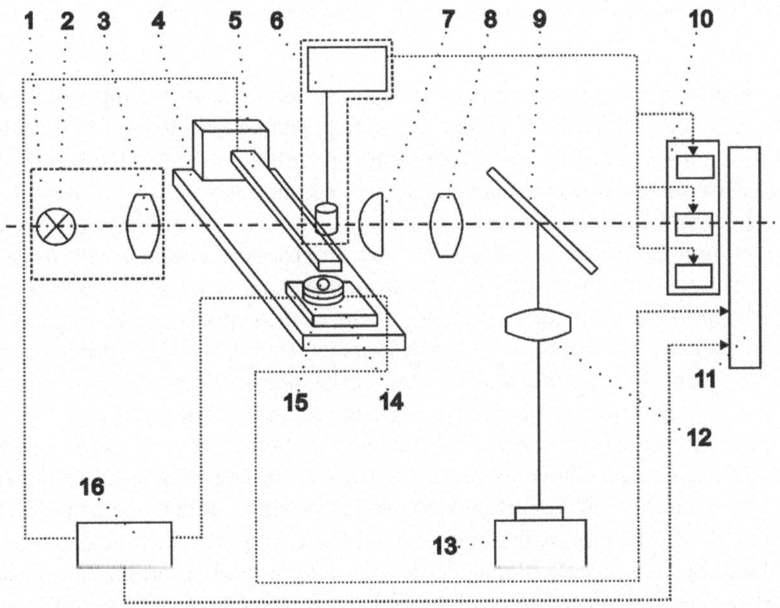

Fig. 4.15 Vibro-impact experimental set-up: The light reflector 1 is composed of a light source 2 and condensing lens 3; the vibro-impact system consists of a cantilever 5 and support 15. The guideways of frame 4 enable one to change the position of the cantilever in the plane of transverse vibrations in two directions perpendicular to each other. The cantilever is shifted by using a mechanical microscrew that enables accurate estimation of the distance between the contact surfaces of the cantilever and the support. Light source 2 exposes the cantilever 5, which is actuated from a statically deflected position by the triggering block 6. A shadow image of the cantilever 5 is enlarged in a transverse direction by the cylindrical lens 7 and 8, then it is projected on the input surface of the photoelectric transducer of displacement 10 by means of the light-refracting screen 9 and the objective 12 on the photo recorder 13

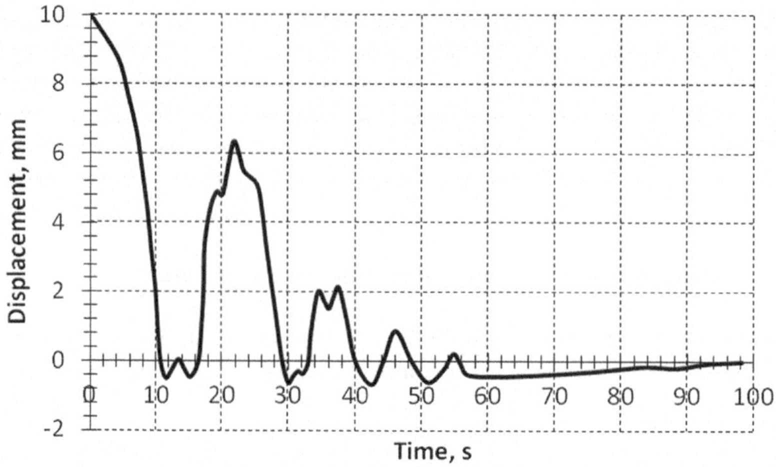

Fig. 4.16 Experimental trajectory of the vibro-impact motion of the cantilever

The working principle of the device is as follows. The triggering attachment *6* actuates the cantilever *5* and runs a block of displacement photoelectric transducers *10*. As the cantilever moves, its several times magnified shadow changes the input light intensity on the displacement transducer *10* and later generates the output of alternating signal, corresponding to the laws of motion of the characteristic points of the vibrating cantilever. At the same time, the triggering attachment *6* runs the photo recorder *13* and the modes of vibration of the cantilever are registered on the film. At the moment when the cantilever *5* bounces the support *15*, the impact sensor *14* measures the contact pressure and transmits the signals to the recorder *11*. At the same time, gauge 16 measures the dynamic resistance between the cantilever *5* and the support *15*. This device has been recognized as an invention [10].

Figure 4.16 features the motion trajectory of the cantilever *5* in case of free impact vibrations when upon the release from the deformed position (initial deflection 10 *mm*), the cantilever makes an impact against the rigid immovable support *15*.

The position of the immovable support could be changed by shifting it in the direction of the cantilever axis. By placing the immovable support in certain fixed positions related to the nodal point positions of the higher vibration modes of the cantilever, the amplitudes of these vibration modes can be markedly increased. Figure 4.17a presents the cantilever vibration momentum T curves after a collision with the support in the third mode nodal point $x/l = 0.87$, where x is the distance of the support from the fixed end of the cantilever and l is the length of the cantilever.

This means that the location of the support in the nodal point of the third vibration mode is beneficial for such a vibro-impact system because these dominating flexural cantilever vibrations are similar to the cantilever vibration at the third natural frequency. The domination of the third mode does not mean that the first two modes are canceled, but rather they are suppressed. A considerable decrease of the third mode amplitudes results in the intensification of the internal energy dissipation in

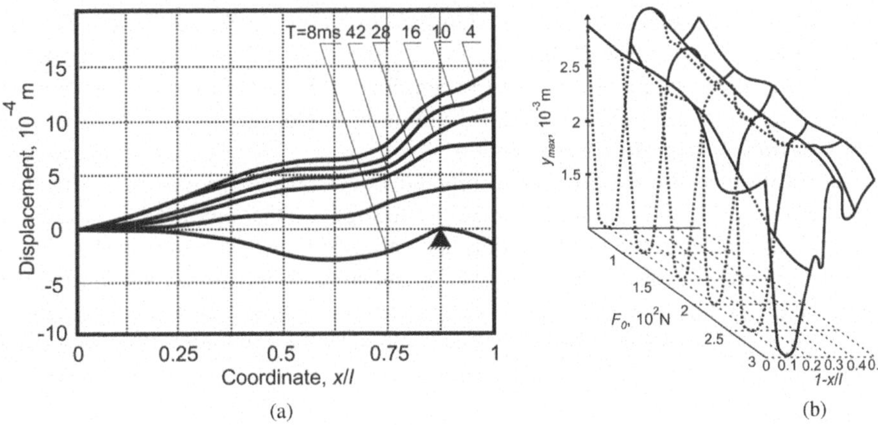

Fig. 4.17 Dependence of the cantilever rebound amplitudes y_{max} on the support position x/l at fixed time moments T after rebound **a** and the pre-stress of the cantilever F_0 on the support **b**

the structure material. This could be confirmed by the drastically decreased cantilever rebound amplitudes when the support was located in the nodal point of the third mode at $x/l = 0.87$ (Fig. 4.17b). The presented explanation is also confirmed by the dependence of vibro-impact motion time on the position of the support x/l and the pre-stress F_0 of the cantilever on the support (Fig. 4.18a).

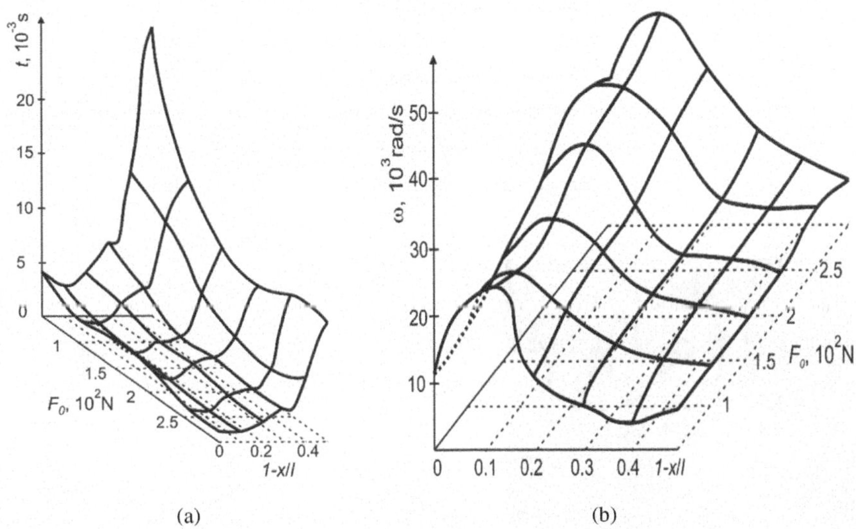

Fig. 4.18 Dependence of vibro-impact motion time **a** and frequency **b** on the position of the support 1-x/l (distance of the support from the free end of the cantilever) and the pre-stress F_0 of the cantilever

The presented explanation is also confirmed by the dependence of vibro-impact motion time on the position of the support *x/l* and the pre-stress F_0 of the cantilever on the support. As the amplitudes of the cantilever rebounds fall when the support is located at the nodal point of the third vibration mode, the time of vibro-impact motion or transitory period decreases too.

The position of the support has a marked effect on the frequency of the free impact vibrations of the cantilever. Figure 4.18b suggests that vibration at maximum frequency is characteristic if the support is positioned in the point $x/l = 0.87$, i.e. in the nodal point of the third eigenmode. The frequencies of free impact vibrations also change at different pre-stress F_0. In the case when there is no pre-stress, the frequency of free impact vibration of the cantilever is equal to the first free transverse vibration frequency.

4.1.1.6 Cantilever Vibration Analysis by Prism Holography System

A Precise real-time instrument for surface measurement (Prism) holography system (Hytec, Los Alamos, NM, USA), whose main characteristics are introduced in Table 4.11 and set-up picture presented in Fig. 4.19a with operation scheme in Fig. 4.19b, was used for the determination and oscillation analysis of the cantilever resonant frequency. The Prism allows completing and processing experimental measurements

Table 4.11 Main characteristics of the Prism system

Measurement sensitivity	<20 nm
Dynamic measurement boundary	100 μm
Measurement boundary	>100 μm
Greatest measurement area	1 m diameter
Distance to the object	>¼ m
Data registration frequency	30 Hz

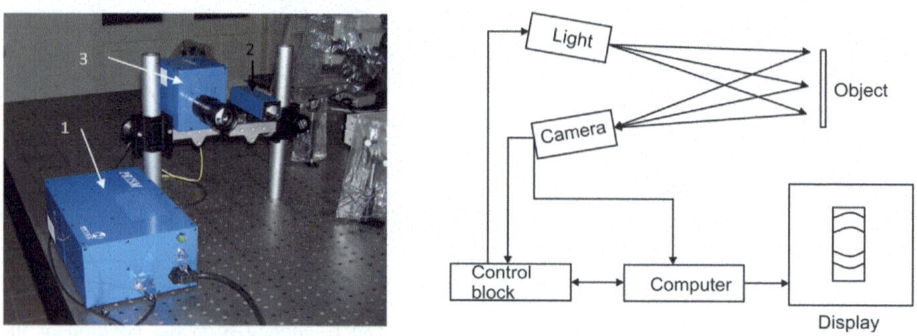

(a) (b)

Fig. 4.19 Holography system prism: **a** set-up picture, **b** operation scheme in less than 5 min as well as capable of determining displacements of less than 20 nm [11].

The main part of the system is a control block, which splits green (532 nm) semiconductor laser beam into two beams—object and reference. Lenses are used to control the object beam and light falling on the object. Light, reflected back from the object, is registered by a camera, which combines object and reference beams, registering the interference pattern (ratio of the object and reference beams may be altered in order to achieve the best definition of interference bands). The interference pattern is transferred to the computer, where it is processed with the Prism-DAQ software, allowing one to monitor real-time dynamic processes occurring at the research object as well as deformations caused by the internal and external forces. For this experiment, an acoustic field was employed to excite the cantilever by ranging the harmonic excitation signal from 10 to 2000 Hz. The shapes and amplitudes of the cantilever surface deformation were determined by analyzing the interference patterns.

The set of experiments was performed with the optimally configured cantilever, whose holograms are presented in Fig. 4.20. For this configuration, the unsupported cantilever was excited at its first resonant frequency at $\omega = 58$ Hz, the second resonance at $\omega = 402$ Hz, and the third one at $\omega = 1368$ Hz. In case of the support located at $x/l \approx 0.78$, the second vibration mode was excited at $\omega = 372$ Hz. Further increasing the vibration frequency, other resonances were observed at $\omega = 390$, 1021, 1341, and 1992 Hz (cantilever flexural and torsional resonance modes). It should be noted that $\omega = 1341$ Hz excited the cantilever torsion to one direction, while at $\omega = 1991$ Hz the cantilever was already turning to the other direction. It may be assumed

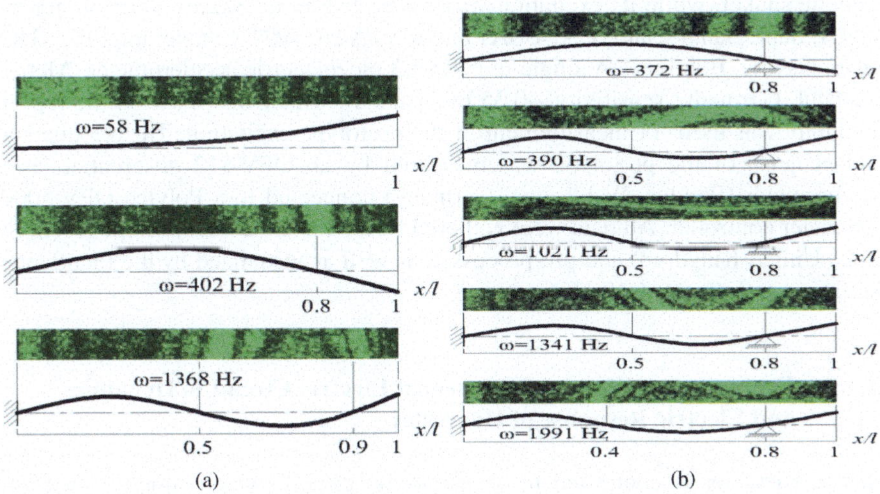

Fig. 4.20 Optimally configured cantilever: natural vibration modes and frequencies: **a** unsupported; **b** rigid support at ($x/l \approx 0.87$)

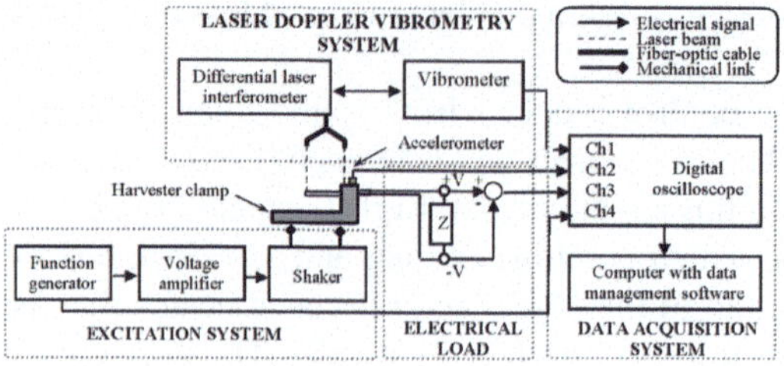

(a)

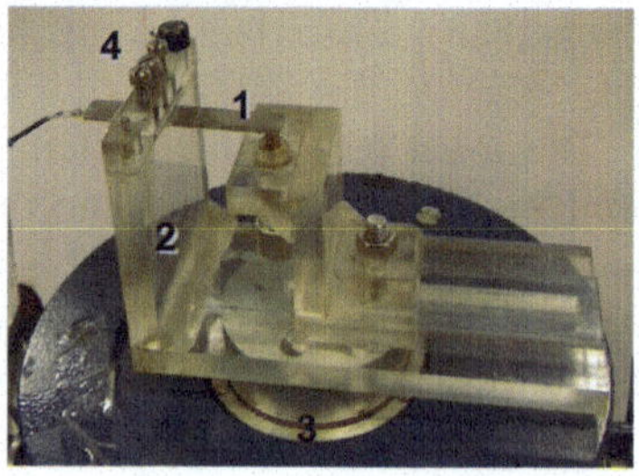

(b)

Fig. 4.21 Scheme of the laser Doppler vibrometry system **a** and experimental set-up for the analysis of the vibration energy harvester prototype response to harmonic excitation **b** 1—vibration energy harvester prototype, 2—harvester clamp, 3—electromagnetic shaker, 4—accelerometer

that the third cantilever resonance occurs at 1341 Hz frequency, while $\omega = 1991$ Hz excites higher cantilever vibration mode.

4.1.1.7 Laser Doppler Vibrometry System

Tip velocity (or displacement) of the harvester prototype in the transverse direction was experimentally analyzed using Doppler vibrometry (Fig. 4.21a). The measurement system consisted of the harvester prototype fixed in a custom-built clamp (Fig. 4.21b) and three main subsystems—excitation system, measurement system, and data gathering system [4]. The harvester prototype was excited by an electromagnetic shaker, while its excitation was controlled by a 33220A function generator (Agilent, Santa Clara, CA, USA) and a VPA2100MN voltage amplifier (HQ Power, Gavere, Belgium). A single-axis KS-93 piezoelectric accelerometer (Metra, Radebeul, Germany; sensitivity—0.35 mV/(m/s^2)), which was fixed on the top of the clamp, was used for measurement of the prototype excitation. Tip velocity (or displacement) of the prototype was measured by an OFV-512 differential laser interferometer (Polytec, Waldbronn, Germany) connected to a Polytec OFV-5000 vibrometer controller. All data were gathered by a 3425 USB oscilloscope (Pico, St Neots, United Kingdom) and pre-processed as well as visualized by the PicoScope Oscilloscope Software 6.10.11.

4.1.1.8 Evaluation of Effects of Connected Electric Circuit to Dynamics and Electric Response of Harvester

When a harvester is connected to an electronic circuit (which should transform harvested energy into electrical energy), it may alter dynamic and electric response [12, 13]. Thus, measurements of harvester frequency, tip displacements, as well as voltage, current, and power were performed by means of laser Doppler vibrometry system. It was important to clarify underlying physical mechanisms that govern coupled dynamic and electrical behavior of harvester prototype when it delivers power to an electrical load (i.e. operates under real-life conditions). Influence of load resistance was considered for mechanical (resonant frequency, quality factor, displacement), and electrical (voltage, current, power) harvester characteristics. It was demonstrated that resistance of connected electrical load significantly influences power generated by operating harvester as well as its other characteristics, thus it must be taken into account when predicting the efficiency of harvesters. The trends in experimentally obtained data were analyzed and the reasons behind them were discussed. The conducted experimental study allowed characterizing coupled mechanical and electrical performance of harvester under variable electrical loading conditions. Moreover, performed measurements allowed extraction of damping and electromechanical coupling coefficients that might be used for numerical modeling of energy harvesters.

A set of resistors with different resistance magnitudes was used in order to create varying electrical loading conditions during frequency response measurements. Effective load resistance acting on harvester was calculated bearing in mind that the input channel of oscilloscope has an impedance of 1 MΩ, which acts in parallel with the load resistor placed across piezoelectric layers. The effective load R_{eff} that is actually exerted on harvester is influenced by this input impedance of oscilloscope and is defined as

$$R_{eff} = \frac{1}{1/R_a + 1/R_{osc}} \tag{4.14}$$

where $R_a$actual resistance of a resistor, R_{osc} input impedance of the oscilloscope.

The load resistances used in this experimental study range from 47 Ω (close to short-circuit conditions) to 1 MΩ (close to open-circuit conditions). Frequency response measurements of tip displacement and voltage output of harvester were performed with the function generator providing swept harmonic excitation within a frequency range of 140–230 Hz and sweep time of 500 s. It was aimed to maintain the same level of acceleration in the course of frequency response measurements, however, acceleration magnitude fluctuated in the vicinity of 1g when sweeping.

The first resonant frequency of the harvester was analyzed in this experimental study. Figure 4.22 illustrates experimentally measured frequency responses of tip displacement and voltage output for varying electrical loading conditions ranging from nearly short circuit (s.c.) to nearly open circuit (o.c.). It was observed that an increase in load resistance leads to a higher voltage output, which correlates with a reduction in tip displacement amplitude at short-circuit resonant frequency (Fig. 4.22a). Yet at the o.c. resonant frequency tip displacement amplitude increases again and one may note that at o.c. resonant frequency both tip displacement amplitude and voltage output (Fig. 4.22b) are amplified as external load resistance increases. Resonant frequency curves in Fig. 4.22a corresponding to low resistance values clearly reveal that the harvester exhibits nonlinear frequency responses with curves being shifted to the left-hand side of the frequency axis.

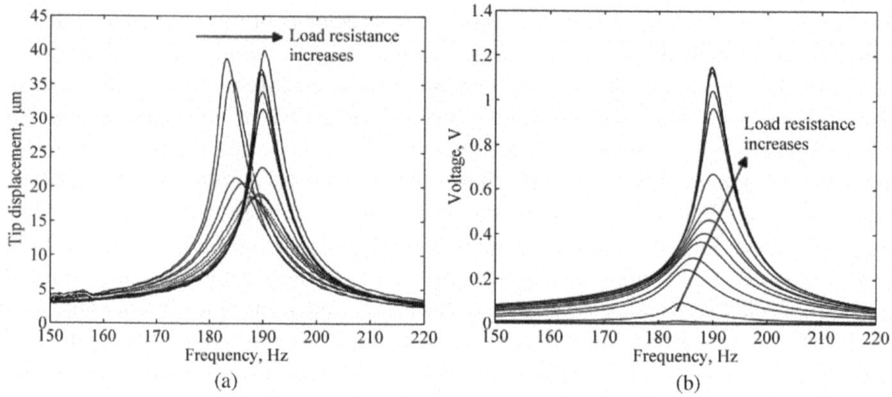

Fig. 4.22 Measured frequency responses of tip displacement (**a**) and voltage output (**b**) for different resistive loads spanning from nearly short circuit to open circuit

The reasons for the observed softening behavior are diverse and are associated with the complex interaction of various effects including but not limited to nonlinear damping (due to air drag and material losses), viscoelectroelasticity, nonlinear electromechanical coupling, dielectric effects, etc. Structural displacements were relatively small in these experiments (Fig. 4.22a), therefore, it was hardly possible that geometric nonlinearities were induced in this case. These experimental results also demonstrated that the observed nonlinear softening response diminishes with larger resistive loads (corresponding to larger electric fields generated inside piezoceramic layers). This, in turn, suggested that electromechanical coupling, which became more prominent with increased electrical loading, counteracted those effects that cause nonlinear softening behavior at lower load resistances (i.e. at weaker electric fields).

Measured tip displacement and voltage output frequency responses in Fig. 4.22 were subsequently used to derive graphs demonstrating variation of harvester resonant frequency and quality factor (Fig. 4.23), power output and tip displacement (Fig. 4.24) as well as voltage and current (Fig. 4.25) as functions of external load resistance.

One of the main characteristics of an operating harvester is the magnitude of generated power, which is directly related to the dynamic response of the piezoelectric transducer. Therefore, it is important to examine the variation of key mechanical characteristics such as resonant frequency, tip displacement, and quality factor during the process of power generation. The resistive load of 47 Ω is very close to the short-circuit (s.c.) conditions for this experimental set-up, therefore, the resonant frequency of 183.2 Hz derived from measured data may be considered as the fundamental short-circuit resonant frequency f^{sc} of the piezoelectric energy harvester. Fundamental open-circuit resonant frequency f^{oc} is measured with a resistive load of 1 MΩ and is equal to 189.9 Hz. The respective 3.7% shift in resonant frequency is obvious in Fig. 4.23. It should be mentioned that the magnitude of the observed frequency

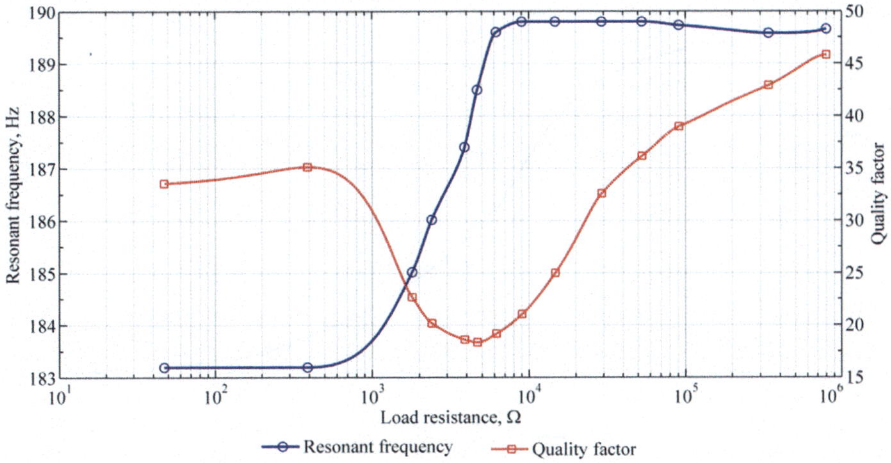

Fig. 4.23 Resonant frequency and quality factor of the harvester as a function of load resistance

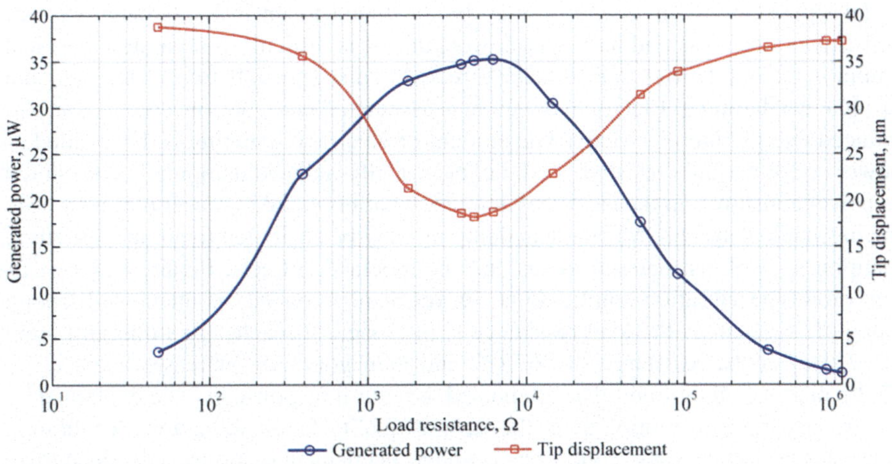

Fig. 4.24 Variation of power output and tip displacement of the harvester as a function of load resistance

shift is directly proportional to the square of the electromechanical coupling coefficient. This shift in resonant frequency is attributed to varying electrical boundary conditions: increase of load resistance from s.c. to open-circuit (o.c.) condition leads to a change in harvester stiffness since elastic modulus of piezoelectric material increases. Figure 4.23 reveals that quality factor, with the initial value of 33.6 at s.c. conditions reaches its minimum value of 18.4 at the electrical load of 4670 Ω and then gradually increases up to 45.8 at the o.c. conditions. The quality factor is explained as a measure of dissipated mechanical energy of a vibrating harvester.

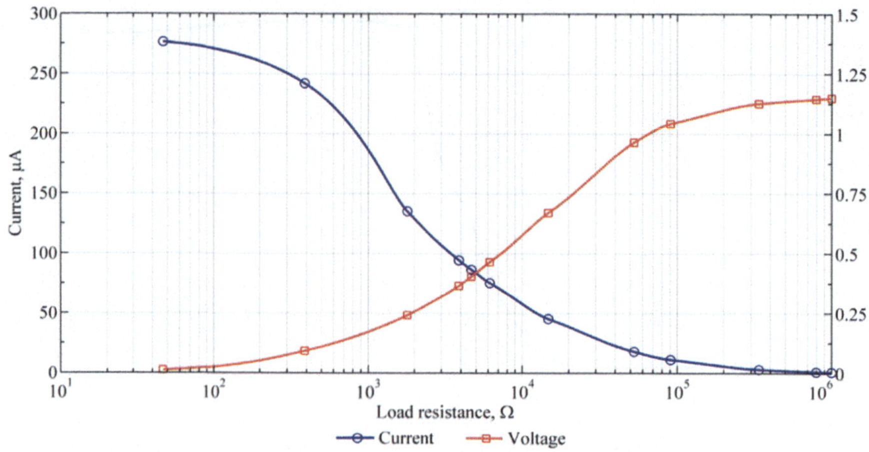

Fig. 4.25 Voltage and current output of the harvester as a function of load resistance

Dissipated mechanical energy may be attributed to intrinsic characteristics of constituent materials and harvester design, thus it is difficult to separate individual damping factors. However, efforts are directed to distinguish between electrical and mechanical damping. Electrical damping is associated with the conversion of strain-induced energy into electricity. The amount of electrical damping is determined as power consumed in a resistive load, i.e. electrically induced damping is considered as power consumed in the electrical domain, which is equal to power removed from the mechanical system. Mechanical damping may be attributed to air and structural damping as well as material losses and thermoelastic effects. Figure 4.24 reveals that prototype vibration amplitude at s.c. resonant frequency is attenuated from a value of 38.8 μm as the load resistance is increased. It reaches its minimum point (18.2 μm) at the resistance of 4670 Ω and then gradually increases again up to 37.2 μm at o.c. conditions. The attenuated structural response may be explained by electrically damped motion as in the case of quality factor reduction. A vibrating harvester prototype at s.c. conditions is under mechanical damping only since there is no electrical power consumed. As resistance is increased, mechanical energy is partially transferred to electrical energy. Harvested electrical energy is considered as electrical damping that sums up mechanical damping, which finally leads to suppression of prototype displacement. It may be concluded that the electrical effect, called backward coupling, manifests in the harvester, i.e. the feedback is sent from the electrical domain to the mechanical one due to power generation, caused by the converse piezoelectric effect. This phenomenon is explained by the theory of piezoelectrics, which are comprised of perovskite crystals with an intrinsic dipole moment. Once these materials are strained, the direction of polarization among neighboring dipoles becomes unified, producing an electric charge on the surface (direct piezoelectric effect). However, when feedback is sent from the electrical domain and electrical energy is applied to the poled piezoelectric material, it distorts the orientation of dipole domains, and overall polarization becomes more random, resulting in mechanical strain (converse piezoelectric effect). Thus, it may be stated that the form of piezoelectric coupling is substantially different from conventional damping mechanisms.

Electrical outputs of the harvester prototype are also analyzed in order to examine variations of generated electrical current, voltage, and power, which subsequently must be considered in the optimization process aiming for maximum power output when the harvester prototype is connected to complicated energy harvesting circuits. Figure 4.24 provides a graph of harvested power as a function of load resistances. It is observed that the maximum power output of 36 μW is delivered at an electrical load of ca. 6160 Ω, which may be considered as an optimal resistance for this harvester. The resistive load of 47 Ω yields a power output of 4 μW at s.c. conditions, whereas the resistive load of 1 MΩ yields 1 μW at the s.c. condition. Results presented in Fig. 4.24 reveal that electrical load resistance delivering the maximum power output (6160 Ω) does not coincide with the load resistance of minimum vibration amplitude (4670 Ω). This phenomenon is attributed to the nature of electromechanical coupling: vibration amplitude of the harvester will not necessarily acquire its lowest value for the magnitude of electrical load corresponding to the maximum power generation. Thus, it is important to note here that harvested power is not considered to be directly influenced by the displacement amplitude of the harvester, but it is also affected by the voltage and external load resistance.

Variations of electric current and voltage generated by harvester for various resistive loads are plotted in Fig. 4.25. Voltage amplitude increases monotonically with increasing load resistance from 0 V to 1.15 V, while current decreases monotonically from 275 μA to 0 μA. Electric current and voltage amplitude curves intersect close to load resistance of 4650 Ω. It should also be noted that the asymptotic character of voltage and current output variation is observed when both curves approach extreme conditions of load resistance.

Measured frequency responses may be used to extract various damping parameters that might be subsequently employed for the development of FE model of the harvester. Mechanical damping ratio ζ and Rayleigh damping parameter β are derived from the measured resonant frequency curves.

Firstly, quality factor Q is calculated as

$$Q = f_r / \Delta f \tag{4.15}$$

where $\Delta f = f_2 - f_1$ is bandwidth, which represents the distance between two points on the frequency axis where the amplitude is equal to $1/\sqrt{2}$ of the maximum amplitude value.

Quality factor Q is used to derive damping ratio ζ

$$\zeta = 1/2Q \tag{4.16}$$

Finally, Rayleigh damping parameter β is calculated as

$$\beta = \frac{1}{2\pi f_r Q} \quad (4.17)$$

Following [14], coupling coefficient k of the system may be determined from the resonant frequencies under open-circuit and short-circuit conditions

$$k^2 = \frac{(f^{oc})^2 - (f^{sc})^2}{(f^{oc})^2} \quad (4.18)$$

The coupling coefficient k for this harvester was determined to be 0.26 which is lower than indicated in the courses equal to 0.72. This reduction is explained by the incorporation of substrate material in harvester configuration as it influences electromechanical coupling of the complete structure. Moreover, backward coupling discussed in the sections before is also thought to be mainly influenced by the thickness and stiffness of harvester substrate material.

4.1.2 Enhanced Harvester Configuration

A piezoelectric bimorph that can transfer mechanical stress to electrical energy is a very simple way to harvest mechanical vibrations. However, this bimorph itself harvests an insignificant amount of energy. The resonant frequency or the frequency at which the piezoelectric bimorph is subjected to the maximum mechanical strain must be adjusted to the target frequencies and generate significant amounts of energy. The piezoelectric bimorph can be adjusted by physically altering the structure of the material by optimizing its shape.

4.1.2.1 Cantilever Beam Optimization

Sometimes there is a need to create a structure whose eigenfrequencies would fit in a certain interval ($\omega_{\min}; \omega_{\max}$). This can be due to many reasons, for example, a vibrating piezoelectric energy harvester generates higher power output values while it is operating at its resonant frequency, coinciding with the ambient source frequency [15]. Figure 4.26 represents a 2D computational scheme of the cantilever with support. The model is made from m linear beam elements located in a single layer. Each finite element has three degrees of freedom at each node. External air pressure forces are neglected in the model, therefore, $Q(t)$ is only a mechanical load that acts in the system. $\overline{K}_i$ and $\overline{C}_i$ are stiffness and viscous damping coefficients of the support, Δ_i—size of the gap between the i-th nodal point of the structure and the surface of the support located at the i-th nodal point.

The target function in such a case is as follows:

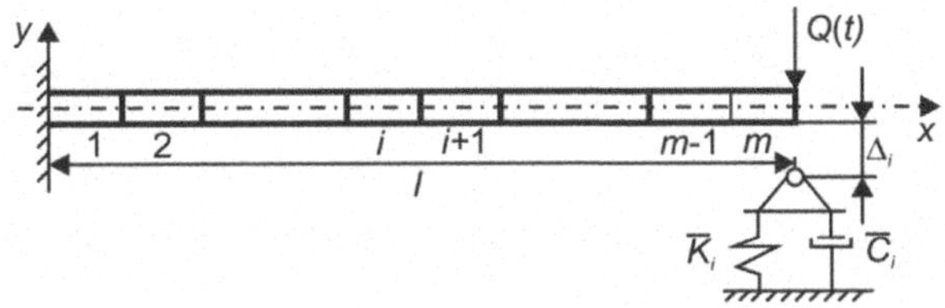

Fig. 4.26 Scheme of the developed 2D FE model of the impacting cantilever

$$\Phi(A) = \min \rho a(A_1 + A_2 + \cdots + A_m) \tag{4.19}$$

where A_i are cross-sections of structure components.

When the method of nonlinear programming, e.g. gradient projection, is used, the unequally-shaped constraints are employed

$$f_m(A) = 1 - A_m/A_i \leq 0 \tag{4.20}$$

$$f_{m+1}(\omega) = 1 - \omega/\omega^* \leq 0 \tag{4.21}$$

where ω^* is the prescribed frequency of structural vibrations.

Optimal substrate configurations obtained with gradient projection method are presented in Fig. 4.27a, which illustrates optimal structures obtained for the operation in prescribed (OPT I$^+$), second (OPT II$^+$), and third (OPT III$^+$) eigenfrequencies. These optimal cantilever structures would attain increased eigenfrequencies if compared to their counterparts with constant cross section. The shapes of the optimal cantilever structures will be symmetrically inverted as the eigenfrequencies are reduced when compared to the analogous structures with a constant crosssection, as shown in Fig. 4.27b for the first (OPT I$^-$), second (OPT II$^-$), and third (OPT III$^-$) eigenfrequencies.

Examination of optimal cantilever substrates in Fig. 4.27 reveals that change of eigenfrequency leads to an increase in the number of cross-sectional minima/maxima along the length of the structure. Moreover, distances from the minimum and maximum cross sections to the clamping site may be easily determined. For example, for the increased second eigenfrequency (OPT II$^+$), the minimum cross-section of the optimised substrate structure from the anchorage point is always 0.24 l for the second eigenfrequency (OPT II$^+$) and 0.15l and 0.5l for the third eigenfrequency (OPT III$^+$), respectively where l is the cantilever length. More detailed analysis of optimal cantilever structures presented in Fig. 4.27 reveals that the recurrence of maximum and minimum cross-sections corresponds to the positions of particular (maximum amplitude and nodal) points of vibration modes (Fig. 4.11a).

The second step of harvester configuration improvement is to adjust its configuration to have two separate piezoelectric material layers. As is it was already discussed in the previous section, harvesters with continuous piezoelectric layers do not perform

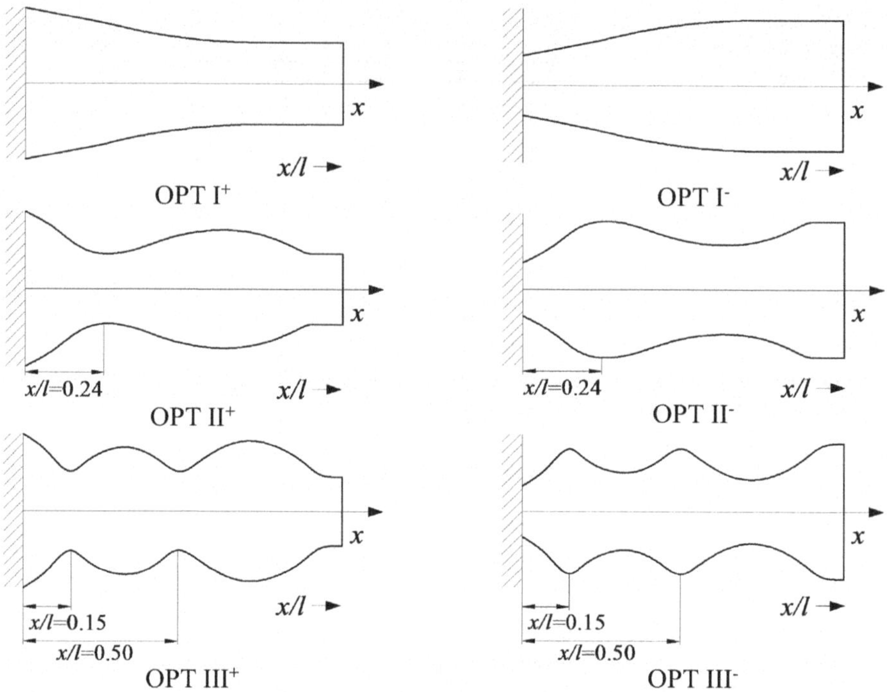

Fig. 4.27 Optimal structures for operation in increased (**a**) and decreased (**b**) eigenfrequency of transverse vibrations (the first (OPT I), the second (OPT II), and the third (OPT III) from top to bottom)

well if excited at greater than the first resonant frequency. Therefore, active piezoelectric material layers must be segmented at the strain nodes of higher vibration modes to avoid undesired charge cancelation effects in the piezoelectric material. Thus, integrating both assumptions described above, rational harvester configuration was developed, a scheme of which is presented in Fig. 4.28. Rational harvester configuration is based on a simplified optimal cantilever structure aimed for operation at decreased second eigenfrequency. Since one of the research objectives was to suggest easily manufacturable harvester configuration, intricate optimal substrate shape (fabrication of which requires sophisticated automated cutting machinery) was replaced by a simplified design of harvester substrate with the hump at $0.24L$. This configuration allows ease of piezoelectric material deposition/attachment and ensures natural segmentation of piezoelectric layer at the strain node of the second eigenmode.

A number of simulations were performed to compare the performance of enhanced configuration harvester (OPT RAT) to its other counterparts (constant cross-section cantilever OPT 0 and with two segmented piezoelectric layers (Segmented, II segments)).

Figure 4.29a presents generated RMS voltage versus relative support location

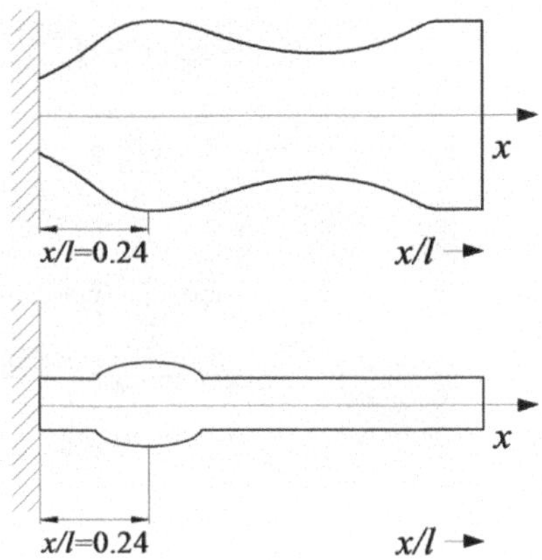

Fig. 4.28 Optimal (top) versus rational (bottom) cantilever configuration

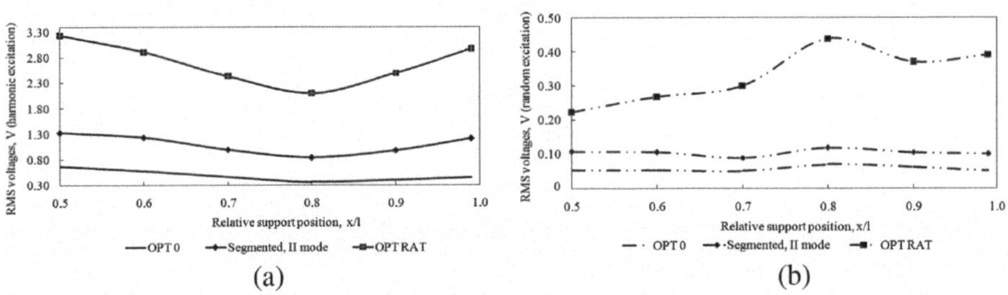

Fig. 4.29 RMS voltages generated by harvesters of different configuration versus relative incorporated support location **a** harmonic and **b** random excitation

graphs for constant cross-section with continuous electrodes (OPT 0), constant cross-section with segmented piezoelectric layer (Segmented, II mode) and rational configuration (OPT RAT); all excited by harmonic signal, meanwhile Fig. 4.29b presents generated RMS voltages for the same harvesters excited by random signal. As one may note from Fig. 4.29b, the rational configuration harvester generates greater RMS voltages (especially prominent if support is located at the nodal point of the third vibration mode ($0.87l$)), if compared to the constant cross-section counterpart with continuous piezoelectric layer. Harvester with piezoelectric layer segmented at the strain node of the second eigenmode (yet constant cross-section) demonstrates higher generated RMS voltages than constant cross-section harvester with continuous piezoelectric layer, yet a few times inferior if compared to the rational configuration. The same trends are noted as harvesters are excited by harmonic excitation signal, with the superior performance of rational configuration harvester, mediocre performance of harvester with segmented piezoelectric layers, and worst performance of constant cross-section harvester with continuous piezoelectric layer. As it was already discussed in the previous sections, incorporated support limits harvester

Fig. 4.30 General view of the high-pressure water jet cutting machine Resato ACM 3060-2

Table 4.12 Main technical characteristics of water jet cutting system Resato ACM 3060-2

Positioning accuracy	±50 μm/m (at 20° C)
Repeatability accuracy	±50 μm/m (at 20° C)
Cutting bed dimensions	3020 × 6020 mm
Angle accuracy	±0.1°
Maximum angle adjustment	55°
Cutting head rotating	from −360° to + 360°

performance if excited by harmonic excitation signal, yet improves the performance once harvesters are excited randomly.

The cantilevers were fabricated by employing a water jet cutting system Resato ACM 3060-2 (PTV, Hostivice, Czech Republic) [16], because it allows cutting steel into intricate shapes without exposing the workpiece to heat during cutting, thus no tension is created in the cut area (Fig. 4.30). In Table 4.12, the main technical characteristics of the water jet cutting system Resato ACM 3060-2 are presented.

4.1.2.2 Investigation of Effect of Piezoelectric Layer Segmentation

Most of the currently explored harvesters do not perform efficiently in real environments, since it is assumed that they are excited only by harmonic signal at their fundamental eigenfrequency. However, if one analyses real environment energy sources, they usually would not consist of a single harmonic, and higher modes of vibration may be excited. Moreover, these higher modes of vibration will be excited if rigid support is incorporated in harvester configuration and the whole system starts operating in vibro-impacting mode. These higher vibration modes of harvester will have strain nodes—i.e. positions on the structure, where bending strain distribution changes sign. It was already demonstrated via numerical simulations that if these strain nodes are covered with continuous piezoelectric layers and the harvester is excited by random excitation signal or higher frequencies, cancelation of harvester electric outputs occur, resulting in the reduction of harvested energy.

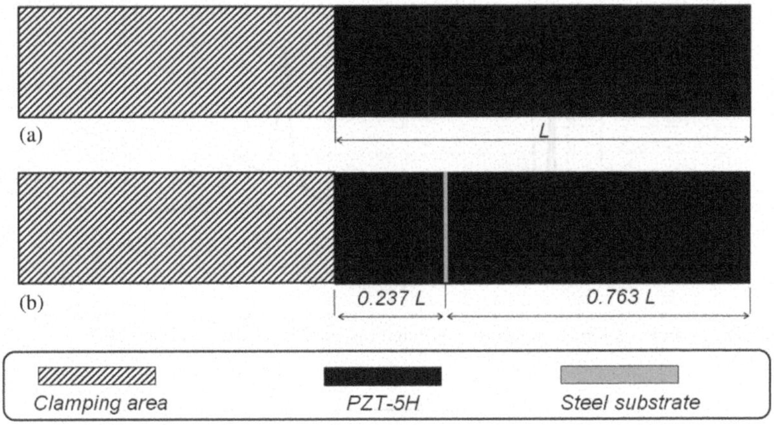

Fig. 4.31 Schemes of fabricated harvester prototypes: **a** non-segmented (i.e. with continuous piezoelectric layer) and **b** segmented at the strain node of the second eigenmode

Thus, this research is devoted to verify numerical simulation results and to ascertain that harvester with piezoelectric layers segmented at the strain node of the second eigenmode would practically demonstrate higher electric outputs as compared to its counterpart with a continuous piezoelectric layer [17]. For the purpose of this research two prototypes of harvester were fabricated. PZT-5H piezoceramic sheets (PiezoSystems, Inc.) covered with nickel electrodes from both sides were cut to the dimensions of 36 × 10 × 10.2 mm by means of an automatic dicing machine and were bonded with epoxy to stainless steel substrates with dimensions of 36 × 11 × 10.4 mm. Wire leads were soldered to harvester electrodes to collect the electrical output. The first harvester prototype (Fig. 4.31a) was designed to have a continuous piezoelectric layer, meanwhile, for the second harvester prototype (Fig. 4.31b), the piezoelectric layer was segmented into two sections (which were confined by mechanically etching nickel electrodes) at the strain node of the second eigenmode.

The measurements were performed by employing Laser Doppler vibrometry system (Fig. 4.21).

The dynamic response of the vibro-impacting harvester was experimentally studied using a prototype with segmented piezoelectric layers (Fig. 4.31b). Firstly, the frequency response function of voltage (Fig. 4.32) was measured (at the arbitrarily selected load resistance of 5000 Ω in order to determine eigenfrequencies of harvester ($\omega_1 = 235$ Hz, $\omega_2 = 1469$ Hz) as well as to analyze resonant frequency bandwidth of operating. Its dependency on the stopgap size was explored: Fig. 4.33 was constructed from a number of frequency response functions, outlaying frequency bandwidth, at which each piezoelectric segment of vibro-impacting harvester prototype generates voltage larger than 0.1 V. For this configuration, the optimal stopgap size is 30 μm since the electrical voltage of more than 0.1 V may be extracted in 50 Hz frequency range. Once the stopgap size is increased up to 110 μm (i.e. until

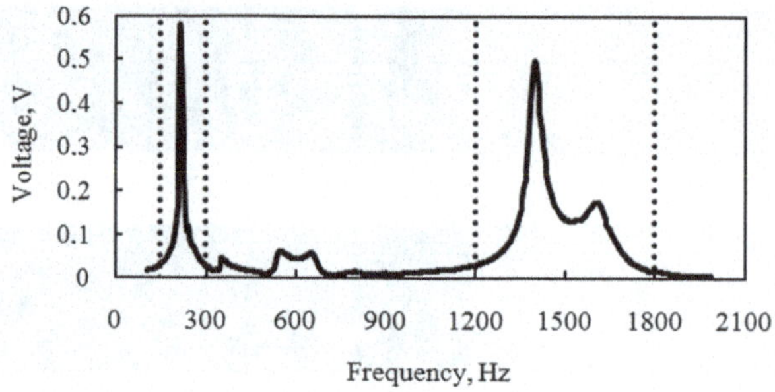

Fig. 4.32 Frequency response function of voltage ($\omega_1 = 235$ Hz, $\omega_2 = 1469$ Hz)

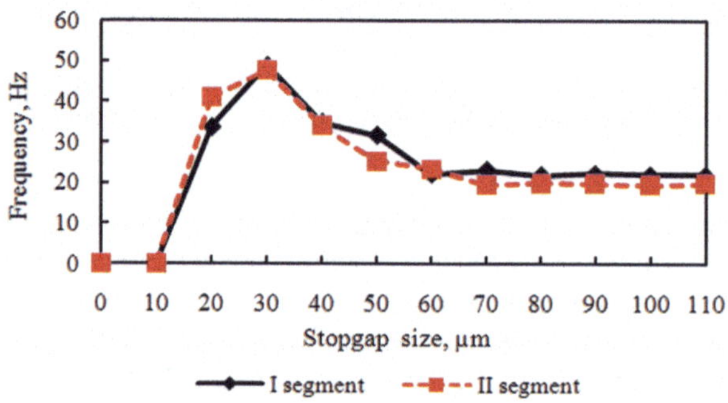

Fig. 4.33 Bandwidth over which each vibro-impacting harvester prototype segment generates voltages larger than 0.1 V

no impact occurs), operating bandwidth decreases by 60%, while maximum voltage increases by 68%.

Measurements of the maximum voltage generated by each segment of the harvester (Fig. 4.34a) confirm the trade-off between a larger piezoelectric energy harvester (PEH) bandwidth and a lower output voltage, as the size of the stopgape

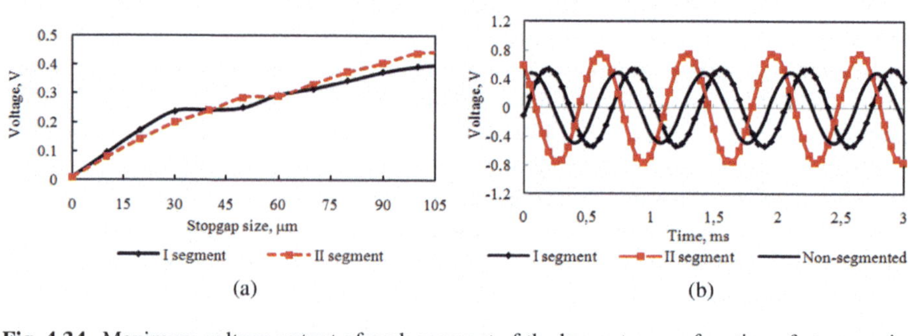

Fig. 4.34 Maximum voltage output of each segment of the harvester as a function of stopgap size (**a**) and comparison of the electrical outputs of segmented and non-segmented PEH prototypes (**b**)

is increased, a higher PEH voltage is generated in each segment. On the other hand, this trade-off could be less pronounced if support is located at the nodal point of the second vibration mode and higher vibration modes of the harvester are excited during the impact. Finally, the maximum voltage generated by segmented and non-segmented prototypes were compared. Separate piezoelectric segments of segmented prototype generate 0.54 V and 0.76 V, respectively, while prototype with continuous piezoelectric layer generates a maximum voltage of 0.5 V (Fig. 4.34b). Implementing a power conditioning circuit that could sum up voltages generated by each piezoelectric segment would result in up to 52% increase in voltage output as compared to PEH prototype with a continuous piezoelectric layer.

Dynamic response of harvester impacting on rigid support was as well evaluated by means of holographic measurement system Prism (Fig. 4.19). Obtained holograms of harvester prototype are presented in Table 4.13: first row—harvester prototype is not impacting on incorporated support; second row—harvester prototype is impacting on rigid support incorporated 29 mm from the clamp or $0.78l$ (i.e. nodal point of second eigenmode at 1469 Hz); and the third row—harvester prototype is impacting on the rigid support incorporated 32 mm from the clamp or $0.87l$ (i.e. nodal point of third eigenmode at 4138 Hz). The rigid support would limit vibration amplitudes at low frequencies (number of black fringes in the hologram decreases), however, it could be used to protect harvester prototypes from rupture at excessive loads and ensure the stability of operation processes at higher frequencies.

The conducted experimental study demonstrates that electrical outputs of vibro-impacting harvester prototype are dependent on incorporated support location. In order to achieve the most favorable trade-off between power output and bandwidth:

- support location should coincide with the nodal point of the second transverse mode or the second nodal point of the third transverse mode in order to intensify amplitudes of these vibration modes during impact and beneficially exploit them for energy harvesting;
- piezoelectric layer of harvester prototype should be segmented in such a way that it does not cover strain nodes of vibrating prototype, thereby avoiding detrimental charge cancelation effects in piezoelectric material and, in turn, reduction in generated voltages.

4.1.3 Investigation of Optimized Cantilever Beam

The main advantage of optimal cantilever substrates relies on the fact that eigenfrequencies of structures are determined by cantilever shape. This allows minimizing the size of the overall system and investigating structures as distributed parameter systems and, moreover, beneficially exploiting their higher vibration modes [18].

For the purpose of dynamic testing, cantilevers of three different configurations were fabricated: the first configuration (OPT 0) possesses constant cross-section; the second (OPT II) and third (OPT III)—optimal cross sections for the operation at the second and third eigenfrequencies of transverse vibrations, respectively. As

Table 4.13 Holograms of harvester vibrations

Support position	Excitation frequency		
	235 Hz	1469 Hz	4138 Hz
Not impacting	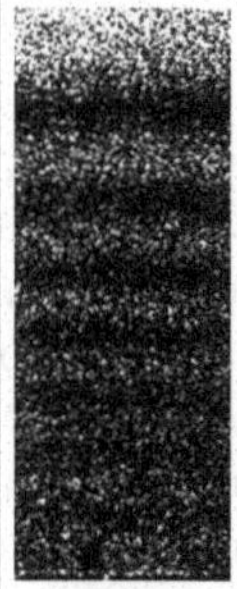		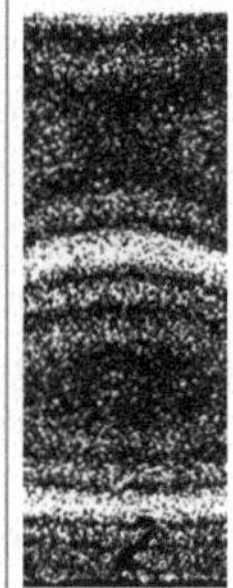
Support located at 0.78*L*	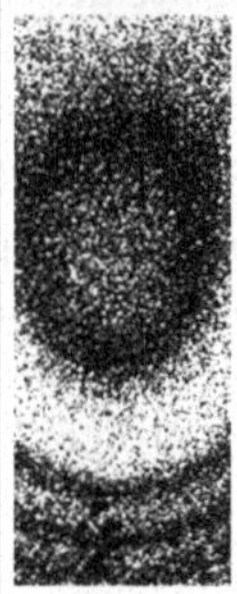	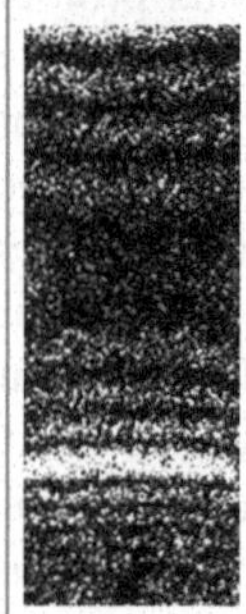	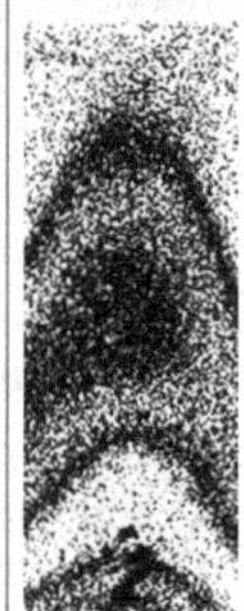
Support located at 0.87*L*	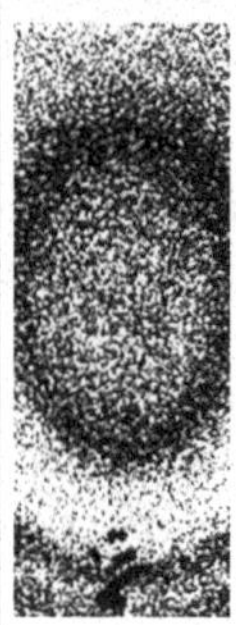	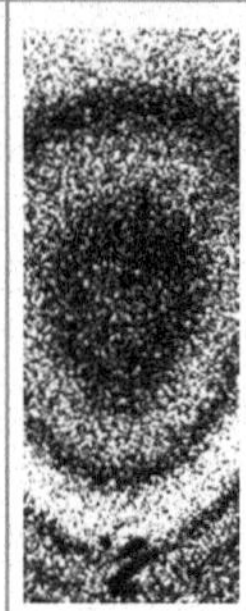	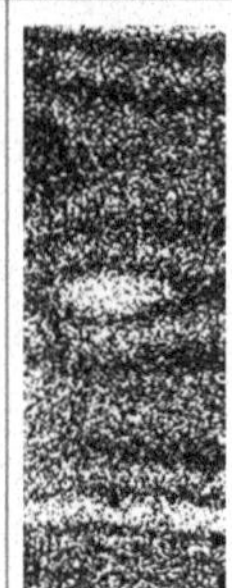

described earlier, optimal cantilever configurations were obtained through optimization with the objective function of cantilever mass minimization in the presence of a constraining equation system. The state of the cantilever was described by a modal analysis equation with bounds (side constraints) imposing that the second and the third natural frequencies of transverse vibrations of the optimal cantilevers must coincide with the corresponding frequencies of the initial cantilever with constant cross-section. Shapes of the optimal cantilevers are depicted in Fig. 4.27. These cantilevers were fabricated from a thin steel sheet by means of water jet cutting machine Resato ACM 3060-2 (Fig. 4.30). This machining method was chosen since it allows accurate cutting of intricate shapes without exposing workpiece to heat,

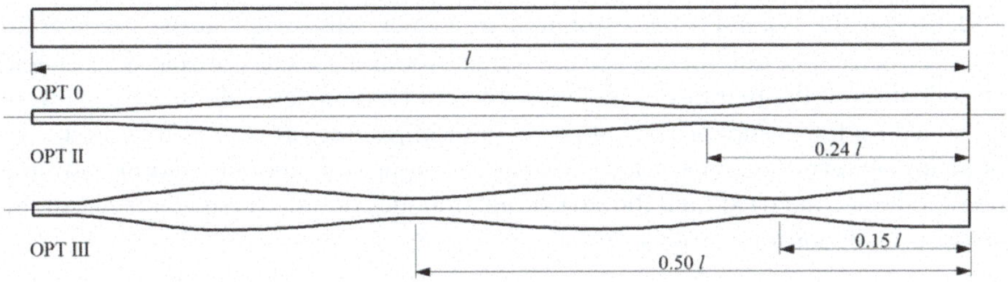

Fig. 4.35 Magnified drawings (side view) of fabricated cantilevers

thereby avoiding thermally induced stresses in the cutting area. Figure 4.35 presents drawings of fabricated cantilevers.

Average measured stiffness values for different configuration cantilevers are presented in Table 4.14 and graphically depicted in Fig. 4.36a after stiffness measurements (Fig. 4.36b). One may note that as cantilever configuration complexity increases, its average stiffness decreases (almost 2.7 times, if comparing OPT III and OPT 0 configurations).

Table 4.14 Average measured stiffness values for different configuration cantilevers

Cantilever configuration	F (N)	y (μm)	c (N/m)	$c_{avergae}$ (N/m)
OPT 0	0.2	467	428.3	430.9
	0.3	692	433.5	
OPT II	0.2	1040	192.3	192.9
	0.3	1550	193.5	
OPT III	0.2	1099	182.0	183.4
	0.3	1623	184.8	

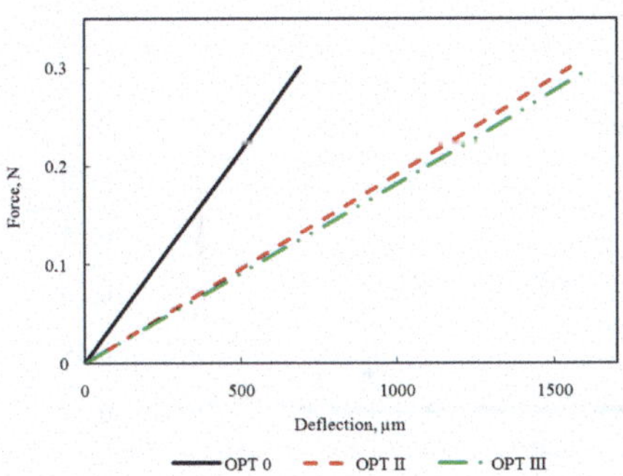

(a)

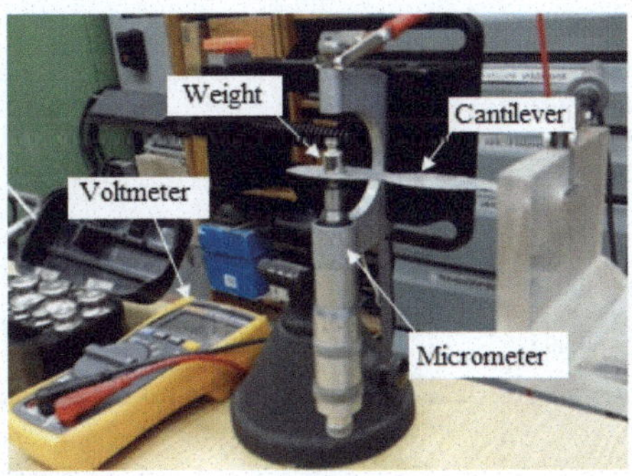

(b)

Fig. 4.36 Stiffness graphs for different configuration cantilevers (**a**) and experimental set-up used to determine the stiffness of harvester prototypes (**b**)

Measuring in dynamics the excitation frequency was swept in the range from 10 to 1500 Hz and amplitudes of transverse vibrations were registered to enable determination of the first three eigenfrequencies of cantilevers. Measurements were performed for the unsupported cantilevers (non-impacting) as well as for cantilevers impacting against the support located at points approximately coinciding with the nodes of the second (0.8*l*) and third (0.9*l*) eigenmodes of transverse vibrations of the cantilever with constant cross section.

4.1.3.1 Investigation by Laser Doppler Vibrometry

For the unsupported cantilever configuration OPT 0 the first transverse vibration mode is registered at 75 Hz, the second at 470 Hz, while the third one was not registered in this case (Fig. 4.37). It is obvious that the resonance peak of the first mode of the unsupported cantilever is significantly greater (up to 10 times) when compared to the corresponding peaks of the impacting cantilevers. As the rigid support is positioned at 0.8*l*, the second transverse mode is recorded at 469 Hz. It should be noted that once the support is placed at the node of the second mode, the amplitude of the corresponding vibration is magnified up to 10 times with respect to the case of the unsupported cantilever. When the rigid support is shifted to 0.9*l*, the first resonance is observed at 46 Hz, while the second one at 410 Hz (with both vibration amplitudes being smaller as compared to the case of 0.8*l*). For the latter two cases of impacting cantilever, one may note amplitude peaks at 135 Hz, which pertains to the fundamental mode of the cantilever clamp and is not considered here.

Frequency responses measured for the cantilever configuration OPT II (Fig. 4.38) reveal that the first eigenfrequency of the unsupported cantilever is excited at 64 Hz and the second one at 490 Hz. In the case of the cantilever impacting the support placed at 0.8 *l*, the second eigenfrequency is recorded at 492 Hz. For the case of support located at 0.9*l*, the first eigenmode is not registered in the considered frequency range, while the second shifts down to 430 Hz. For the configuration

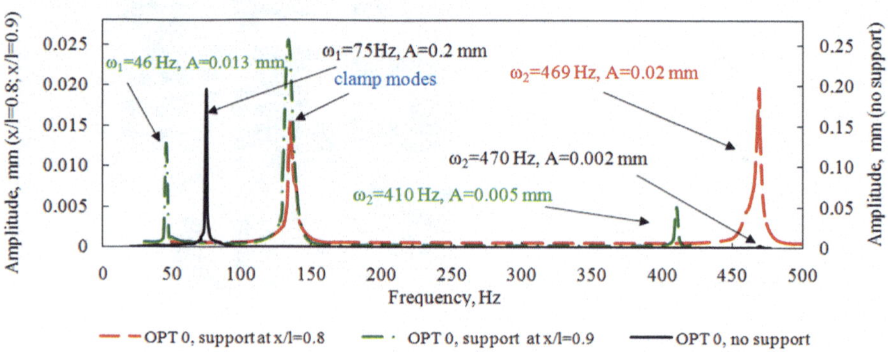

Fig. 4.37 Amplitude-frequency characteristics for the cantilever configuration OPT 0

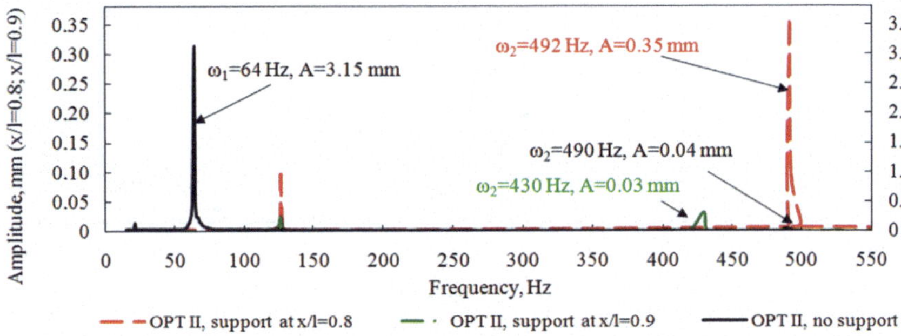

Fig. 4.38 Amplitude—frequency characteristics for the cantilever configuration OPT II

with support located at 0.8*l*, the higher transverse vibration eigenmodes were registered as well, indicating that the free end of the cantilever vibrates more easily. As expected, the unsupported cantilever is characterized by the highest amplitudes of the first mode, meanwhile amplitudes of the second eigenmode increase nearly 10 times (compared to the unsupported cantilever case) once the support position is juxtaposed with the node of the second transverse eigenmode.

Finally, Fig. 4.39 provides frequency responses for the optimal cantilever configuration OPT III: for the unsupported cantilever, the first eigenfrequency is registered at 58 Hz, the second and the third one at 396 and 1336 Hz, respectively. The amplitude of the first eigenmode is highest for the unsupported cantilever. When the support is introduced at 0.8*l*, the second eigenmode is observed at 395 Hz and the third one at 1249 Hz (the amplitude of the second eigenmode reaches the highest value for this particular case as already noted for OPT 0 and OPT II cases). As the support is moved to 0.9 *l*, the second eigenmode is reduced to 350 Hz, while the third one is observed at 1325 Hz with vibration amplitudes being considerably smaller when compared to the other cases.

Figure 4.40 illustrates the shift of natural frequency for different unsupported cantilever configurations. The plots indicate that cantilever configuration OPT 0 is

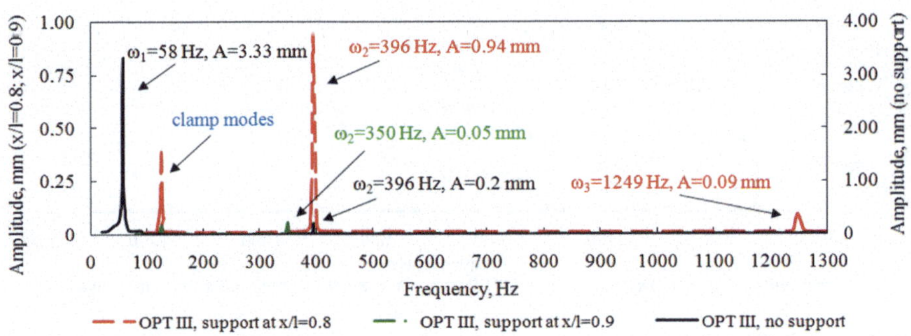

Fig. 4.39 Amplitude-frequency characteristics for the cantilever configuration OPT III

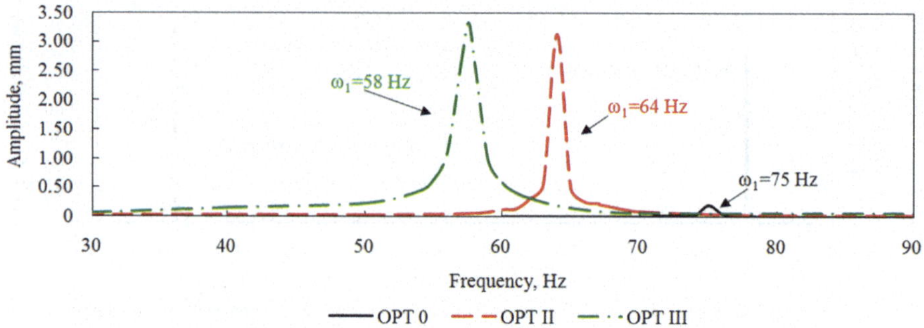

Fig. 4.40 Shift of the first natural frequency for different cantilever configurations

the stiffest one as it exhibits the highest first eigenfrequency (75 Hz) accompanied by the lowest amplitude peak. Optimal cantilever OPT III is characterized by the lowest first eigenfrequency of 58 Hz, which constitutes 23% reduction with respect to OPT 0, while the amplitude peak is higher more than 6 times. Cantilever OPT II is somewhat in the middle between the other configurations in terms of first eigenfrequency and modal amplitude. These small discrepancies between the frequency values confirm that the chosen cutting method with water jet was sufficiently accurate in realizing optimal cantilever structures with the second and third eigenfrequencies as close as possible to the corresponding frequencies of the cantilever with constant cross section.

Figure 4.41 combines frequency responses of different cantilever configurations for the case when the rigid support is located at $0.8l$ (node of the second transverse vibration eigenmode). The plots reveal that cantilever OPT 0 exhibits the lowest amplitudes of the second eigenmode, while OPT III is characterized by the highest vibration amplitudes for second and third eigenmodes with the lowest second eigenfrequency and highest third eigenfrequency. Cantilever OPT II acquires frequency and amplitude values in-between other configurations.

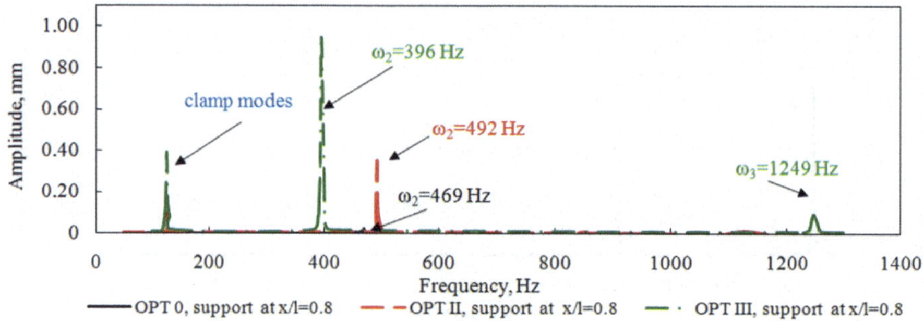

Fig. 4.41 Amplitude-frequency response characteristics for different configuration cantilevers with the support located at $0.8l$

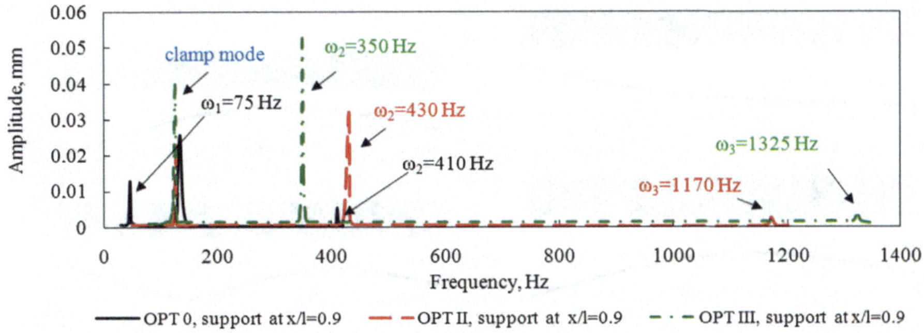

Fig. 4.42 Amplitude-frequency response characteristics for different configuration cantilevers with the support located at $0.9l$

Similarly, Fig. 4.42 provides frequency responses of different cantilever configurations for the case when support is located at $0.8l$ (second node of the third transverse eigenmode). Analogously to the preceding case, cantilever OPT III exhibits the highest vibration amplitudes of the second and third eigenmodes, demonstrating the lowest second eigenfrequency and the highest third eigenfrequency.

4.1.3.2 Investigation by Holographic Measurement

Further on, cantilever beams were analyzed by employing a holographic measurement system Prism. An acoustic field was employed to excite the cantilevers by varying the harmonic driving signal from 10 to 2000 Hz. Eigenmode shapes of the cantilevers with the corresponding displacement amplitudes and eigenfrequencies were determined via digital analysis of the registered interference patterns.

Figure 4.43 provides visualization of the measured eigenmode shapes of cantilever OPT 0 including a schematic representation of tested configurations, both the unsupported and the impacting one with the support placed approximately at $0.8l$. Results for the unsupported cantilever indicate that the first transverse eigenmode is excited at 77 Hz, the second and third eigenmodes at 480 and 1335 Hz, respectively. Both transverse and torsional displacements are observed for the third eigenmode. In the presence of the rigid support, the second transverse eigenmode is detected at 465 Hz, while the third at 1186 Hz. Further increase of excitation frequency leads to an eigenmode at 1402 Hz, where both transverse and torsional components are visible.

Similarly, Fig. 4.44 visualizes eigenmodes shapes for the optimal cantilever OPT II characterized both in the unsupported and impacting configurations. For the former, the first transverse eigenmode is excited at 63 Hz, the second and third at 490 and 1158 Hz, respectively. For the latter case, the second transverse eigenmode is detected at 479 Hz, while the third at 1133 Hz. Further increase of excitation frequency induces flexural-torsional mode at 1660 Hz.

Finally, Fig. 4.45 presents measured eigenmode shapes for the optimal cantilever

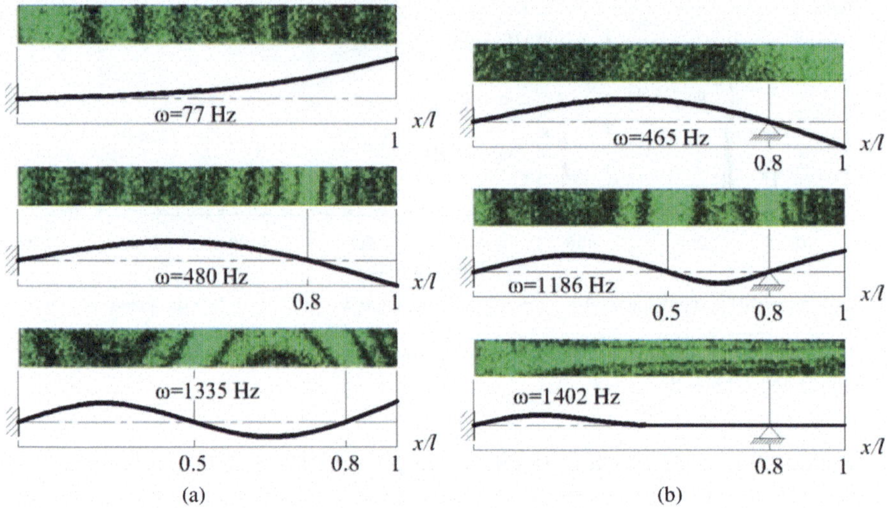

Fig. 4.43 Measured natural vibration modes and frequencies for cantilever OPT 0 (**a**) unsupported and (**b**) rigid support located at ~0.8*l* (mode number increases from top to bottom)

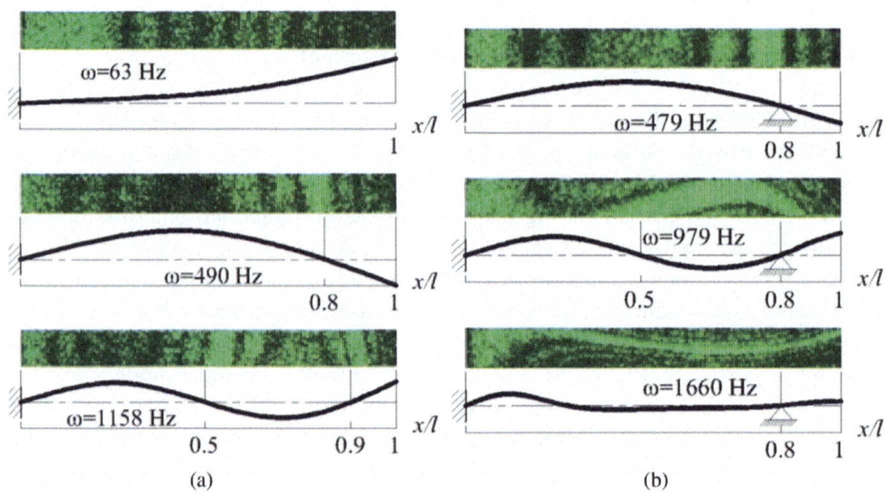

Fig. 4.44 Measured eigenvibration modes and frequencies for cantilever OPT II **a** unsupported and **b** rigid support located at ~0.8*l*

OPT III. For the unsupported case, the first transverse eigenmode is excited at 58 Hz, the second and third at 402 and 1386 Hz, respectively. In the presence of support at 0.8*l*, the second eigenmode is observed at 372 Hz. Further increase of excitation frequency yields a number of higher eigenmodes at 390 Hz, 1021 Hz, 1341 Hz, and 1992 Hz, which are characterized by both transverse and torsional

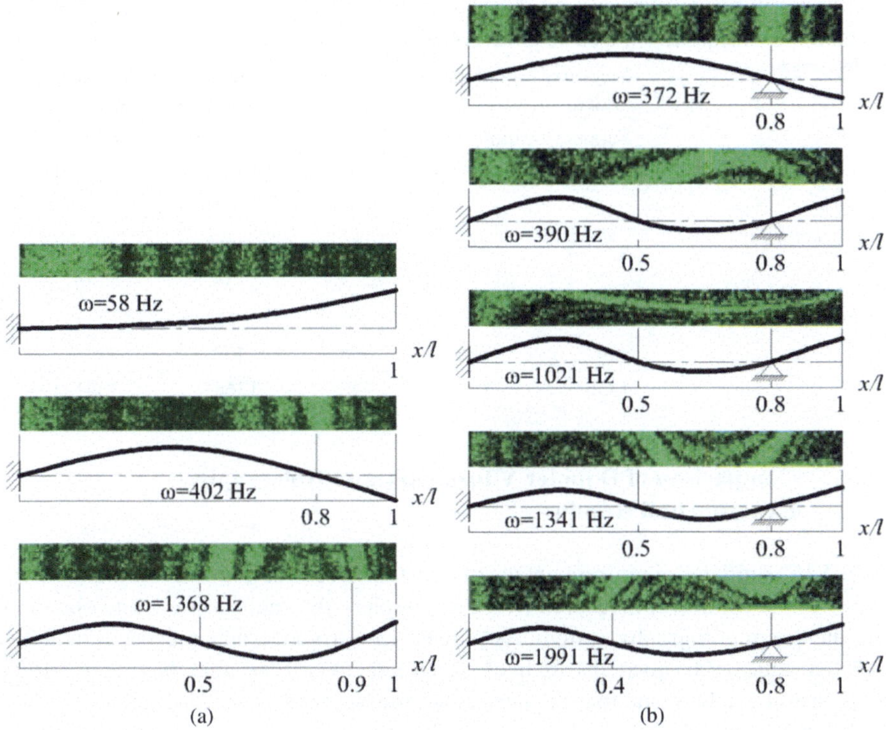

Fig. 4.45 Measured eigenmodes and frequencies for cantilever OPT III: **a** unsupported, **b** rigid support located at ~0.8l

displacement components. The holographic image of the mode shape at 1341 Hz reveals that the cantilever undergoes torsional oscillations in one direction, while at 1991 Hz, the torsion is observed in another direction. One may assume that the third cantilever resonance occurs at 1341 Hz frequency while higher vibration modes are excited at 1991 Hz. Table 4.15 lists eigenfrequencies derived from Dopler and holographic interferometry measurements. The values of eigenfrequencies obtained using different optical measurement techniques are in close agreement (discrepancies are in the range of 2–3%). Moreover, it could be reiterated that this study attempted to optimize and then fabricate optimally shaped cantilevers (OPT II and OPT III) so as their second and third eigenfrequencies of transverse vibrations would be equal to the corresponding frequencies of the cantilever with constant cross section OPT 0 (at 480 Hz and 1335 Hz, respectively). Presented experimental results indicate that for the cantilever OPT II, the second eigenfrequency is equal to 490 Hz, while for OPT III, the third eigenfrequency is 1368 Hz, which amounts to 2% error with respect to the values of corresponding frequencies obtained for cantilever OPT 0.

Table 4.15 Summary of eigenfrequencies for different cantilever configurations measured with the Doppler vibrometry system and the Prism holography system

Cantilever configuration	Frequency	Eigen frequencies (Hz)				
		Measured with doppler vibrometry system			Measured with prism holography system	
		Support location			Support location	
		Unsupported	$x/l \approx 0.78$	$x/l \approx 0.87$	Unsupported	$x/l \approx 0.87$
Constant cross section	ω_1	75	–	46	77	–
	ω_2	470	469	410	480	465
	ω_3	–	–	–	1335	1186
Optimal	ω_1	58	–	–	58	–
	ω_2	396	395	350	402	372
	ω_3	1336	1249	1325	1368	1341, 1991

4.1.3.3 Comparison of Doppler Vibrometry and Holography Experimental Results

Table 4.15 compares eigenfrequencies of constant cross-section and optimal configuration of the cantilever measured with the Doppler vibrometry system and obtained with the Prism holography system. As may be seen from the data presented in Table 4.15, the values of frequencies are in close agreement (2–3% value fluctuation), thus a conclusion can be made that the performed measurements are qualitative.

The dynamic responses of cantilever beam structures of non-optimized (i.e. constant cross-section, OPT 0), optimized for operation at increased second eigenfrequency (OPT II) and optimized for operation at increased third eigenfrequency (OPT III) show, that shape of the cantilever has a significant influence on its dynamic responses. Insights on the dynamic response of optimized cantilever beam structures presented in this section may serve as a basis for enhanced harvester designs, allowing controlling resonant frequencies as well as device amplitudes without the addition of concentrated masses at the tip of the cantilever. In such a way, the device still possesses all features of distributed parameter system, and higher vibration modes may also be exploited for energy harvesting as opposed to the typical single degree of freedom systems (cantilevers with concentrated masses at the tip), which utilizes only its first resonant frequency.

4.1.4 Appropriate Way to Extract the Low-Frequency Vibration Energy

The most appropriate way to transform the energy of low-frequency oscillations into electricity is associated with the excitation of the third eigenmode. For this purpose, an optimally configured cantilever (Fig. 4.46) with an optimized cross-section for

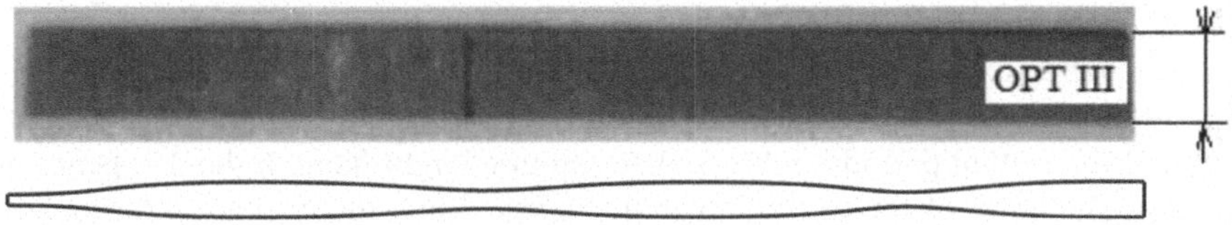

Fig. 4.46 Technical drawing of the fabricated cantilever

the operation at the third eigenfrequency was fabricated and analyzed. The optimal cantilever configuration was obtained to solve the optimization problem as per the above-described theory with the aim of mass minimization (yet the diminution of the cantilever was limited) in the presence of a constraining equation system. The state of the system is described by the mode analysis equation, which requires that the eigenfrequency of the third flexural vibration mode coincides with the increased values of the third eigenfrequency of the constant cross-section cantilever.

From the sketch in Fig. 4.46, the distances from the minimum and maximum cross-sections to the fixing site of the structure are easy to measure. The minimum cross-section of the structure that is optimal at the third value of transverse vibration frequency ω_3 has located the distances of 0.15 l and 0.5 l from the fixed (right) end of the cantilever, where l is its length.

4.1.4.1 Frequency Response Analysis

The peculiarities of a cantilever beam vibrating in the third mode are related to the significant increase of the level of deformations capable of extracting significant additional amounts of energy compared to the conventional harvester vibrating in the first mode. To show the possibility of effectively using a cantilever vibrating in the third mode for energy harvesting, its FE model was developed with the Comsol multiphysics software. During the simulation, various vibro-impact regimes of the optimally configured cantilever according to the given third eigenfrequency were determined (Fig. 4.47).

Under the sine excitation Q, the optimally shaped cantilever hits the support which is placed at its free end and starts the vibro-impact motion. On the right side of

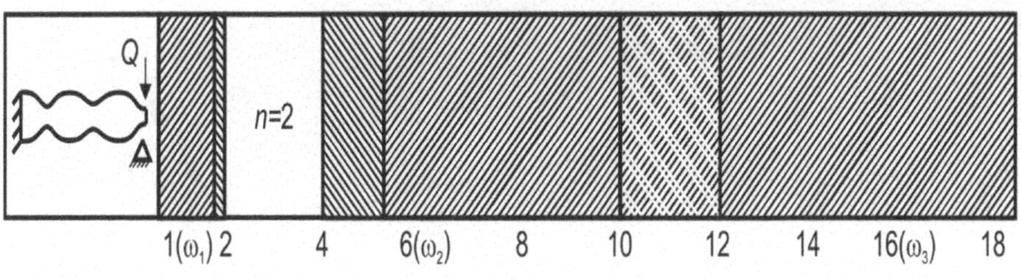

Fig. 4.47 Periodically excited vibro-impact motion laws of the optimal cantilever

Fig. 4.47, the particularly settled areas are presented at different excitation frequencies ω/ω_1. When excitation frequency ω is equal to the first natural frequency ω_1 of the cantilever, the first resonance of such a vibro-impact system has a place and the area of the settled periodic vibro-impact motion (right-hand dashes) appears in the narrow diapason of the excitation frequencies. When the excitation frequency increases, the vibro-impact motion becomes close to the settled one (points and dashes), with its areas alternating with unsettled or chaotic recurrent vibration laws (left-hand dashes). Between these two chaotic regimes there is a $n = 2$ non-dashed area, characterised by settled vibro-impact motion, where the optimal cantilever hits the support only once during the two periods of sine excitation. When the excitation frequency coincides with the second natural frequency of the cantilever, the second resonance is characterized by the settled vibro-impact motion. However, the most beneficial is the third resonance characterized by a very large diapason of the settled vibro-impact motion. This phenomenon could be useful for increasing energy harvesting efficiency, as well as for creating high-speed vibro-impact mechanisms.

4.1.4.2 Design of Energy Harvester Prototype

Two types of prototype piezoelectric vibration energy harvester devices were analysed. The first without electrode segmentation and the second with segmentation at the deformation nodes of the third vibration mode to achieve efficient operation at the third eigenfrequency. The resonant frequencies, eigenmodes as well as the preliminary location of the deformation nodes of the vibrating structure have been determined in the simulations [9]. Figure 4.48 presents a scheme of the FE model which is based on a stainless steel rectangular cantilever beam ($l_s = 37$ mm, $T_s =$

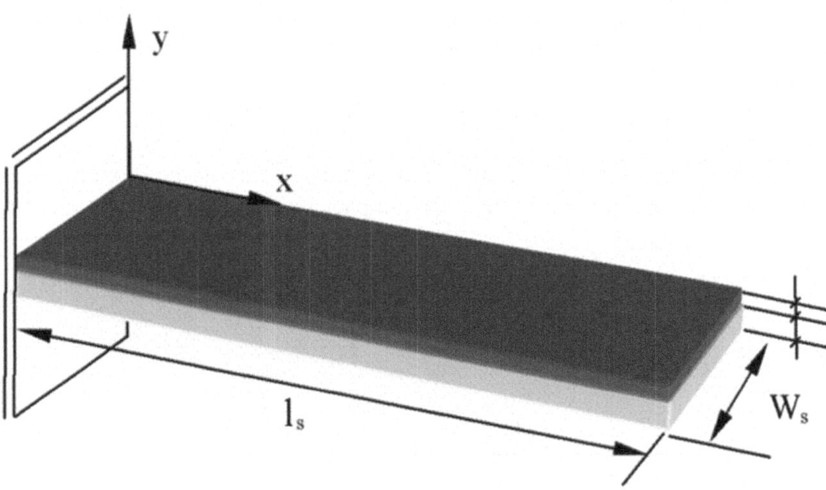

Fig. 4.48 Scheme of the simulated energy harvester prototype

0.36 mm, $W_s = 11.05$ mm) with one T107-H4E-602 piezoelectric layer bonded on its top surface ($l_p = 37$ mm, $T_p = 0.191$ mm, $W_p = 10$ mm).

Simulation results are presented in Fig. 4.49. It shows the deformed shape and von Misses stress distribution in the third vibration mode of the piezoelectric vibrating energy harvester prototype.

Charge cancelation in piezoelectric materials does not appear in the first eigenmode (strain distribution function does not change the sign when the harvester is operating at its first eigenfrequency), thus higher vibration in the third vibration mode (Fig. 4.49a). As the von Misses stress distribution plot in Fig. 4.49b suggests, the simulated locations of the strain nodes in the third eigenmode are 6 mm and 19 mm from the clamping point, respectively.

Figure 4.50 presents drawings of the harvester prototypes built for the experimental study. Both prototypes were built from T107-H4E-602 plate (Piezo Systems, Inc., Woburn, MA, USA) covered with conductive layers. The first harvester

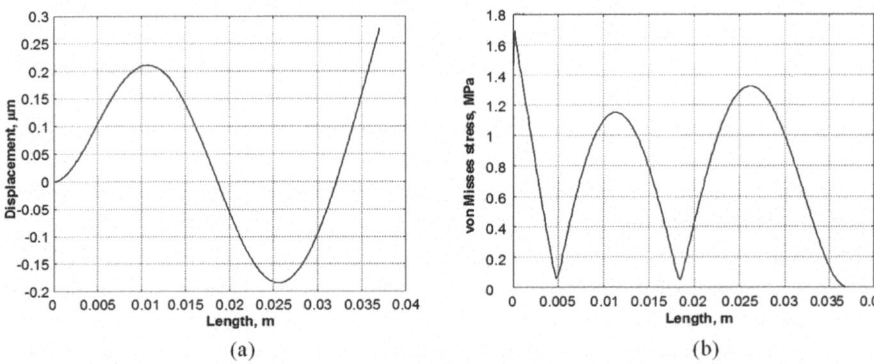

Fig. 4.49 Deformed shape **a** and von Misses stress distribution and **b** for the third eigenmode of the energy harvester prototype

Fig. 4.50 Schemes of elaborated piezoelectric energy harvester prototypes: **a** non-segmented; **b** segmented for operation in the third eigenmode

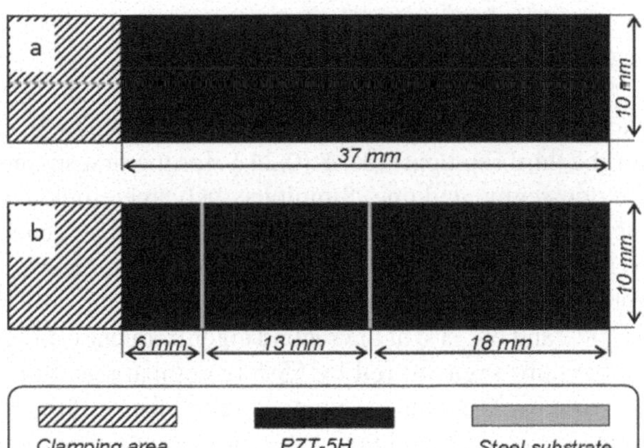

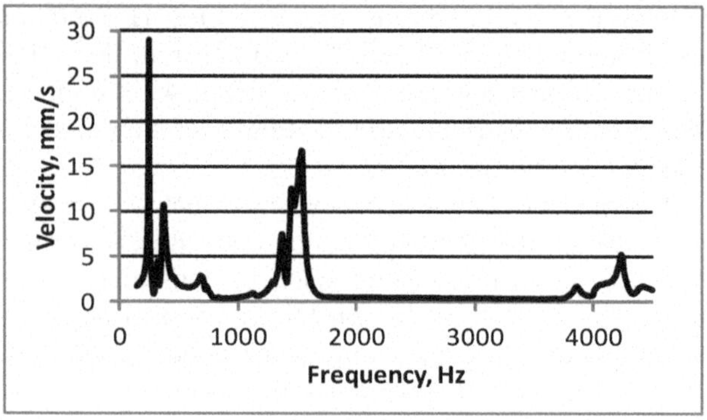

Fig. 4.51 Measured frequency response of tip velocity of the harvester prototype

prototype (Fig. 4.50a) had no electrode segmentation, while the second prototype (Fig. 4.50b) featured electrode segmentation configured for a device operating at the third resonant frequency. The electrode of the piezoelectric material was divided into three parts at the strain nodes of the third vibration mode in order to boost its output voltage. If these strain nodes were covered by continuous electrodes, cancelation of electric outputs occurred, resulting in an overall reduction of the harvested energy.

Figure 4.51 shows the measured frequency response of tip velocity of the piezoelectric energy harvester (the energy harvester was excited by a 3*g* acceleration). It was found that the first resonant frequency of the harvester is 252 Hz, the second is 1452 Hz, and the third one is 4231 Hz.

The measured output electrical potential (electrical load resistance of 4700 Ω, arbitrarily selected) of the non-segmented energy harvester (Fig. 4.52d) at the first eigenfrequency was 0.69 V, at the second—0.37 V, and at the third—0.05 V, respectively. These results show that the best operating frequencies for a non-segmented energy harvester are 252 and 1452 Hz, while at higher frequencies, this harvester is useless. The segmented energy harvester has three electrodes: 6 mm, 13 mm, and 18 mm length that are located between the strain nodes of the harvester vibrating at the third eigenmode. Figure 4.52 provides the measured frequency responses of output electrical potential of the first (6 mm), the second (13 mm), and the third segment (18 mm) of the energy harvester.

The graphs show that voltage, generated by any segment of the energy harvester excited at the third resonant frequency (Fig. 4.52a, b, c), increased from 3.4 (0.17 V for the third segment) to 4.8 (0.24 V for the first segment) times in comparison with the non-segmented one. Simultaneously, the efficiency of the energy harvester also increased at lower eigenfrequencies. The voltage generated by the first segment rose by 14% in comparison with the non-segmented harvester and by 57% at the second eigenfrequency. As follows, the voltage generated by the second segment increased by 28% and by 33% at the second eigenfrequency. In fact, only the voltage generated by the third segment fell by 55% in comparison with the non-segmented harvester and rose by 27% at the second eigenfrequency. If the voltages generated by separate segments could be added electrically, then the segmented energy harvester would generate 2.9 times higher voltage than the non-segmented one at the first, 4.2 times

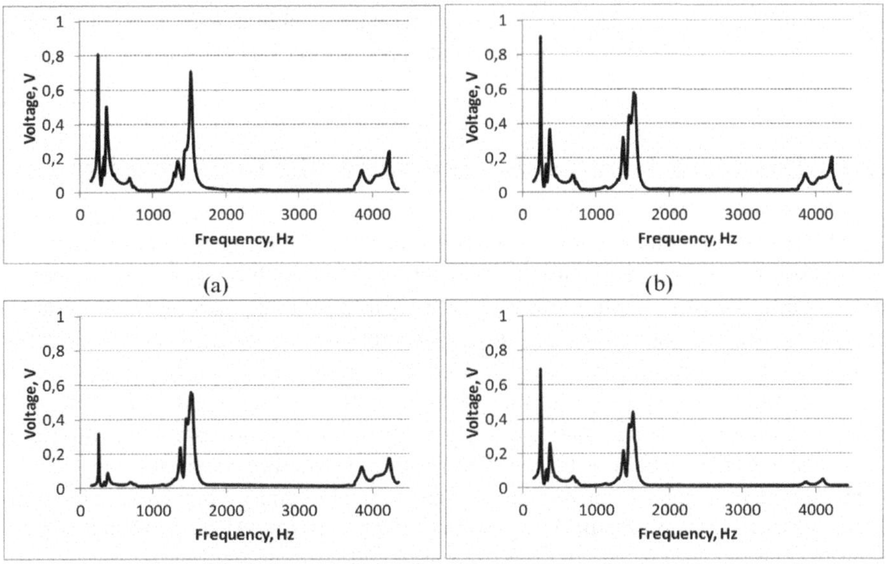

Fig. 4.52 Measured frequency response of output voltage of the first 16 mm (**a**), second 13 mm (**b**), and third 18 mm (**c**) segments generated by the energy harvester in comparison with the frequency response of output voltage of the non-segmented energy harvester (**d**)

at the second, and 12.2 times at the third eigenfrequency. These experiments show that a greater area of the piezoelectric layer on the energy harvester does not necessarily result in higher generated voltages, i.e. optimal location and size of the piezoelectric layer on the substrate should also be defined when designing a piezoelectric energy harvester.

On the basis of the obtained research results, patents for two inventions were obtained [19, 20].

4.2 The New Principles of Energy Harvesting in Micro Level

The vast majority of reported harvester designs are based on elastic structures covered with piezoelectric layers, commonly configured as uni- or bi-morphs. Literature review reveals that relatively little research work has been performed on structural optimization of microcantilever-type vibro-impacting harvesters with thorough dynamic response studies being scarce. Investigation of dynamics-related performance parameters of vibro-impact systems such as operation speed, stability, reliability, longevity is a high priority topic among the other research work conducted in the field of harvesters. Designing a commercially viable device is possible only through in-depth understanding and accurate prediction of its vibrational behavior.

4.2.1 Enhanced Vibration Energy Harvesting Configuration

The initial step of enhancement of micro-cantilevered vibration energy harvester configuration was associated with the shape optimization of harvester substrate structure [21, 22]. The aim of the optimization was to select such geometrical parameters that would correspond to technical characteristics of the considered dynamic system and give a minimum value to a certain quality function (a very common example would be system mass minimization with the constraint of a prescribed vibration frequency). It was also important to distinguish geometrical and structural performance constraints in order to avoid irrational structure configurations.

The target function of optimization of the vibration energy harvester microcantilever substrate and constraints were given in Eqs. (4.19)–(4.21) and optimal configurations are presented in Fig. 4.27. As the obtained optimal configurations are difficult to implement, especially at the micro level, an original way of simplifying these configurations by replacing them with rational ones in terms of manufacturing has been proposed. Harvesters with continuous piezoelectric layers do not perform well if excited at greater than the first eigenfrequency. Therefore, active piezoelectric material layers must be segmented at the strain nodes of higher eigenmodes to avoid undesired charge cancelation effects in the piezoelectric material. The rational harvester configuration, scheme of which was presented in Fig. 4.28, is based on a simplified optimal cantilever structure aimed for operation at decreased second eigenfrequency. Since one of the research objectives was to suggest easily manufacturable harvester configuration, intricate optimal substrate shape (fabrication of which requires sophisticated automated cutting machinery) was replaced by a simplified design of harvester substrate with the hump at $0.24l$. This configuration allows ease of piezoelectric material deposition/attachment and ensures natural segmentation of piezoelectric layer at the strain node of the second eigenmode.

4.2.1.1 Physical Twin Frequency Response Measurement

In order to verify the above improvement of the vibration energy harvester design by dynamic testing, three different configurations of microcantilevers were fabricated from nickel using the Sigma Laser Micromachining System (Amada Miyachi America, USA) with a precision of ± 3 μm: the cross-section of the 1st configuration (OPT 0) is constant; the cross-section of the 2nd (OPT II) and the 3rd (OPT III) configurations are optimum (Fig. 4.53) to operate at the 2nd and 3rd transverse vibration eigenfrequencies, respectively. The length of all cantilevers is $l = 5$ mm, width is $0.1l$, thickness varies from $0.01l$ in the thickest part to $0.005l$—in the thinnest

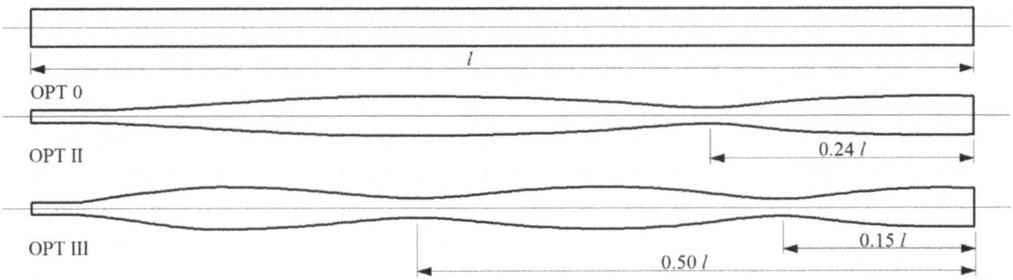

Fig. 4.53 Drawings of optimal cantilevers that were fabricated by means of three different configurations: OPT 0, OPT II, and OPT III

part. All dimensions and measurement results are presented in dimensionless form, which allows comparing results obtained in different scales.

Frequency response measurements of the fabricated cantilevers were performed using Polytec scanning laser Doppler vibrometer (Fig. 4.54). The tested cantilever was fixed on the platform attached to the piezoelectric actuator PSt 150/4/20VS9 (Piezomechanik GmbH, Germany) excited using a periodical signal generated by

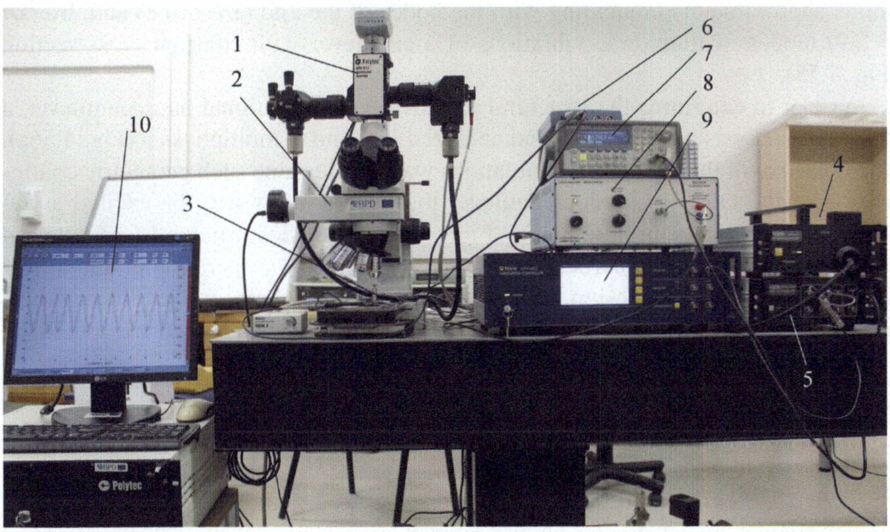

Fig. 4.54 Photo of experiment set-up: 1—Polytec OFV-072 Microscope adapter with Polytec OFV-073 Microscope scanner unit and Polytec OFV-071 Microscope manual positioner, 2—Nikon microscope Eclipse LV100 with digital video camera Pixelink PL-A662, 3—piezoelectric actuator PSt 150/4/20VS9 (Piezomechanik GmbH, Germany) with a researched object, 4—Polytec OFV-512 fiber-optic interferometer, 5—Polytec MSV-Z-40 Scanner controller, 6—PC oscilloscope PicoScope 3424 (PicoTechnology Ltd.,GB), 7—function waveform generator Agilent 33220A, 8—linear amplifier EPA-104 (PiezoSystems Inc., USA), 9—Polytec OFV-5000 Vibrometer controller, 10—Polytec Vibrascan DAQ PC

function waveform generator Agilent 33220A and amplified by linear amplifier EPA-104 (PiezoSystems Inc., USA). Interference optical laser signal generated and registered by Polytec OFV-512 fiber-optic interferometer was transformed into electrical using Polytec OFV-5000 vibrometer controller via MSV-Z-40 Scanner controller and transmitted to Polytec Vibroscan DAQ PC for the analysis. For the laser beam and specimen adjustment, Nikon microscope Eclipse LV100 with digital video camera Pixelink PL-A662 was used. The laser beam spot was positioned at the end of the cantilever. The electrical signal generated by the piezoelectric layer of the energy harvester was collected by oscilloscope PicoScope 3424 (PicoTechnology Ltd, GB).

4.2.1.2 Physical Twin Test Results

The excitation frequency was swept in the range from 0 to 20 $\omega/\omega_{1\ OPT0}$ (ω—excitation frequency, $\omega_{1\ OPT0}$—the 1st eigenfrequency of optimal microcantilever OPT 0) and amplitudes of transverse vibrations were registered to enable determination of the first three natural frequencies of the microcantilever. Measurements were performed with the freely-vibrating microcantilevers (no impacting) as well as for vibro-impacting ones, i.e. microcantilevers impacting against a stopper located at points approximately coinciding with the nodes of the 2nd ($x/l \approx 0.8$) and 3rd ($x/l \approx 0.9$) modes of transverse vibrations of a cantilever with constant cross section (Fig. 4.55).

In order to determine the structural parameters of the rational microcantilever, a 2D FE model was implemented and analyzed in Comsol multiphysics (Fig. 4.56a). The objective of the conducted simulations was to determine such geometric configuration of the OPT RAT microcantilever that would allow it to be excited at the predetermined 2nd natural frequency. Figure 4.56b provides structural configuration of the modeled rational microcantilever that is intended for operation in transversal (d_{31}) mode. A rational design approach that is adopted here implies that an optimally shaped zone of increased cross section (with center located at $x/l = 0.24$) in

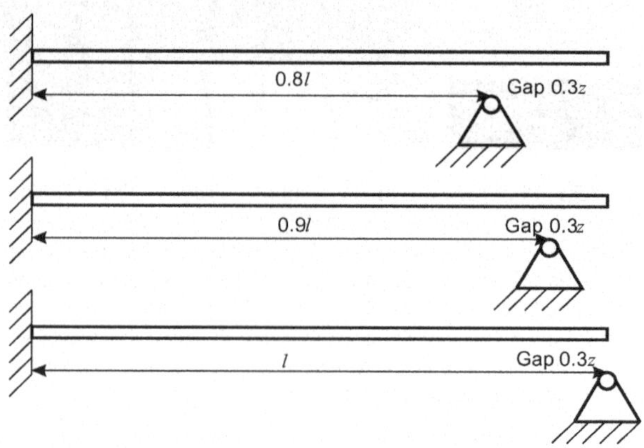

Fig. 4.55 Support location at points approximately coinciding with the nodes of the 2nd ($x/l \approx 0.8$), 3rd ($x/l \approx 0.9$) eigenmodes of transverse vibrations, and at the end of the cantilever ($x/l = 1$)

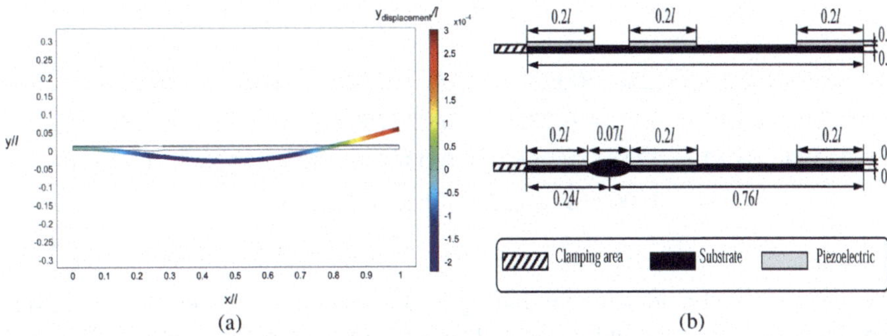

Fig. 4.56 FE model (**a**) and schematics of the fabricated piezoelectric harvester prototypes (**b**): OPT 0 (top), OPT RAT (bottom)

the optimal microcantilever OPT II$^-$ (Fig. 4.28) is replaced by a hump-like zone, length of which varies from $0.01l$ to $0.07l$, while sections of the structure outside the hump retain constant cross section (overall length and width of the microcantilever do not change with respect to the initial microcantilever of constant cross-section).

The response of randomly excited rational microcantilever OPT RAT was determined numerically and compared with cantilever OPT 0. Both OPT RAT and OPT 0 models were subjected to random base excitations, which were defined as vertically acting body load. In order to introduce a realistic random excitation signal to the FE model, transient acceleration signal (Fig. 4.57) was measured using single-axis piezoelectric accelerometer Metra KS-93 that was mounted on an operating milling machine. A registered signal was approximated using Matlab code, which resulted in several mathematical expressions of the excitation signal, qualitatively defined by the coefficient of determination R^2. The mathematical expression of the approximated random signal with the coefficient of determination of $R^2 = 0.7869$ is presented

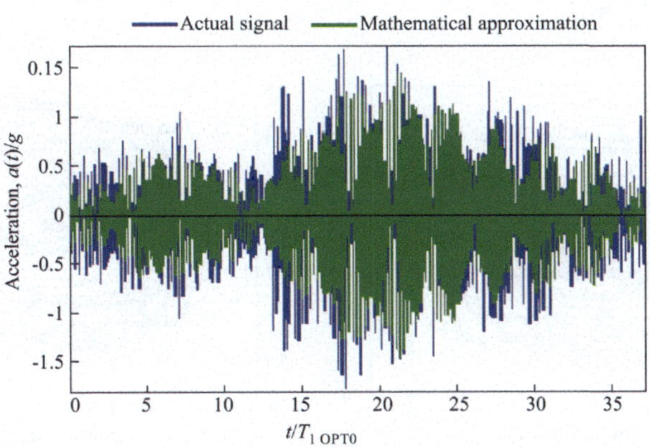

Fig. 4.57 Comparison or real (blue) and approximated (green) random excitation signal ($R^2 = 0.7869$)

below

$$a(t)/g = 0.01322 \cdot \cos(2\pi \cdot 300t - 1.6153) + 0.01164 \cdot \cos(2\pi \cdot 302t + 0.235) + \\ 0.02796 \cdot \cos(2\pi \cdot 320t - 0.2856) + 0.03476 \cdot \cos(2\pi \cdot 322t + 2.8952) + \\ 0.03383 \cdot \cos(2\pi \cdot 324t - 0.9567) + 0.0344 \cdot \cos(2\pi \cdot 3326t + 1.3189) + \\ 0.00898 \cdot \cos(2\pi \cdot 384t - 2.2426). \tag{4.22}$$

An approximated random excitation signal and its comparison to the actual excitation signal are presented in Fig. 4.57. This signal was introduced in the FE model as a vertically acting body load.

Modeling results of the response of OPT RAT and OPT 0 microcantilevers to random excitation are presented in Figs. 4.58 and 4.59. The tip of the rational

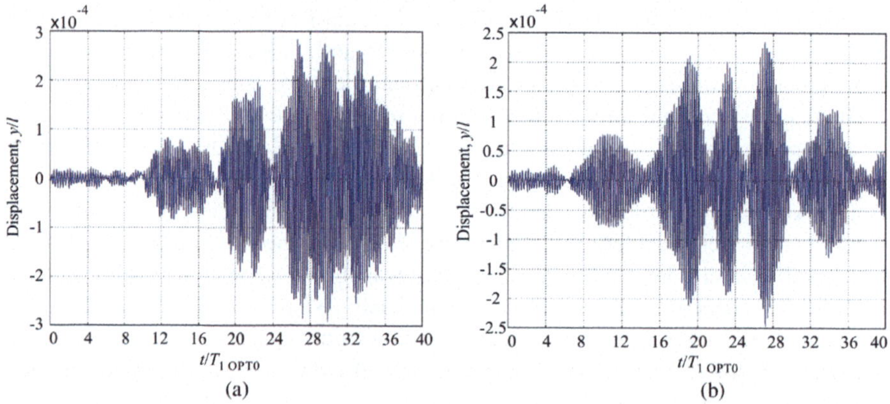

Fig. 4.58 Response of the tip of randomly excited **a** microcantilever OPT 0 and **b** rational microcantilever OPT RAT

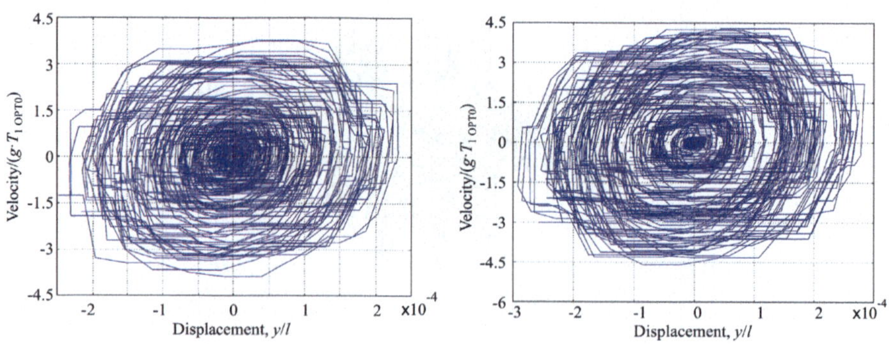

Fig. 4.59 Phase diagram of the tip of randomly excited **a** microcantilever OPT 0 and **b** rational microcantilever OPT RAT

cantilever OPT RAT vibrates with a higher amplitude in comparison to microcantilever OPT 0. It means that microcantilever OPT RAT falls quicker into the 2nd resonant vibration shape. In addition, the vibration amplitude of microcantilever OPT RAT remains constant for the time interval that is 5 times longer with respect to OPT 0 case.

Stability comparison of these two microcantilevers is very clearly represented using phase diagrams (Fig. 4.59). Curves of the phase diagram of microcantilever OPT 0 are mainly concentrated in the center, which means lower displacement and velocity as well as lower generated mechanical energy that could be converted into electrical energy.

In addition, simulations were performed with the freely-vibrating and vibro-impacting cantilevers (stopper located at $x/l = 1$). In the case of freely-vibrating cantilever, the results of the numerical modal analysis indicate that natural frequencies of OPT RAT and OPT 0 models are as follows: 1st—0.77 $\omega/\omega_{1\ OPT0}$, 2nd—4.52 $\omega/\omega_{1\ OPT0}$, 3rd—13.17 $\omega/\omega_{1\ OPT0}$ and 1st—1.16 $\omega/\omega_{1\ OPT0}$, 2nd—7.15 $\omega/\omega_{1\ OPT0}$, 3rd—19.9 $\omega/\omega_{1\ OPT0}$, respectively.

These microcantilevers were also tested experimentally. Geometric parameters of the OPT RAT model ($l = 1$ mm, width 0.1 mm) were used to fabricate a prototype of the piezoelectric harvester by employing the proposed rationally-shaped substrate made of nickel that was covered with three segments of screen-printed piezocomposite layer (Polyvinyl butyral 20% mixed with PZT nanopowder) with electrodes.

Microfabrication technology used for the realization of microcantilever OPT RAT is presented in Fig. 4.60. For cantilever formation, a sacrificial layer (oxide) was deposited onto a silicon wafer using the plasma enhanced chemical vapor deposition technique (Fig. 4.60a). The photoresist layer for the patterning of a sacrificial layer was deposited using a spin coating technique and exposed using Ultraviolet (UV)

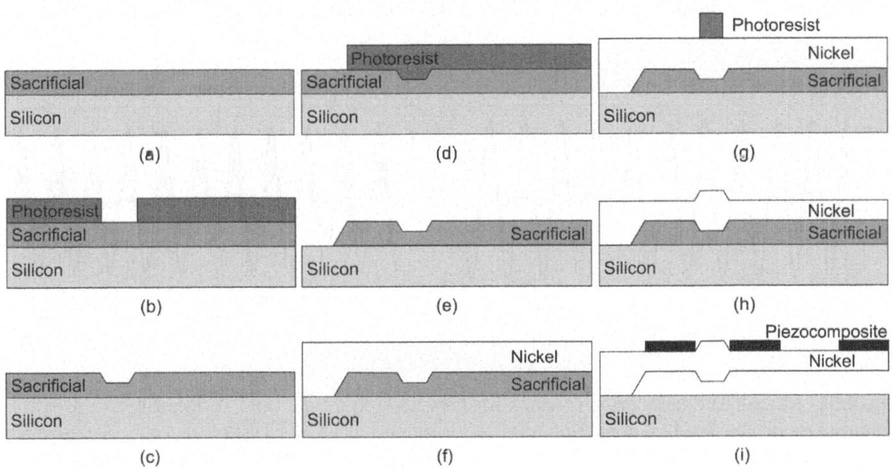

Fig. 4.60 Microfabrication of microcantilever OPT RAT

lithography (Fig. 4.60b). The exposed region was treated and used as a mask for the formation of the underlying sacrificial layer. A *V*-shaped groove was formed using isotropic reactive-ion etching and photoresist was removed (Fig. 4.60c). The new photoresist layer for the patterning of a sacrificial layer in the region of the cantilevers' base was deposited, exposed, and treated (Fig. 4.32d). The sacrificial layer was etched using isotropic reactive-ion etching (Fig. 4.32e). Subsequently, a thick nickel layer was deposited (Fig. 4.60f) and coated by a photoresist (Fig. 4.60g). The photoresist layer was exposed using UV lithography and treated. The top surface of the cantilever was formed using isotropic reactive-ion etching (Fig. 4.60h). Lastly, three segments of piezocomposite were screen-printed and the temporary sacrificial layer was removed (Fig. 4.60i) and the nickel cantilever OPT RAT was free to move.

Eigen frequencies of transverse vibrations of the fabricated rationally-shaped vibration energy harvester were measured to be (freely-vibrating mode): 1st—0.64 $\omega/\omega_{1\ OPT0}$, 2nd—3.61 $\omega/\omega_{1\ OPT0}$, 3rd—9.95 $\omega/\omega_{1\ OPT0}$. Discrepancy between the measured and simulated frequency values is attributed to non-ideal clamping of the fabricated device, which contributes to the decrease of the measured natural frequencies. A vibration energy harvester prototype OPT 0 was also fabricated following the same procedure but using nickel microcantilever of constant cross section as a substrate. Both vibration energy harvester prototypes were subjected to electrical characterization in order to compare their energy harvesting performance. Generated open-circuit voltages were measured for two cases: (i) freely-vibrating cantilever, (ii) vibro-impacting cantilever with stopper located at the free end ($x/l = 1$). Figure 4.61 presents plots of voltage signals collected by three piezocomposite segments of OPT 0 prototype, which is subjected to harmonic excitation at its 1st natural frequency when the microcantilever of the constant cross section is vibrating without impacts (Fig. 4.33a) and with impacts (Fig. 4.61b). Meanwhile, Fig. 4.62 provides analogous plots for the case of OPT RAT prototype (excitation signal frequency–0.64 $\omega/\omega_{1\ OPT0}$). Comparison of voltage responses generated by both prototypes operating in vibro-impact mode (Figs. 4.61b and 4.62b) indicates markedly larger content of

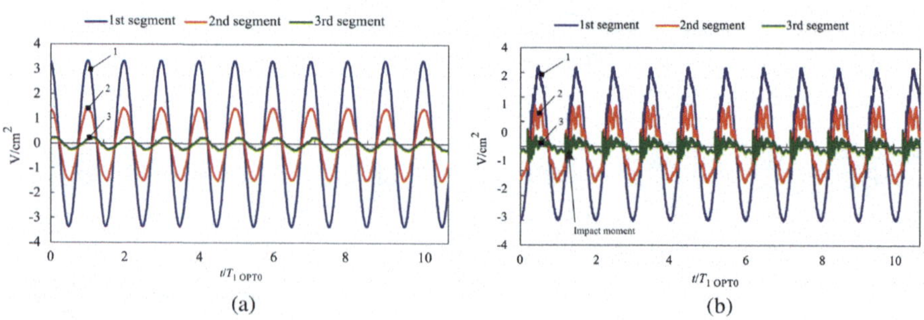

Fig. 4.61 Measured open-circuit voltages for harvester prototype OPT 0 operating in: **a** freely-vibrating mode and **b** vibro-impacting mode with support at $x/l = 1$ (1st piezocomposite segment (near the clamped part of the microcantilever)—blue line (1), 2nd segment (in the middle)—red line (2), 3rd (at the free end)—green line (3))

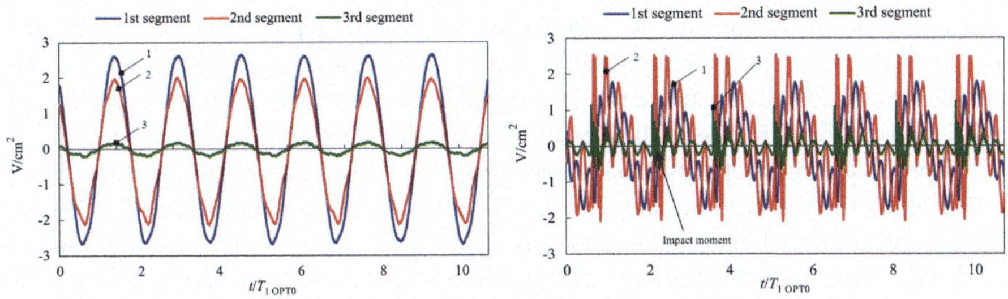

(a) (b)

Fig. 4.62 Measured open-circuit voltages for vibration energy harvester prototype OPT RAT operating in: **a** freely- vibrating mode, **b** vibro-impacting mode with a stopper at $x/l = 1$ (1st piezocomposite segment (near the clamped part of the microcantilever)—blue line (1), 2nd segment (in the middle)—red line (2), 3rd (at the free end)—green line (3))

higher-order harmonics in the case of OPT RAT response, though its vibration amplitudes are comparable or larger with respect to the OPT0 case. It demonstrates that OPT RAT prototype undergoes self-excitation at higher vibration modes with the dominant 2nd mode of transverse vibrations. Thus, base excitation of the proposed rationally-shaped vibro-impacting energy harvester at low 1st eigenfrequency (0.64 $\omega/\omega_{1\ OPT0}$) leads to self-excitation of the vigorous vibrations at much higher frequencies (>3.6 $\omega/\omega_{1\ OPT0}$), which translates into higher deflection velocities and strain rates. As a result, the observed amplification of the higher-order mode response in OPT RAT device would lead to higher power output since it is strongly dependent on the strain rate in the piezoelectric material. In addition, it would also improve the efficiency of mechanical-to-electrical energy conversion, which can be explained by means of the following relationships:

$$\eta = \frac{\zeta_e}{\zeta_T} = \frac{\zeta_e}{\zeta_m + \zeta_e} \tag{4.23}$$

and

$$\zeta_e = \frac{(C_p \omega_n R_L) k_e^2}{\left(C_p \omega_n R_L \left(\omega_{OP}/\omega_n\right) + \pi/2\right)^2} \tag{4.24}$$

where ω_n and ω_{op} are the eigen and operational (excitation) frequencies, respectively; C_p and k_e^2 are the capacitance and alternative electromechanical coupling coefficient of the piezoelectric transducer, respectively; ξ_T, ξ_m, ξ_e are total, mechanical, and electrical damping ratios, respectively.

Equations (4.23) and (4.24) indicate that increase in vibration frequency of the piezoelectric transducer leads to stronger electrical damping (higher ξ_e), which, in turn, improves the efficiency. It means that with higher vibration frequencies, more mechanical energy is removed from the vibration energy harvester during the energy harvesting process. Higher vibration frequencies of the piezoelectric transducer result in lower matched load resistance ($R_{ML} = 1/\omega_n C_p$), which enhances the average power generated by vibration energy harvester ($P_{av} = (V_{rms})^2/R_{ML}$).

The proposed rationally-shaped vibro-impacting vibration energy harvester would be able to not only accommodate wide variations in excitation magnitude, but would also deliver improved energy harvesting performance owing to the amplified higher-order mode responses, resulting in improved energy conversion efficiency due to increased electrical damping and higher power output due to larger response velocities of the piezoelectric transducer. The adopted design approach is referred to here as rational since the near-optimal shape of the piezoelectric transducer is derived on the basis of optimal cantilever structures. In other words, the rationally-designed microcantilever effectively reproduces modal behavior that is characteristic to the optimal cantilever structures, i.e. structures with the distribution of cross-sectional areas, which was determined through dynamic shape optimization. The proposed nonlinear vibration energy harvester constitutes an adaptive micro-power generator, which could provide enhanced energy harvesting performance under varying real-life excitation conditions. Due to rational design modifications, the device is amenable to conventional macro-scale processing methods and advanced microfabrication methods, thereby representing a cost-effective alternative to energy harvesters that are based on complex-shaped optimal structures.

References

1. Zizys G, Gaidys R, Dauksevicius R, Ostasevicius V (2013) Segmentation of piezoelectric layers based on the numerical study of normal distributions in bimorph cantilevers vibrating in the second transverse mode. Mechanika 4:451–458
2. Zizys D, Gaidys R, Ostasevicius V (2016) Electric power output maximization for piezoelectric energy harvester by optimizing resistive load. In: Mechanika 2016: Proceedings of 21st International Science Conference, pp 314–316
3. Zizys D, Gaidys R, Ostasevicius V, Narijauskaite B (2017) Vibro-shock dynamics analysis of a tandem low frequency resonator-high frequency piezoelectric energy harvester. Sensors 17(5):1–21
4. Milasauskaitė I, Dauksevicius R, Ostasevicius V, Gaidys R, Janusas G (2014) Influence of contact point location on dynamical and electrical responses of impact-type vibration energy harvester based on piezoelectric transduction. Zeitsch Angew Math Mech = J App Math Mech Weinheim: Wiley-VCH Verlag 94(11):898–903
5. IEEE Ultrasonics, Ferroelectrics, and Frequency Control Society, 176-1987—IEEE Standard on Piezoelectricity (1988)
6. Catalogue of Ferroperm Piezoceramics A/S. https://www.ferroprm-piezo.com
7. Erturk A, Tarazaga PA, Farmer JR, Inman DJ (2009) Effect of strain nodes and electrode configuration on piezoelectric energy harvesting form cantilevered beams. Vibr Acous 11:1–11
8. Kim S, Clark WW, Wang QM (2005) Piezoelectric energy harvesting with a clamped circular plate: experimental study. Intel Mat Sys Str 16:847–854
9. Ostasevicius V, Janusas G, Milasauskaitė I, Zilys M, Kizauskiene L (2015) Peculiarities of the third natural frequency vibrations of a cantilever for the improvement of energy harvesting. Sensors 15(6):12594–12612

10. Ostasevicius V et al (1986) Device for semi-natural modeling of vibroimpact mechanical systems. SU Invention Nr 1251116:5
11. Ostasevicius V, Gitis N, Palevicius A, Ragulskis MK (2005) Tamulevicius S (2006) Hybrid experimental-numerical full-field displacement evaluation for characterization of micro-scale components of mechatronic systems. Sol St Phen: Mech Syst Mater MSM 113:73–78
12. Dauksevicius R, Milasauskaite I, Ostasevicius V, Jurenas V, Mikuckyte S (2012) Experimental study of coupled dynamic and electric characteristics of piezoelectric energy harvester under variable resistive load. J Vibroeng 14(3):1435–1443
13. Dauksevicius R, Milasauskaite I, Ostasevicius V, Gaidys R (2012) Investigation of piezoelectric bending actuator for application in kinetic energy harvesting. In: Mechanika 2012: Proceedings of 17th International Conference, pp 48–51
14. Lesieutre GA, Davis L (1997) Can a coupling coefficient of a piezoelectric device be higher than those of its active material? J Smart Str Mat: Smart Str Integ Sys 3041:281–292
15. Zizys D, Gaidys R, Dauksevicius R, Ostasevicius V, Daniulaitis V (2016) Segmentation of a vibro-shock cantilever-type piezoelectric energy harvester operating in higher transverse vibration modes. Sensors 16(1):14
16. Resato, Waterjet technology, brochure. http://www.resato.com/waterjet/fileadmin/resato/Waterjet-cutting/Brochures_PDF/Waterjet_UK.pdf
17. Janusas G, Milasauskaite I, Ostasevicius V, Dauksevicius R (2014) Efficiency improvement of energy harvester at higher frequencies. J Vibroeng 16(3):1326–1333
18. Dauksevicius R, Kulvietis G, Ostasevicius V, Milasauskaite I (2010) Finite element analysis of piezoelectric microgenerator towards optimal configuration. Proc Eng Eurosens 5(4):1312–1315
19. Ostasevicius V, Gaidys R, Dauksevicius R, Milasauskaite I (2016) Piezoelectric generator of high frequency vibrations for converting into electricity. Patent LT 6329 B: 6
20. Ostasevicius V, Gaidys R, Dauksevicius R (2016) Piezoelectric generator converting low frequency vibration into electrical energy. Patent LT 6330 B: 6
21. Ostasevicius V, Dauksevicius R, Gaidys R, Palevicius A (2007) Numerical analysis of fluid-structure interaction effects on vibrations of cantilever microstructure. J Sound Vibr 308(3–5):660–673
22. Migliniene I, Ostasevicius V, Gaidys R, Dauksevicius R, Janusas G, Jurenas V, Krasauskas P (2017) Rational design approach for enhancing higher-mode response of a micro cantilever in vibro-impacting mode. Sensors MDPI ag 17(12):1–15

Chapter 5
Digital Twin-Driven Technological Process Monitoring for Edge Computing and Cloud Manufacturing Applications

5.1 Edge Computing-Enabled Wireless Vibration Sensor Node

Edge computing is one part of a larger IoT system designed to address latency issues and barriers arising from existing cloud architecture. IoT edge computing enables on-site data processing, allowing continuous monitoring, analysis, and connection. By eliminating the need to reach the cloud to gain time-sensitive insights and make real-time decisions when needed, edge computing reduces the distance between the user and the server. When using edge computing, the goal is to make the data available in the shortest amount of time. This means that companies define the benefits themselves, taking into account the use cases, goals, and problems they are trying to solve. IoT edge computing in a production environment may be limited to one building.

5.1.1 Use Case of the Non-rotating Tool

With the rapid increase in the number of wireless sensor nodes and their networks, it is necessary to develop continuously powered systems in order to avoid the most expensive maintenance issue—periodic battery replacement or charging. This requires the use of the energy of the vibrations generated during cutting, which is transformed into electricity. A force acting on a non-rotating tool between the tool surface and the chip causes the tool to vibrate. In order to make efficient use of tool vibrations for energy harvesting, the eigenfrequency of the piezoelectric transducers should be close to the eigenfrequency of the tool, which allows the piezoelectric transducer to be excited in resonant mode and to harvest maximum energy.

V. Ostaševičius, *Digital Twins in Manufacturing*, Springer Series in Advanced Manufacturing, https://doi.org/10.1007/978-3-030-98275-1_5

5.1.1.1 Energy Harvesting from Turning Tool Vibration

Machine tool chatter is a self-excited vibration problem occurring in large rates of material removal, resulting from the unavoidable flexibility between the cutting tool and workpiece. It is important to mention the turning process and turning tool, which is a flexible structure vibrating during cutting process at several eigenmodes [1]. The source of vibration or chatter is related with cutting force action dependent on the regimes of manufacturing as well as on the tool wear. This means that it is impossible to stop vibrations of the turning tool structure. That's why it could be useful for the energy generation. Exploiting the vibrations of machine units is one of the effective ways, allowing harnessing ambient energy for autonomous systems powering at the point of placement without the power supply cable or batteries. Integrating a harvesting system into the cutting tool structure (Fig. 5.1a), the electrical energy can be generated from mechanical vibrations. The accelerometer KD91 was used for cutting tool vibration measurements. Some world-wide producers (Morgan Inc., Noliac A/S, Piezo system Jena GmbH, etc.) offer piezo-electric cantilevers for transmitting mechanical energy into electrical energy. Under the vibration or a change in motion (acceleration), the piezoelectric material "squeezes" and produces an electrical charge. Furthermore, the maximum amount of energy could be produced if industrial processing machine tools and piezo-electric energy generator's resonant frequencies coincided. According to the research work [2], considering several modes of elastic turning tool vibrations, it was identified that the turning tool vibrates on the main mode, depending on the clamping, i.e. the length of the tool hanging from the claim, in a range from 2.5 kHz to 5.6 kHz. It means that commercial piezoelectric cantilevers are not available for energy harvesting because of the low resonant frequency. For this purpose, the circular piezo transducer bimorph assuring the resonant frequency in the required diapason (2.5…5.6 kHz) was chosen. Turning tool impact test results presented in Fig. 5.1b, when tool shank length is L = 55 mm, confirm that turning tool vibration period presented in time domain of the tool shank is ~2,2 ms, which corresponds to 4.5 kHz.

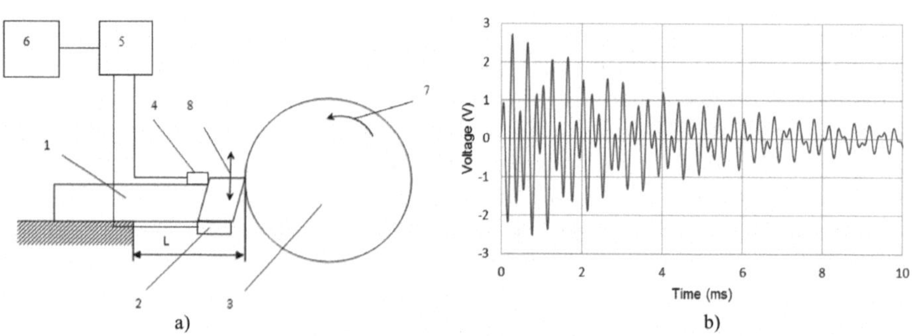

Fig. 5.1 Turning experiment set-up (**a**): 1-turning tool, 2-energy harvester element, 3-workpiece, 4-KD91accelerometer, 5-Pico Scope 3424, 6-PC, 7-cutting speed direction, 8-excited turning tool vibration and impact test, when tool shank length is L = 55 mm (**b**)

During the cutting process, the piezoelectric harvester fixed on the tool tip removes energy in the form of electricity. During the research, the disk-shaped piezo transducer bimorph (model AB1541, dimensions ø15.0 × 0.15 mm, resonant frequency ~4 kHz.) was used for mechanical to electrical energy conversion. Under uniform pressure q_0, the disk deflection W can be expressed as follows:

$$W = \frac{q_0}{64D}\left(R^2 - r^2\right)^2 \tag{5.1}$$

where R—disk radius; $D = Eh^3/12\left(1 - \mu^2\right)$—flexural rigidity; E—Young's modulus; h—thickness of the plate; μ—Poisson's ratio.

Next, the relation between strain and displacement of the circular plate with axisymmetric vibrations has been calculated in such a manner:

$$\varepsilon_r = -zd^2/dr^2 \tag{5.2}$$

where ε_r—radial strain, $w(r)$—out-of-plane transverse deflection of the middle surface.

$$\varepsilon_r = \frac{q_0}{64D}\left(-4z\left(3r^2 - R^2\right)\right). \tag{5.3}$$

Equations (5.4) and (5.5) give the expression of the strain energy U of the plate caused by bending the axisymmetric circular plate with an initial deflection:

$$U = \frac{q_0}{64D}\int_0^R \left(\frac{\partial^2 w}{\partial r^2}\right)^2 rdr \tag{5.4}$$

or

$$U = \pi \frac{q_0}{64D} 8R^6. \tag{5.5}$$

The first natural frequency of the circular plate with clamped edge is as follows:

$$\omega = \lambda^2 \sqrt{\frac{D}{\rho h}} \tag{5.6}$$

where $\lambda^2 = 10.215$ andρ—mass density.

Figure 5.2 shows that the dependence of the amount of the generated energy in the disk plane on the width of the disk is linear, while the dependence of the generated energy on the disk radius is non-linear. In both cases, the biggest amount of energy is generated by the disk with larger geometrical dimensions.

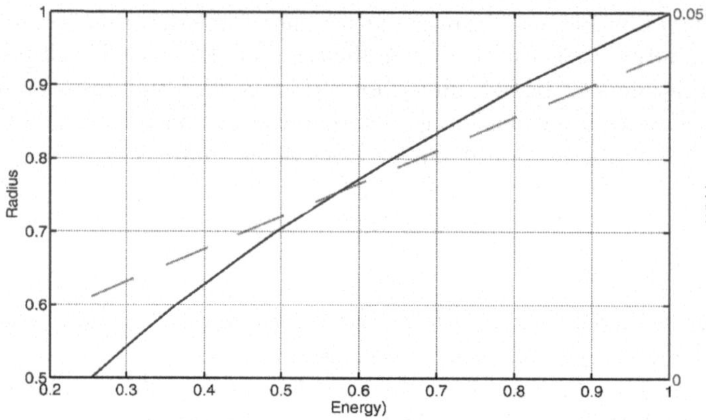

Fig. 5.2 The dependence of the amount of generated energy on the geometric parameters of the disk-shaped piezo harvester: radius (continuous line) and width (dashed line), under the fixed natural frequency

Commercial Noliac CMBP piezoelectric transducers are not manufactured for a non-rotating tool because of very low resonant frequency. Therefore, circular piezo-transducer bimorphs that are made of a piezoelement and a bronze disk and have the necessary resonance frequency diapason were used. A mathematical model was created to analyze the possibility to apply frequency tuning to the piezoelectric transducer by using Comsol multiphysics as a FE tool [3]. Hence, as Fig. 5.3a shows, the model of the piezoelectric transducer bimorphs is composed of a brass disc (diameter—15 mm, width—0.1 mm) and a piezoelectric disk (diameter—10 mm, width—0.1 mm). During the simulation, the fixture was made like a spring foundation having damping losses and the electrical borders like 10 m ohm load of resistor. Then, the model was linked with loose tetrahedral elements of brass base and loose triangular swept mesh of piezoelectric part. It is really complicated to use the capability of tuning the natural frequency, but in order to attain this effect, the suggested method was the relocation of the circumferential fixture with the one having a smaller diameter. Thus, the fixture was made into a ring-shaped spring foundation of 0.5 mm width on a foundation of the transducer. Mode frequency analysis (Fig. 5.3b) reveals that the resonant frequency initially increases because of the fixture radius reduction, but then the frequency begins to drop approximately at 0.7 r/R of size ratio. If the excitation frequency is defined earlier, it is possible to choose thoroughly the geometry and dimensions of a conventional linear harvester that fit to its excitation frequency and resonant frequency. The application of the circular piezoelectric bimorph, including the option of approaching the frequency of piezoelectric bimorph to the frequency of the cutting tool, helped to develop a self-powered wireless node of the sensor for the cutting tool. The developed advanced wireless sensor architecture with separate ultralow-power voltage detector and elaborated algorithm could reach only 100 $\div$ 150 nA triggering current after capacitor charged. State-of-art nano-watt triggering power consumption allows accumulate energy from very low

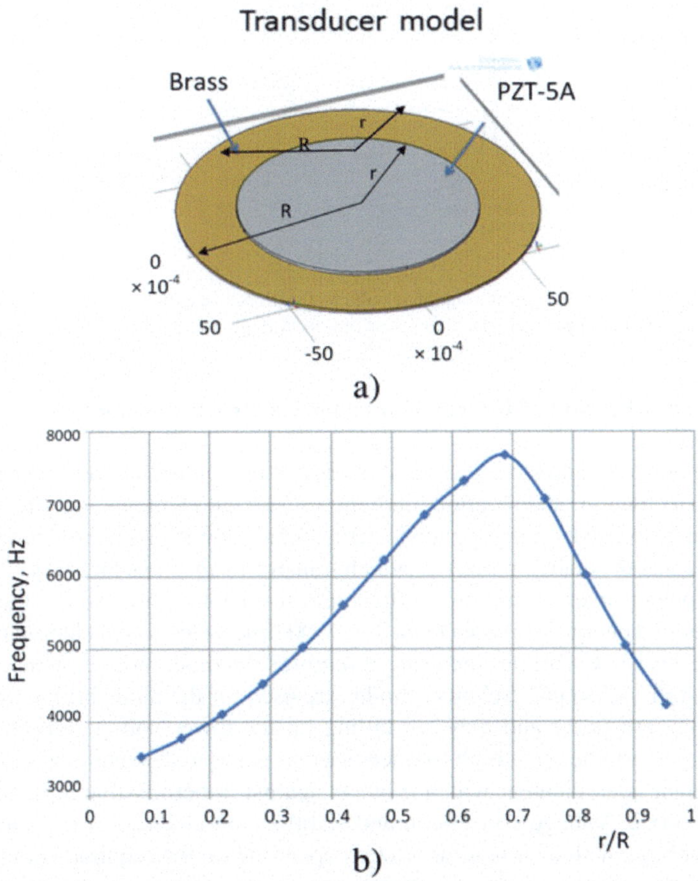

Fig. 5.3 Diagram of the circular piezo transducer bimorph (**a**) and a chart of dependence of natural frequency on fixture size ratio r/R (**b**)

energy sources, wake up embedded microsystem, and transfer sensor status without additional energy source.

The voltage-frequency characteristics for five circular piezo transducers were received vibrating on the electrodynamical stand (Fig. 5.4a). It can be seen that there is a significant frequency distribution of the energy harvester resonance in the 3.9 ÷ 4.5 kHz range.

For efficient energy harvesting, the first resonance of the turning tool should be close to the resonant frequency of the piezo generator. The FE model of circular piezoelectric transducer was developed with Comsol multiphysics software which confirms that ultimate voltage is harvested at resonant frequency of 4.2 kHz (Fig. 5.4b).

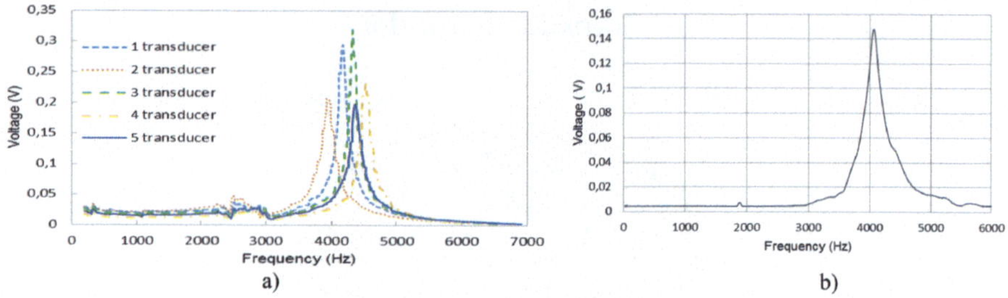

Fig. 5.4 **a** Piezo-electric harvester generated voltage-frequency response under acceleration of 1 **g** and **b** ultimate simulated piezoelectric harvester output voltage at 4.2 kHz resonant frequency

5.1.1.2 Identification of Energy Harvester Electrical Parameters

In order to generate and store electrical energy, it is required to have energy generation, transformation, stabilization, and energy storage elements together with the elements from excessive surge protection (Fig. 5.5a). Power inverter or energy generating element is the energy converter which transforms mechanical energy into electrical. All power inverters are most efficient at resonant regime, where the amount of the generated energy increases from 2 $\div$ 100 times. Generated electric power by power inverter is AC current; therefore it requires the diode bridge, which changes the power from AC to DC. Schottky diodes are used for the diode bridge for several reasons: they are faster and they are falling on the lower voltage (approximately 0.3 V) and perform better at high frequencies compared to conventional diodes. The guard element is an element which protects against impermissible high voltage. It is a Zener diode. Voltage conversion and stabilization electronics is a chip which increases voltage with a voltage stabilizer depending on the required stabilized DC voltage which is used to power the electronics and power inverter generated voltage amplitude. Energy storage element denotes an energy storage capacitor as well as powered electronics denotes the microcontroller.

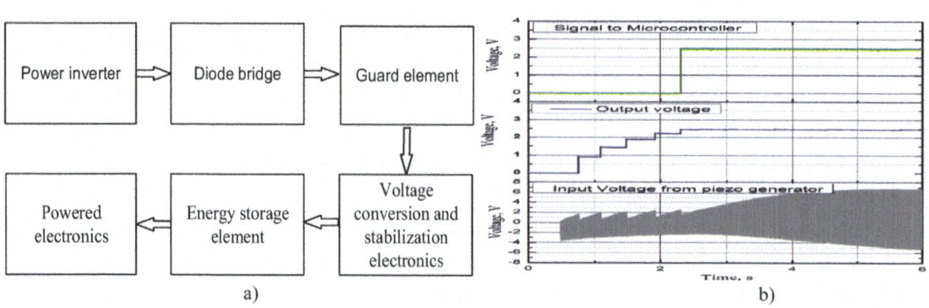

Fig. 5.5 **a** Components of energy recovery and storage device and **b** power transducer sub-system signals charging 100 μF capacitor

The subsystem of the electrical energy conversion was realized to convert electrical energy, generated by energy transducer, to microcontroller voltage level. The primary transducer the piezo generator converts mechanical vibrations' energy into electrical energy, which is variable and must be of sufficiently large amplitude. The created subsystem is a secondary transducer, which converts the variable voltage into direct constant voltage, which can be modified. This transducer is designed using Linear Technology LTC3588-1 microchip, which integrates diode bridge, Zener diode, and AC/DC voltage decreasing converters.

Power conversion subsystem performance is shown with removed signals, charging 100 μF capacitor (Fig. 5.5b). When output voltage reaches the target voltage of 2.5 V, the output induces that stabilized voltage does not go out of the constrained boundaries. This signal is fed to the measuring electronics. As it is seen from Fig. 5.5b, the output voltage reaches 6 V without load and 2.5 V with load. Experimental results have showed instability of high amplitude tool vibrations. In order to increase the efficiency of energy collection from piezo transducer, the system was improved and Schottky diode bridge was replaced with input voltage quadrupler.

5.1.1.3 Wireless Sensor Node Architecture

In order to increase the efficiency of energy collection from piezo transducer, the energy harvesting system was improved: the guard Zener diode was changed to Mosfet in order to minimize the reverse current. Possible wireless sensor node architectures, embedded into smart tool, are presented in Fig. 5.6.

As wireless sensor nodes operate on a tough power budget, Ultra-Low Power (ULP) Microcontroller Units (MCU) are required for processing and power management. A typical MCU, like the Texas instruments MCU-MSP430, is ideal for energy harvesting for its characteristics: it has the standby current of less than 1 μA and the active current of 160 μA/MHz, a quick wakeup time of less than 1 μs, the temperature sensor inside, besides, it operates at the range of 1.8 V to 3.6 V. After the

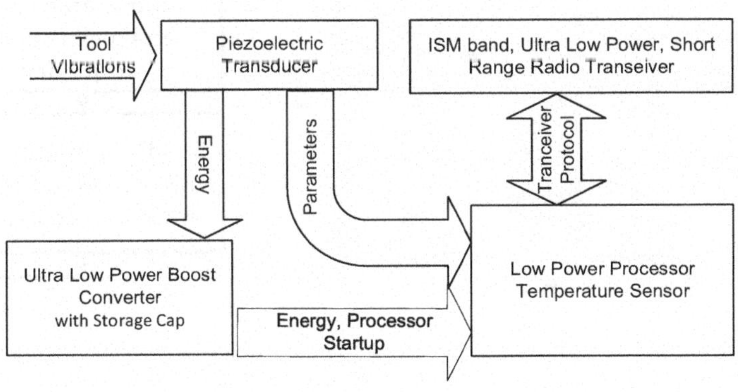

Fig. 5.6 Wireless sensor node design of smart tool-sensor with separate low power boost converter

experimental tests, the developed sensor has showed that at the beginning of the charging process of the sensor subsystem (energy converter or processor and radio transmitter) it goes to an uncertain state—power consumption strongly exceeds the current generated by the piezo transducer. This stops the charging process, and the sensor is not able to switch to the measuring mode.

The wireless sensor node architecture (Fig. 5.7a), proposed advanced sensor architecture with separate ULP charge detection (Fig. 5.7b), enabled reducing the energy consumption during MCU start-up process. The system only uses the energy for input voltage until the capacitor is fully charged. The algorithm of WSN is presented in Fig. 5.7c. The MCU inside temperature sensor for tool temperature monitoring is used.

Eight different devices of charge detection were tested and experimental results are presented in Fig. 5.8 Extremely low energy consumption, more than 100 $\div$

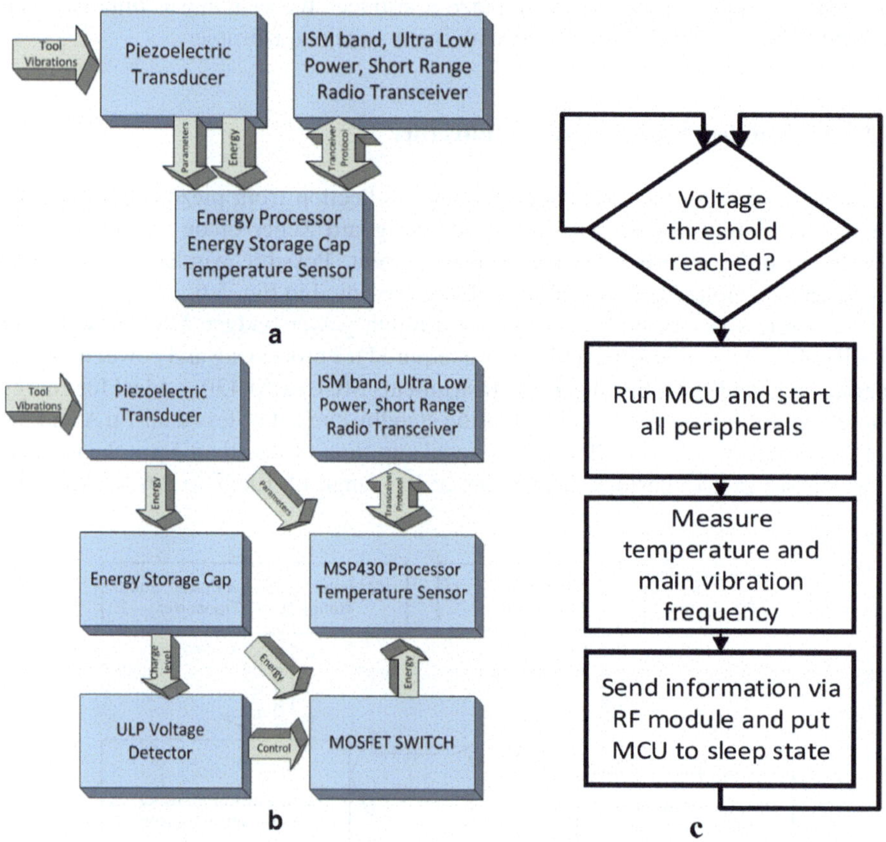

Fig. 5.7 Wireless sensor node architecture (**a**), advanced wireless sensor architecture with separated ultra-low power voltage detector (**b**) and algorithm of created wireless sensor node (**c**)

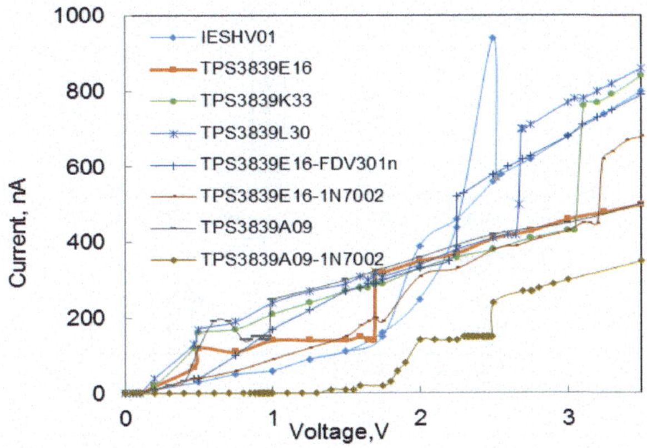

Fig. 5.8 Wireless sensor node energy consumption using eight different modes of charge detection

150 nA current consumption could be reached only with IESV01 and TPS3839A09 1N7002 of charging detectors during capacitor charging. This method achieves the accumulation of the required energy even at low level of tool vibrations.

The proposed sensor schematics with separate ULP charge detection is shown in Fig. 5.9.

The proposed architecture for wireless sensors' nodes is presented in Fig. 5.10.

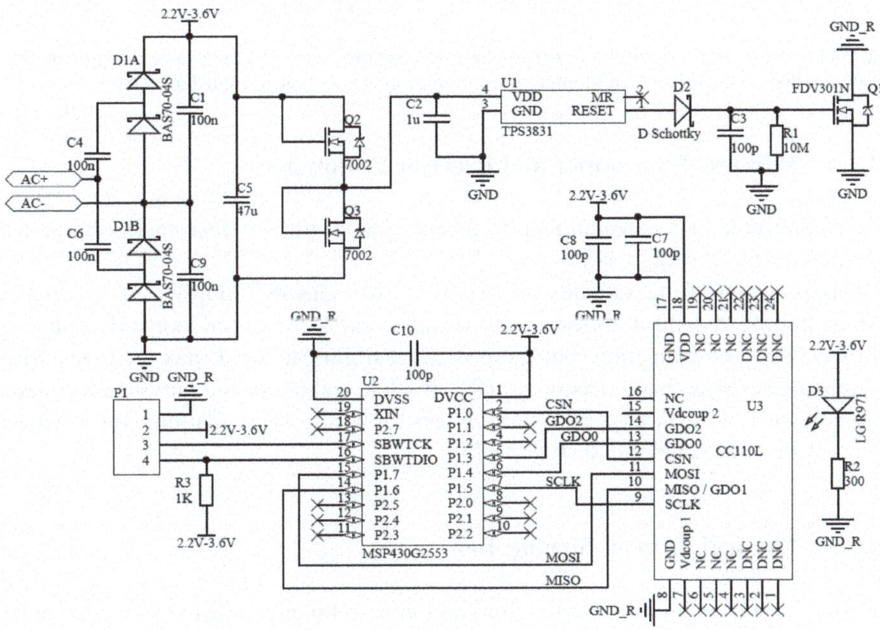

Fig. 5.9 Schematics of the created wireless sensor node

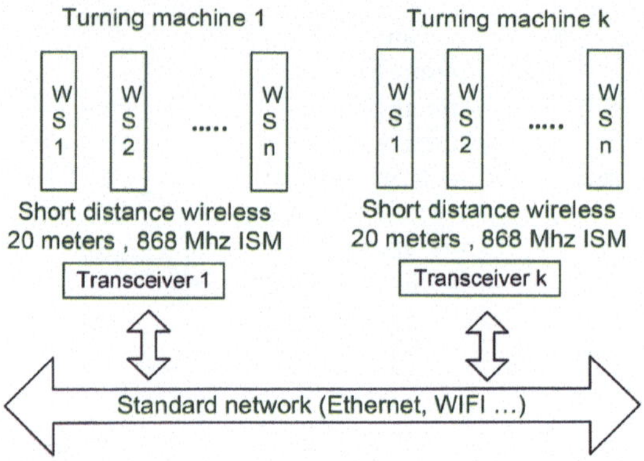

Fig. 5.10 Wireless sensors' nodes' network architecture

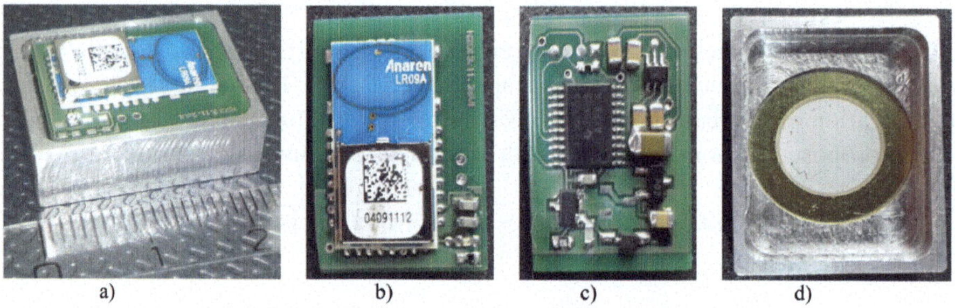

Fig. 5.11 A prototype of wireless energy harvester: general view (**a**), subsystem of information wireless transfer (**b**), subsystem of energy accumulation (**c**), and piezo transducer (**d**)

5.1.1.4 Wireless Transmitter and Receiver Prototypes

According to the experimental results, a prototype of the wireless energy harvester was manufactured (Fig. 5.11).

This prototype fully satisfies energy needs for sensors' supply and is capable of transmitting the information at the distance of 20 m for indoor environment. 10 bytes are sent each time, but there is a possibility to send max 64 bytes with accumulated energy. Radio frequency (RF) module wakes up, and initialization takes approximately 3 ms. Two types of transceivers for wirelessly transmitted information from sensors were elaborated (Fig. 5.12).

5.1.1.5 Physical Twin of Turning Tool

Installation of a monitoring system into an industrial manufacturing process can be necessary for various reasons. Due to redundant wear of cutting tools, distortions of dimensions of the produced components arise and this often increases scrapped

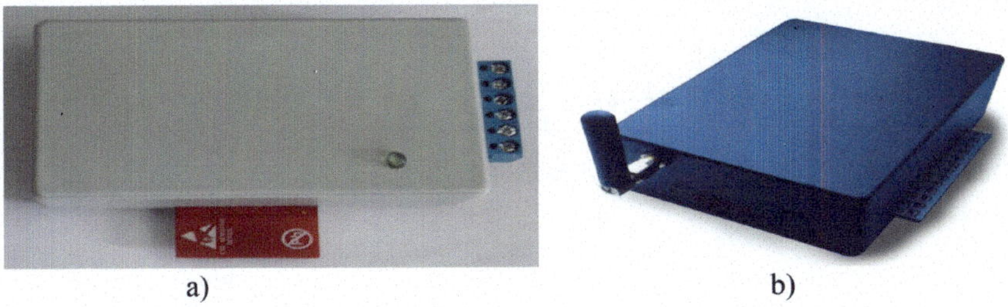

Fig. 5.12 Wireless sensor node transceiver prototypes: surface mount antenna (**a**) and separate antenna (**b**)

levels, sustaining extra expenses. Signal components of the parameters of the cutting process which can be treated as suitable to detect and monitor tool-wear can be recognized using the data which was derived from a number of experimental test cuts. The analyses of the signal clearly show correlation between the chosen signals and the cutting tool wear. It is possible to detect the changes of sensor signals in the vibration signatures in the course of the machining operation process.

Cutting vibration energy harvester concept was demonstrated for steel turning operation on CNC lathe Rayo 165. The depth of cut was 0.5 mm. The experimental energy harvesting set-up is shown in Fig. 5.13. The created wireless sensor node is mounted on the CNC turning tool. Anaren RF module LR09A and the transceiver with surface-mount antenna were used in wireless sensor node. During the turning operation, the piezo harvester is excited by the high frequency vibrations. The harvested energy is accumulated by 100 μF capacitor.

The created wireless sensor node was mounted on the CNC turning tool 2, and the harvested energy was accumulated by a 100 μF capacitor. The aim of the experiment was to reveal the influence of the tool wear on the tool vibrations amplitudes. A

Fig. 5.13 General view of cutting energy harvesting set-up: 1-personal computer, 2-wireless receiver, 3-CNC machine tool RAYO 165

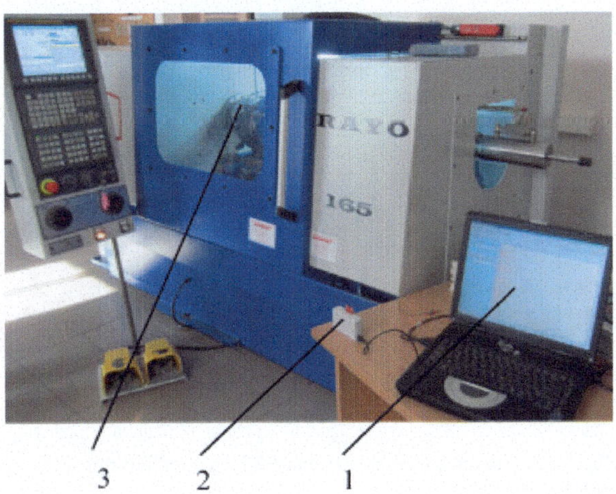

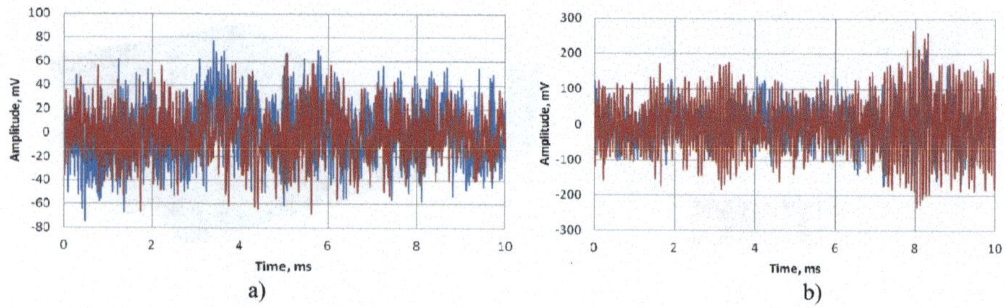

Fig. 5.14 Turning tool vibrations cutting steel with the new **a** and the worn **b** tools (blue color—axial vibrations, red color—transversal vibrations)

correlation was detected between the vibration acceleration amplitude and the tool wear. The analysis of vibration signature in the course of the cutting process is proved as useful method for the detection of the tool wear. Thus, the analysis of the measurements revealed (Fig. 5.14) the possibility to use the analysis of vibration acceleration amplitudes directly for the assessment of the tool wear. When the cutting time increases, the signal of acceleration differs, but this was not visible on the plots of time domain because bigger frequency components hid the smaller ones, but the full spectra could be observed in the frequency domain. The time domain also clearly demonstrated that the signals reacted in a clearly accidental way and showed dynamic properties. The results of the measurements show that vibration amplitudes of the worn-out tool increased 3 times in axial and 5 times in transversal directions.

To assess the vibration signatures and the signals of dynamic cutting force, the measured signals had to be transformed from time to frequency domain using Fast Fourier Transform (FFT). Figure 5.15 shows the turning tool vibration FFT and frequency distribution cutting steel with the new tool.

This method enabled to reveal the properties or specific information of the cutting process covered up into the time domain and to start the analyzed and produced

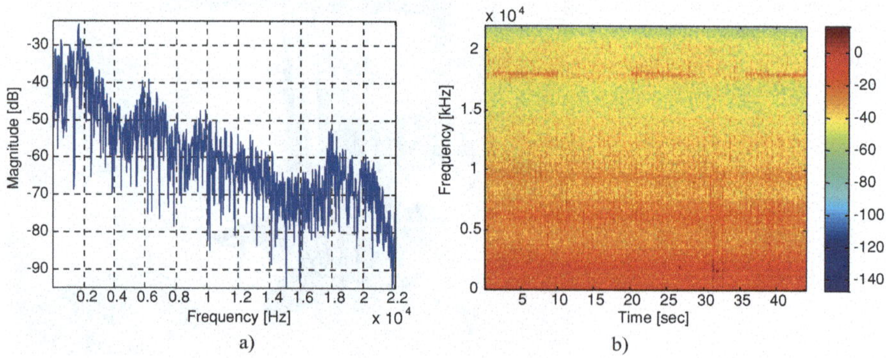

Fig. 5.15 Turning tool vibrations FFT (**a**) and vibration frequency distribution (**b**) cutting steel with the new tool

spectra with properties characterizing the tool at the initial contact (i.e. with a new and sharp tool). Then it was finished by defining the spectra of properties of an apparently worn tool (or the random failure/chipping modes). The results of the spectra enabled to detect a distinctive difference of signal characteristics once the cutting process was carried out using a new insert in comparison to the other one done with a damaging degree of wear. The specific amplitude growth of the principal resonance frequency peak of the intense force signal was referred to the periodic stick–slip effect within the contact area. The physics of this effect can be explained by the sharp edges of a new tool that reduced the contact between the tool and the workpiece. The growing degree of wear caused a raise of the contact area as a result of deteriorated cutting edges. High signal pulses that are developed by the friction transition from steady to floating influence the growth of the principal resonant peak magnitude, and the emergence as secondary peaks at the band region of about 10 kHz. Having reached a particular wear value of the flank, the principal frequency peak becomes damped when the contact area evenly expands, partially because of the growth of contact friction, its plastic deformation and the growth of the further nose wear. Thus, it is possible to demonstrate how the development of the even nose wear leads to a drop in the resonant peak magnitude. The control of the shaped chips showed that the high burst was caused by a growth of the lamination frequency (from 0.1 ÷ 10 kHz approximately). The growth of the lamination frequency was caused by the alteration of the material deformation behavior on the shear zone. It is possible that the decrease in the peak amplitude of resonant frequency due to a strong wear of the tool is caused by the growth of the rake angle from the friction effect of the rough chip.

5.1.1.6 Tool Wear Influence on Energy Harvesting

The purpose of this experiment is to show the influence of the tool wear on the frequency of wireless transmission of the information to the transceiver (Fig. 5.16).

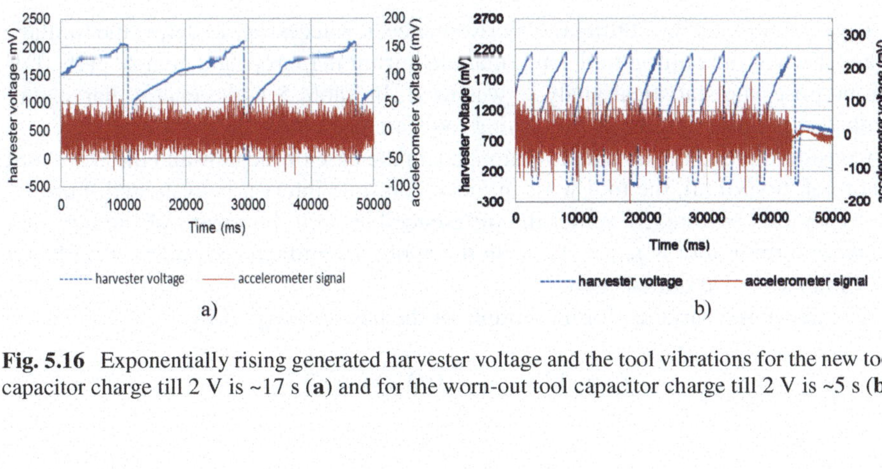

Fig. 5.16 Exponentially rising generated harvester voltage and the tool vibrations for the new tool capacitor charge till 2 V is ~17 s (**a**) and for the worn-out tool capacitor charge till 2 V is ~5 s (**b**)

The maximum output voltage of the vibration harvester, when cutting by the new tool is 2 V (Fig. 5.16a). The generated harvester voltage is exponentially rising till the capacitor is fully charged and wireless signal is sent to the receiver. The signal sending time intervals are ~17 s. The accelerometer signal is shown in red color. At the end of tool life, when the tool cutting edge is worn out, the accelerometer's signal amplitude increases (Fig. 5.16b). The frequency of wireless transmission of the measured results increases too, and the time interval is decreased approximately 4 times till ~5 s.

During the exploitation, the resonant frequencies of the tool and the harvester are not changing and are very close, what is the necessary condition for the energy accumulation efficiency. With the deterioration of the cutting tool technical state, the vibrations' level increases shortening the charging time of the capacitor. The capacitor's charging time is strictly proportional to the tool's vibration intensity and could be useful for the cutting tool technical state characterization. Such statistical procedure was used for the evaluation of the experimental results: samples from the population were taken by occasional sampling method. Considering that the capacitor charging time coincides with the normal distribution law, the samples of capacitor charging for the new and the worn tool are independent. Thus, for the verification of the research results, the Student's t-test was used to check the statistical hypothesis for unknown variances (case $\sigma_1^2 = \sigma_2^2 = \sigma$).

$$t_0 = \frac{\overline{x}_1 - \overline{x}_2 - \mu_D}{S_p\sqrt{\frac{1}{n_1} + \frac{1}{n_2}}} \tag{5.7}$$

where t_0—test statistics, n_1—number of samples for the new tool, n_2—number of samples for the worn tool.

Rating of dispersions:

$$S_p^2 = \frac{n_1 - 1}{n_1 + n_2 - 2} S_1^2 + \frac{n_2 - 1}{n_1 + n_2 - 2} S_2^2. \tag{5.8}$$

For evaluation of the cutting tool performance, it's necessary to define the limitary moment when the cutting tool starts manufacturing inappropriate quality parts. For this purpose, the statistical analysis was used. In Table 5.1, the energy harvesting results for the new and the worn turning tools are presented.

Considering that the data distribution law in the case of the new and the worn tool is normal, first of all, the tool wear influence on time interval is evaluated. Further, the hypothesis is tested, whether the difference between the means of the new and the worn turning tool is $\mu_D = 12$, with the applied significance level $\alpha = 0.05$ and the variances are assumed equal.

The mean and variance for the sample of the new turning tool is

$$\overline{x}_1 = 17.1,\ S_1 = 0.97,\ n_1 = 21.$$

Table 5.1 Energy harvesting signal formation period (s)

New tool	Worn-out tool	New tool	Worn-out tool	New tool	Worn-out tool
18.842	4.428	15.638	6.492	15.384	4.752
17.824	5.552	17.892	4.488	18.324	5.546
16.486	6.488	15.145	5.826	19.026	4.996
16.308	7.154	15.848	4.566	18.878	6.218
18.384	4.202	15.566	5.582	16.845	5.286
17.226	5.536	16.559	5.606	17.546	4.678

The mean and variance for the sample of the new turning tool is

$$x_2 = 5.4,\ S_2 = 0.9,\ n_2 = 21,$$

where n_1 and n_2 are the sample sizes.

Then, we verify if $\mu_1 - \mu_2 = \mu_D$ (where $\mu_D = 12$): μ_1 and μ_2 denotes the population mean of the new and the worn-out tool:

$$H_0 : \mu_1 - \mu_2 = 12$$
$$H_a : \mu_1 - \mu_2 \neq 12 \text{ (alternating hypothesis)}.$$

The statistics of the t-test is marked t_0. The null hypothesis H_0 is rejected if

$$nt_0 > t_{0,05.40} = 1.684,$$

or if

$$t_0 < -t_{0,05.40} = -1.684.$$

At $S_p = 0.9$ is obvious, that the statistics of the test $t_0 = -1.1$. The t value with 40 degrees of freedom having an area 0.05 to the right in t distribution is $t_{0,05.40} = 1.684$. As $-1.684 < t_0 = -1.1 < 1.684$, the null hypothesis could not be rejected. That's why in the case of new turning tool (with importance level 0.05) it‘s possible to assume that the harvester signal formation interval is by 12 s longer than in the case of the worn turning tool.

5.1.1.7 Turning Process Acoustic Emission Data

The Acoustic Emission (AE) belongs to the group of phenomena that develops transient elastic waves using the rapid release of energy from a fixed source or sources in a material, or the transient elastic wave(s) generated in this way. The device for measuring AE (indicated by *1* in Fig. 5.17) was elaborated for evaluating the

Fig. 5.17 Experimental set-up for the diagnostics of acoustical signal during turning: 1-microphone of the acoustic measurement device, 2-zone of turning process, 3-controller of acoustic signal, 4-PC with the acoustic signal analysis program

reliability of vibrational analysis.

Using AE for controlling tool condition is the reason that *AE* signal frequency range is considerably greater than that of the environmental noises and machine vibrations, and does not interrupt the operation of cutting. According to the results of the research, AE of the stress waves, derived from the abrupt release of energy in deforming materials, was effectively applied in laboratory tests to trace tool wear during the operations of the single point turning.

On the basis of the analysis of AE signal sources, AE generated from metal turning comprises the permanent and transient signals having completely diverse characteristics. Permanent signals are related to shearing within the fundamental zone and wear on the flank and tool face, whereas transient signals or burst derive from chip fracture or tool breakage (Fig. 5.18).

An AE signal is non-fixed and frequently consists of overlapping transients with unknown waveforms and times of arrival. The biggest problem of processing an AE

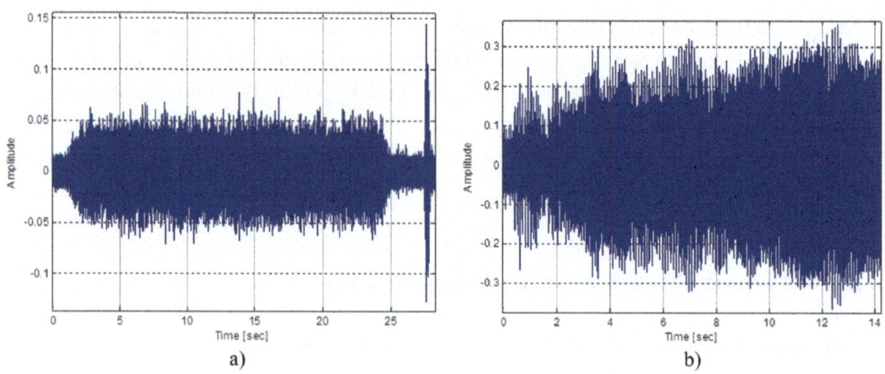

Fig. 5.18 Typical AE signals in turning for a new **a** and worn out **b** tool

signal is to obtain proper physical parameters, like tool wear, when these parameters include variations of frequency as well as time. FFT was used to analyze signals of AE (Fig. 5.19) and to identify frequency diapasons (Fig. 5.20).

The graphs in Fig. 5.20 show that the turning process using the new tool with an acoustic signal is characterized by low vibration amplitudes (0.05) and an even frequency spectrum. In contrast, the worn tool raises the noise level by 5 ÷ 6 times and develops higher eigenmodes at 10 and 15 kHz in the frequency spectrum beyond the first mode at 5 kHz.

Distribution of acoustical signal in turning operation of steel with the new (Fig. 5.20a) tool could be characterized by the low amplitude (approximately 0.05 mm) and quite smooth frequency spectrum as well as in turning with the worn tool (Fig. 5.20b) the vibration amplitude becomes by 5 ÷ 6 times higher and the first

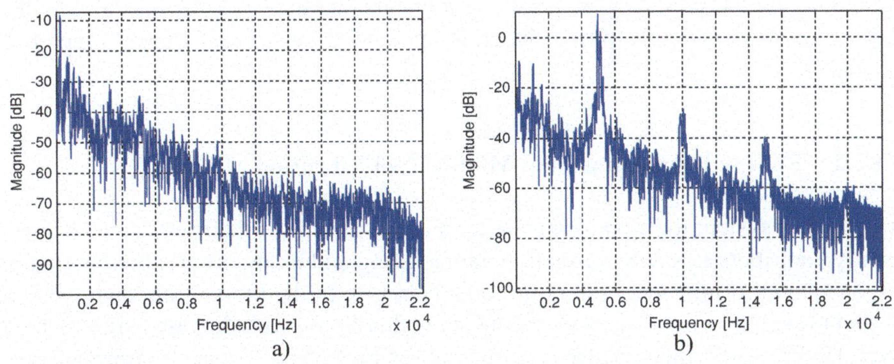

Fig. 5.19 FFT applied to AE signal of the turning tool cutting steel with the new **a** and worn out **b** tool

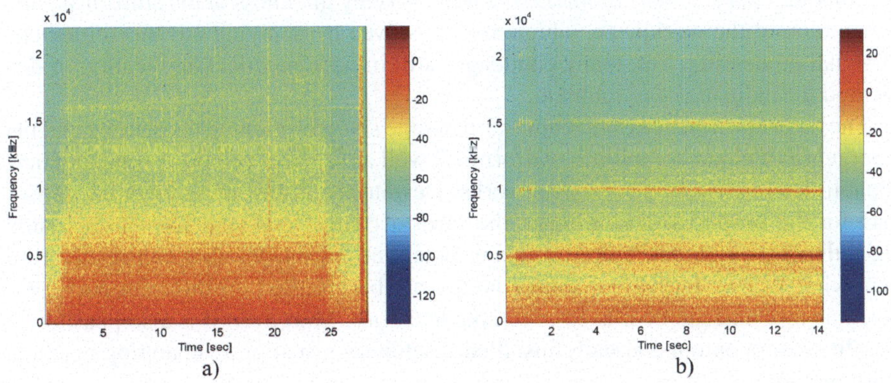

Fig. 5.20 Distribution of acoustical signal in turning operation of steel with the new **a** and worn out **b** tool

(approximately at 5 kHz) and higher (at 10 and 15 kHz) modes could be recognized from the frequency spectrum.

5.1.2 Use Case of the Rotating Tool

Unlike a non-rotating, the rotating tools have a series of cutting teeth that gradually attack the workpiece surface by removing the chip. Therefore, energy is successfully extracted due to tool vibrations when the cutting tooth strikes the workpiece. In this case, the decisive factor is the vibrations of the tool's response to the impact and the frequency of repetition of these shocks. Therefore, the eigenfrequency of the selected piezo transducer are depending on the number of teeth cutting the tool and its rotation speed. When the frequency of tool cuts into the workpiece is close to the eigenfrequency of the piezo transducer, the maximum amount of energy generated is reached.

5.1.2.1 Energy Harvesting from Milling Tool Vibration

An effective system for monitoring the wear of the machine tool inserts could significantly contribute to saving costs in manufacturing and many different methods for effective monitoring have been proposed so far. One of the most recent and popular of them revolves around the use of sensing technologies to indirectly estimate tool wear. However, this method also presents a major challenge as sensory information is difficult to collect from machine tools due to the extremely poor signal-to-noise ratio of the relevant tool wear-related information. This is caused by the milling operation itself as it is of interrupted nature since the workpiece is in contact with the tool edge several times per second. Another issue is the varying thickness of the chip during the penetration of the workpiece, which may be solved by applying advanced methods of signal processing. Yet, many challenges still impede practical application of this method for industrial environment.

The recent prevalence of machine-to-machine service-oriented computing and the emergence of cloud computing platforms as well as services provide promising new capabilities for wireless sensor networks. Despite the issues in aggregating sensor data for the purposes of sharing in the form of "Big Data" and for sensor fusion algorithm development, there are many potential advantages that can result from deploying WSNs. Individual sensor nodes can be installed on different parts of a machine tool and can be placed in remote locations. Moreover, it is also possible to retrofit sensors onto the already-installed machinery just slightly changing machine configuration.

Drilling, turning, and milling are three main machining operations, while all the other processes are ascribed to various categories. In milling and drilling processes, the cutting edges are brought to bear against the workpiece due to the rotating cutting tool. To evaluate the possibility to harvest energy from the rotating tool, it was

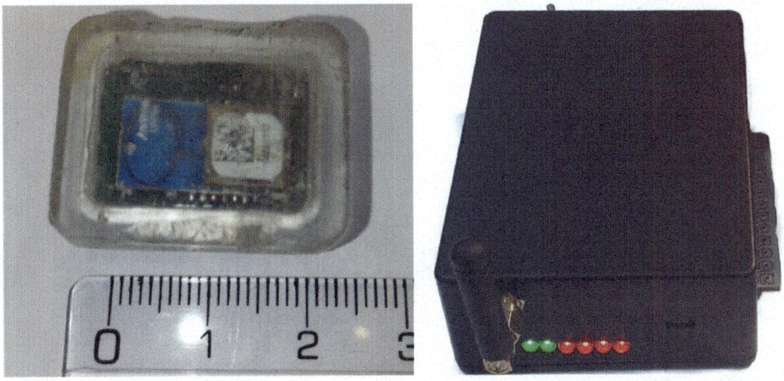

Fig. 5.21 General view of the wireless rotating tool acceleration measurement device (**a**) and the receiver (**b**)

necessary to develop a special alimented from battery device with an accelerometer and wireless transmission (Fig. 5.21a) as well as a receiver (Fig. 5.21b).

The angular vibrations of the milling tool holder were measured directly on the rotating tool [4]. A pronounced periodicity was present in the acceleration measurements carried out during the milling operations. This periodicity can be directly associated with the harmonics of the tool structure, the number of cutting edges as well as the spindle speed n. Increasing the cutting depth resulted in increasing cutting forces, and thus also in greater excitation levels. When changing the feed speed (feed rate), no significant changes were observed, although a small increase in the vibration level with increasing speed was noticed. The main purpose of using acceleration sensors was to discover any angular motion relating to torsional modes of the milling tool. Milling can be described as an interrupted cutting process because during each revolution of the cutter, its teeth enter and exit the workpiece and in this way they are subjected to an impact force cycle. It can be seen from Fig. 5.22 that the 150 Hz frequency dominates as the result of mill tooth and workpiece material collision. During this impact, the vibrations of the cutting tool are excited.

The amplitudes of the vibrations between the interrupted cutting periods are considerably lower than during the impact and are characterized by lowered amplitudes till the new tooth comes in contact with the workpiece. These amplitudes are not sufficient for energy generation, and the most favourable are vibro-impact accelerations whose sequence intervals depend on cutting regimes. This means that higher effectiveness is demonstrated by the rotating cutting tool accelerations, mostly related to the tool speed and the number of cutting edges, rather than the cutting tool vibration modes. It is also claimed that energy harvesters for rotating and non-rotating cutting tools should be designed by using a few different types of piezoelectric transducers whose natural frequencies differ more than 10 times. In such a case, due to the force resulting from any changes in acceleration or motion, the mass squeezes the piezoelectric material, and it, in turn, generates an electrical charge corresponding

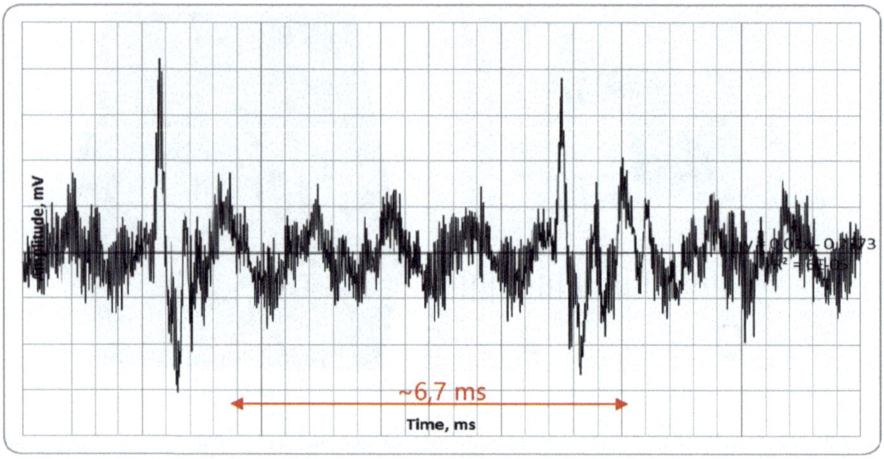

Fig. 5.22 End milling tool vibrations under 3200 rpm of 4 tooth mill

to the force it was affected by. The initial conditions for energy harvester design are related to the necessity to find an appropriately shaped piezoelectric transducer for high and for sufficiently low frequencies. The first will be used for energy extraction from a non-rotating tool, while the second from a rotating one.

One of the objectives is to find a shape of the piezoelectric harvester characterized by the given eigenfrequency and generating maximum power [5]. As the first mode shape of a fixed-free beam is close to the deflection curve subjected to a uniform load, such a piezoelectric cantilever load was chosen. In the case of linear deformations, the energy generated by piezo layers is proportional to the amount of deformation in this layer [6]. The equation of the deflection curve for a cantilever beam subjected to a uniform load of intensity q is as follows:

$$\upsilon = -\frac{qx^2}{24EI}\left(6L^2 - 4Lx + x^2\right) \tag{5.9}$$

where E—Young's modulus, I—moment of inertia bending about neutral axis, L—length of the cantilever, v—deflection.

The deformations caused by the strains εx are actual, thus strain component εx is analyzed.

$$\varepsilon_x = \frac{d^2 v(x)}{dx^2} = -z\left(\frac{qx^2}{24EI}\right)\left(12L^2 - 24xL + 12x^2\right). \tag{5.10}$$

The plane strain energy is determined by integrating over the upper plane of the beam:

$$W = \frac{q^2}{(24EI)^2}\frac{EW_b}{2}\frac{H^2}{4}\int_0^L \varepsilon^2 dx \tag{5.11}$$

where W_b—the width and H—height of the beam and uniform load q.

$$W = \frac{q^2}{(24EI)^2}\frac{EW_b}{2}\frac{H^2}{4}\frac{144L^5}{5}. \tag{5.12}$$

The first eigenfrequency of the cantilever can be expressed as follows:

$$\omega = \frac{\lambda^2}{2\pi}\sqrt{\frac{E}{12\rho}\frac{H}{L^2}} \tag{5.13}$$

where $\lambda = 1.8751$.

To choose the geometrical parameters of the cantilever, the optimization task can be expressed as

$$\begin{aligned} &\text{Max } W(L, WB, H) \\ &\omega(H, L) = \omega^0, \\ &L_{\min} < L < L_{\max}, \\ &W_{\text{bmin}} < W_{\text{b}} < W_{\text{bmax}}. \end{aligned} \tag{5.14}$$

From the frequency equation, H expression is inserted into the equation for the calculation of energy and thereby the harvested energy will be calculated for the cantilever of the given natural frequency. The analysis was performed on a rectangular cross-section cantilever with the following dimensions: $L = 0.0485$ m, $W_{\text{b}} =$ 0.0078 m, $H = 0.0007$ m, $E = 7.5\ \text{e}^{10}$ N/m^2, $\rho = 7750$ kg/m^3, and natural frequency $\omega^0 = 149{,}5$ Hz. The coefficient of stiffness of the cantilever can be calculated as $K = 3EI/L^3$ and the equivalent distributed charge is $q = q_0/L^4$, where $q_0 = 1$. The variation of the cantilever dimension diapasons is 1e−2 m < L < 4e−2 m, 2e−3 m < W_{b} < 1.2e−3 m. As can be seen from Fig. 5.23a, the biggest strain energy is generated by the cantilever whose width and length proportion gains the upper limits of constraints [7].

Figure 5.23a demonstrates that the amount of energy from the cantilever length is non-linear as compared to width-linear, thus the non-linearity of wider cantilevers is bigger. Figure 5.23b presents the dependence of energy field values on the length and width of the cantilever, which allows properly selecting the dimensions of the cantilever assuring maximum amount of the delivered energy. It is clear that the thicknesses of the cantilevers are different, but the frequencies are equal. As the search for the shape of the piezoelectric transducer of the rotating tool indicates the structure of the cantilever, the Noliac piezoelectric transducer CMBP04 was chosen.

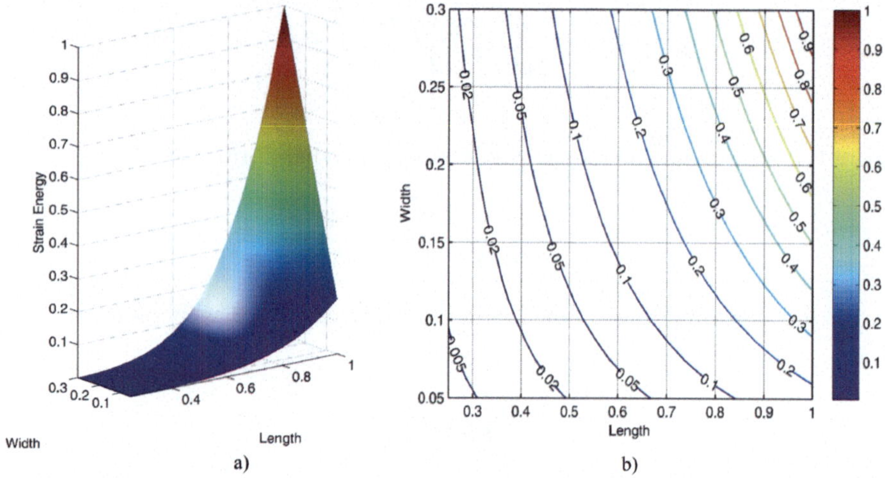

Fig. 5.23 **a** The dependence of resilient energy on the length and width of the cantilever when natural frequency is fixed and normalized for (0.1) interval and **b** when the length and energy are normalized for (0.1) interval widths according to length

It can be described as a co-fired multi-layered piezoelectric transducer requiring low voltage, capable of converting a bending movement into an electrical output, and normally applied when a large displacement can be caused by a low force. Figure 5.24 shows the full architecture of the wireless energy generator and its components [8].

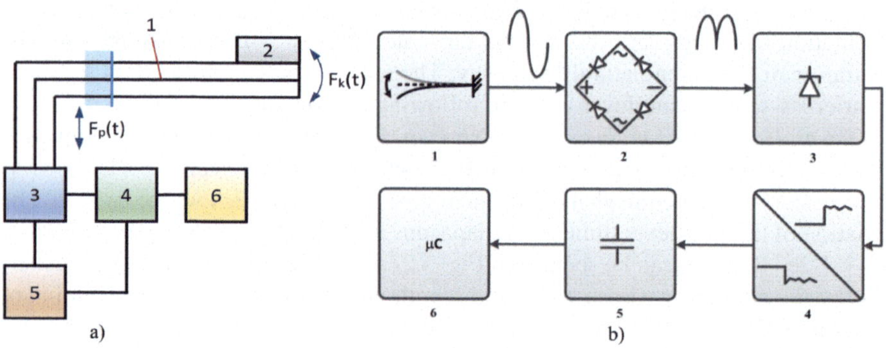

Fig. 5.24 **a** Full architecture of wireless vibration energy generator: 1-cantilever type Noliac CMBP04 piezoelectric transducer, 2-concentrated inertial mass for the tuning of resonant frequency, 3-connection unit, 4-controller, 5-electrical energy accumulator, 6-wireless transmitter, (F_k(t)–dynamic force, generated by the piezo transducer function, F_p(t)–dynamic force, generated by the cutting tool function) and **b** components of the energy recovery and storage device: 1-power inverter, 2-diode bridge, 3-guard element, 4-voltage conversion and stabilization electronics, 5-energy storage element, 6-powered electronics

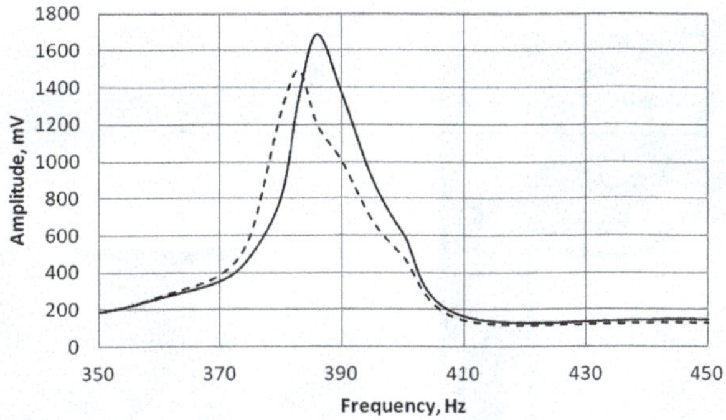

Fig. 5.25 Amplitude-frequency characteristics of the electric signal of the cantilever without concentrated inertial mass piezoelectric transducer when shunt resistance is infinite (open circuit—continuous line) and when one layer is shunted (short circuit—dotted line)

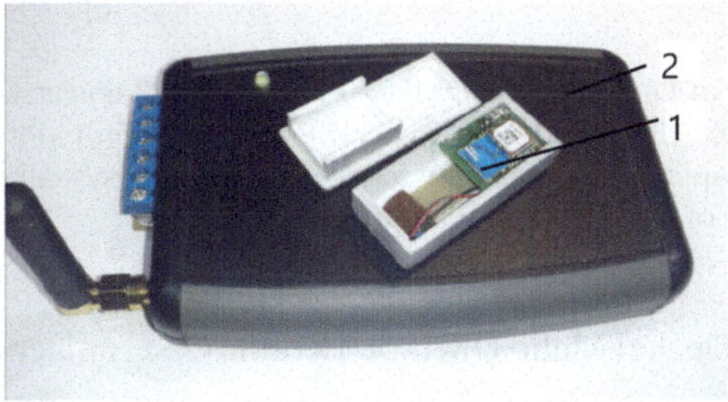

Fig. 5.26 General view of the electrical energy harvester with the energy accumulation, wireless transmitter 1 and diagnostic signal receiver 2

To increase the effectiveness of energy harvesting and sensitivity of the piezoelectric transducer 1, the resonant frequency of transverse vibrations F_k(t) is tuned by changing the concentrated inertial mass 2 and approaching it to the frequency of 150 Hz excited by cutting forces Fp(t). As Noliac CMBP piezoelectric transducer is composed of two layers, the resonant frequency of the cantilever type piezoelectric transducer can be changed by connecting the appropriate shunt resistance to one of these layers (Fig. 5.25).

As Fig. 5.21 suggests, it is possible to change the resistance slightly and the amplitudes of generated voltage considerably. The general view of the implemented prototype of the wireless energy harvester together with the diagnostic signal receiver is presented in Fig. 5.26.

5.1.2.2 Physical Twin of Milling Tool

The developed wireless piezoelectric energy harvester with a subsystem of wireless information transfer and a subsystem of energy accumulation was applied for steel

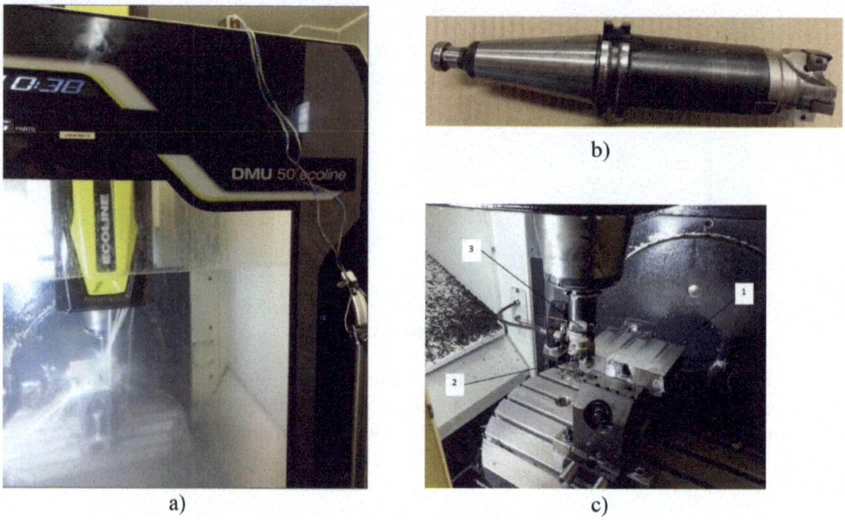

Fig. 5.27 General view of Deckel Maho DMU 50 Ecoline CNC universal machining centre (**a**), 4 teeth end mill (diameter 40 **mm**) (**b**) and general view of end mill vibration harvesting (**c**): 1-workpiece, 2-end mill, 3-piezoelectric energy harvester with a subsystem of wireless information transfer and a subsystem of energy accumulation

milling operation on Deckel Maho DMU 50 Ecoline CNC universal machining centre shown in Fig. 5.27.

The experiment was done on a steel workpiece with end mill working regimes: mill speed 3200 rpm; feed 0.1 mm/rev; depth of the cut 1 mm. Two kinds of mills were chosen: one was a new mill and the other was worn out. These two mills were expected to show different results: predictably, the worn mill would have higher resonance frequencies due to higher vibrations. A simple 1 mm depth, 5 mm milling edge of the tool on the workpiece, with the feed rate of 600 mm/min and 3200 rpm was chosen and a "zigzag" milling principle was applied. *FFT* of mills vibration characteristic is presented in Fig. 5.28.

As Fig. 5.24 shows, the dominant frequencies of the end mill are 150 $\div$ 212 Hz and its multiples.

A further experiment was organized to show the influence of the tool wear on the frequency of wireless information transmission to the receiver (Fig. 5.29a). During milling operation, piezo harvester was excited by the mill vibrations and the harvested energy was accumulated by a 100 μF capacitor. Figure 5.29 suggests that the maximum output voltage of the vibration harvester when cutting by the new tool is 2 V. The generated voltage of the harvester was exponentially rising till the capacitor was fully charged and wireless signal sent to the receiver. The signal sending time intervals amounted to ~28 s with new (Fig. 5.29a) and worn-out (Fig. 5.29b) milling tool for a steel workpiece manufacturing.

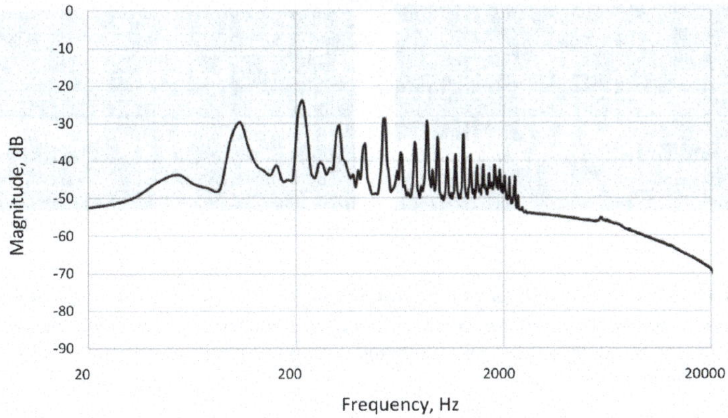

Fig. 5.28 Amplitude-frequency characteristic of end mill vibrations at 3200 rpm

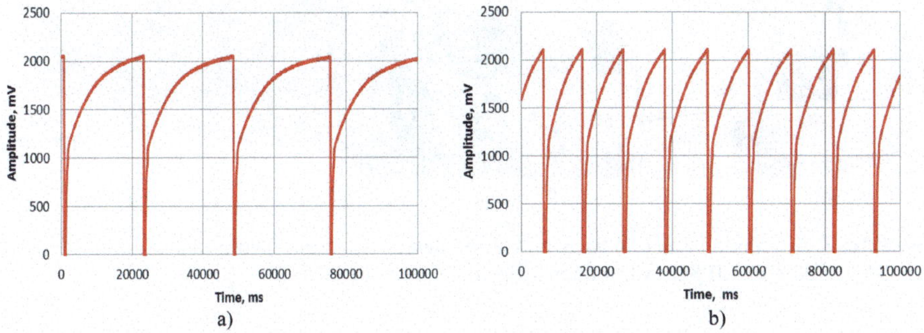

Fig. 5.29 Exponentially rising generated voltage of the harvester when milling with new **a** and worn out **b** milling tool for steel workpiece manufacturing: the time of capacitor charge till 2 V is ~28 s (**a**) and 10 **s** (**b**) accordingly

When the tool life comes to an end and the cutting edge of the tool is worn, the frequency of wireless transmission of the capacitor discharge increases, while the time interval is decreased by approximately $2 \div 3$ times.

5.1.2.3 Milling Process Acoustic Emission Data

The device for measuring acoustic emission was elaborated for evaluating the reliability of vibrational analysis. *AE* can be described as non-fixed signals that very often consist of overlapping transients with unknown waveforms as well as the arrival time. The biggest problem of processing an *AE* signal is obtaining proper physical characteristics, such as tool wear, when they include variations of frequency and time. *FFT* was used to analyze *AE* signals and identify frequency diapasons (Fig. 5.30).

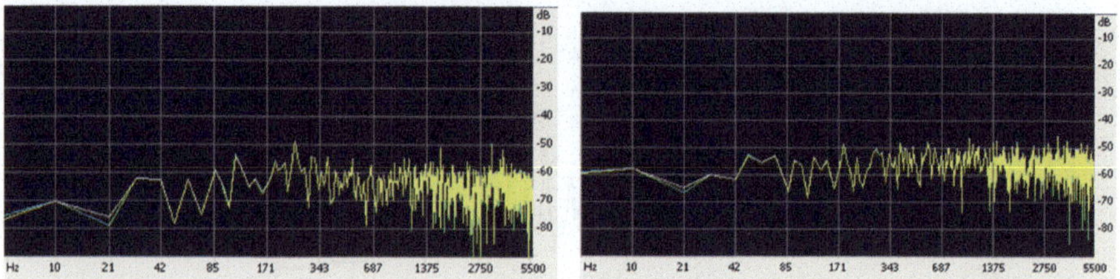

a) b)

Fig. 5.30 In case of the new mill (**a**), the average noise level is -65 dB, the frequency of the mill tooth and workpiece material collision is 120 Hz; in case of the worn-out mill (**b**), the average noise level is −55 dB, the frequency of the mill tooth and workpiece material collision is inhibited by the noise of higher frequencies

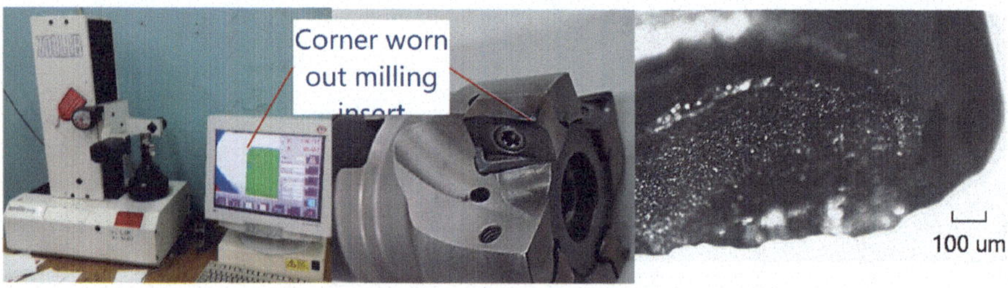

a) b) c)

Fig. 5.31 Zoller Smile v300 tool wear measuring device (**a**), corner worn-out end milling insert (**b**) and its Nikon ECLIPSE LV150 microscope digital picture (×5) (**c**)

Comparing the noise spectrums of the new and worn mills (Fig. 5.30), registered by the microphone placed near the cutting zone, a 10 dB level difference is evident, which coincides with the 3 times difference in the amplitudes. For the identification of the wear level of the milling tool, the industrial wear measuring device Zoller Smile V300 was used (Fig. 5.31).

As a result, the dynamic interaction between the workpiece and the cutting tool, self-excited vibrations distort the quality of the surface and also deteriorate the lifetime of the tool. Moreover, machining process may become less effective due to the occurring vibrations which reduce the capacity of chip removal of the tool. A typical type of vibrations occurring in the process is regenerative chatter where the varied thickness of the chip causes excitation in the machine-tool-workpiece system. Figure 5.32 illustrates the vibrations of the workpiece measured by an accelerometer attached to it.

This figure clearly demonstrates the increase of workpiece excitation level under the worn-out mill cutting.

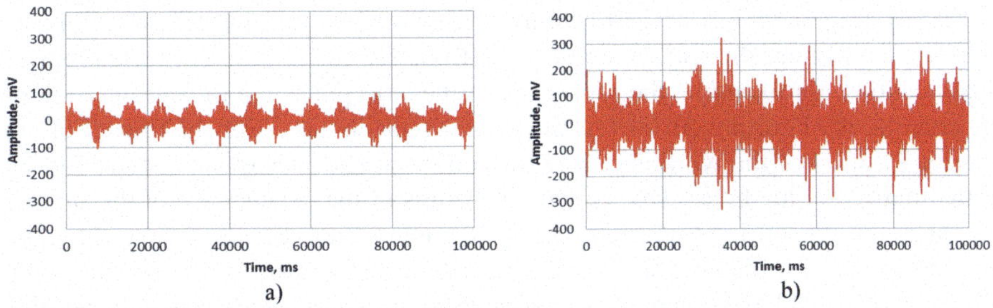

Fig. 5.32 Vibration signal of the workpiece measured by an accelerometer KD 91 when milling with a new (**a**) and worn-out cutting tool (**b**)

5.1.2.4 Application of Wireless Acoustic Emission Sensor in Industrial Environment

The results of ultrasound measurements were registered by self-made sensor of acoustic emission, whose general view and structure are presented in Fig. 5.33. The sensor of acoustic emission (Fig. 5.33a) is composed of the housing 1, which has a disc-shaped piezoelectric actuator 3, electrodes 5, 6 and wires 7, 8 by which the measured acoustic emission signal is transmitted to the measuring device. From the direction of the acoustic signal, the piezoelectric actuator is armoured by a ceramic disc 2 and a diaphragm-screen 4, characterized by low damping. From the backside, the acoustic signal damping compound 9 is placed ensuring the sensing direction. The frequency diapason of the acoustic sensor is from 10 Hz to 2 MHz, and its overall dimensions are equal to $\varnothing 16 \times 22$ mm.

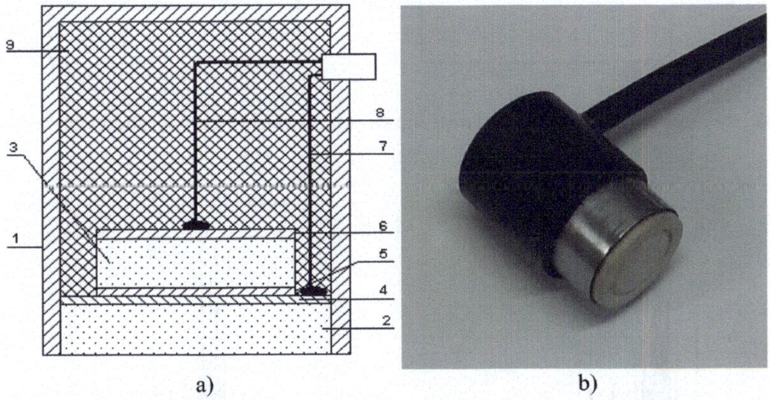

Fig. 5.33 Structure **a** and general view **b** of acoustic emission sensor: 1-housing, 2-element of acoustic contact, 3-piezoelectric element, 4-diaphragm-screen, 5 and 6-electrodes of piezo element, 7 and 8-cables, 9-compound for the acoustic signal suppression

Surface roughness has a significant influence on how the completed components perform. In milling, the surface quality itself is affected by different technological specifications, for example, the properties of the workpiece and the cutting tool, or the conditions of cutting. However, literary sources do not cover the impact of the relative location of the workpiece and tool on the surface roughness of the former or the lifetime of the latter. That is why the results of the dependence of the surface roughness on the milling tool position could be interesting. During face milling, the measurements of roughness were realized in *1*, *2*, and *3* positions according to the scheme in Fig. 5.34 and compared to the acoustic emission measured for the relative mill and workpiece behavior in the indicated positions (Fig. 5.35).

The analysis of acoustic emission signals excited by the new (Fig. 5.35) and the worn-out (Fig. 5.36) milling tools clearly shows that the most favourable position from the point of view of possible chatter excitation is related to the relative tool-workpiece location *2* which assures a higher workpiece stiffness under the action of the milling forces. The test results suggest that the amplitude of the acoustic signal

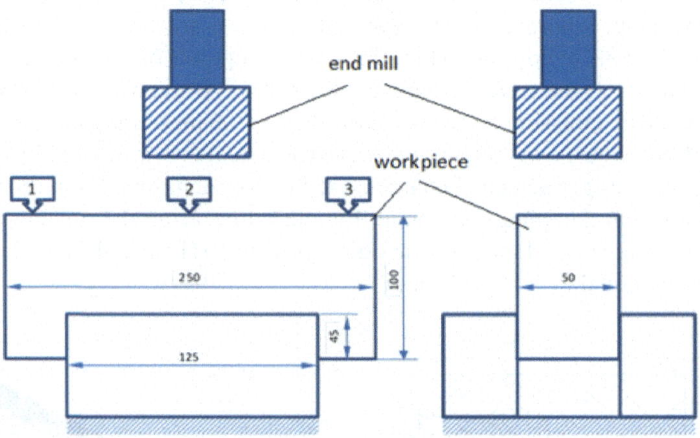

Fig. 5.34 Relative positions of the end mill and workpiece during the measurements

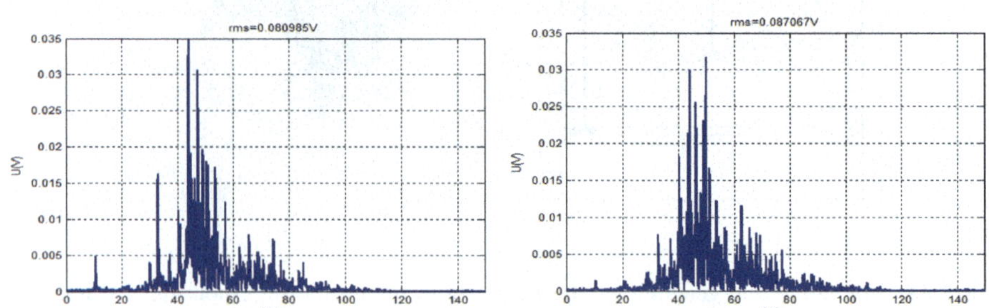

a) b)

Fig. 5.35 Acoustic emission signal in the workpiece for the new mill in 1 and 3 (**a**), 2 (**b**) positions

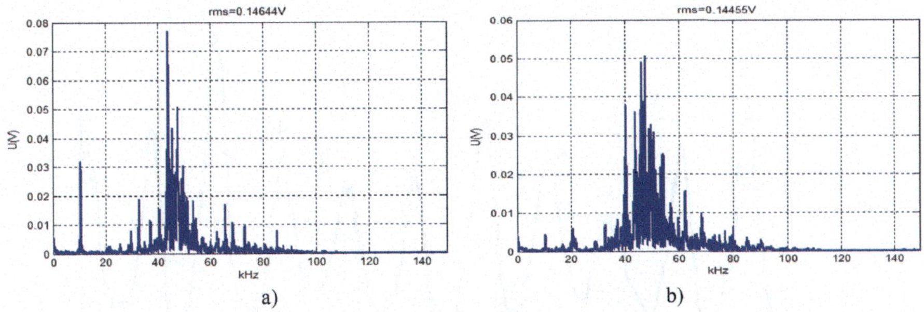

Fig. 5.36 Acoustic emission signal in the workpiece for the worn mill in 1 and 3 (**a**), 2 (**b**) positions

generated in the workpiece material is twice lower for the new mill than for the worn one. However, the diapason of the acoustic signal is wider and comprises from 20 to 100 kHz for the former (Fig. 5.35a), while in case of the worn tool, this diapason is narrowed to 80 kHz (Fig. 5.36a). It is evident that the cutting forces of the worn tool excite more intensive vibrations of the workpiece, which have resonance frequency equal to 10 kHz (Fig. 5.36a). According to the fixing scheme of the workpiece in Fig. 5.34, when the cutting starts (1 position) or ends (3 position), these parts of the workpiece could be considered as cantilever beams, the stiffness of which is lower than in the case of the 2 position.

This statement was confirmed by measuring the surface roughness in the indicated points (Fig. 5.37).

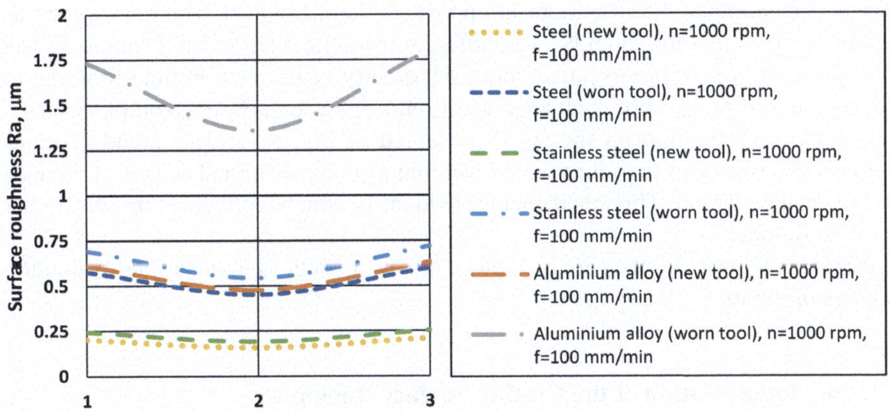

Fig. 5.37 Dependence of the surface roughness measured at different relative positions of the cutting tool-workpiece

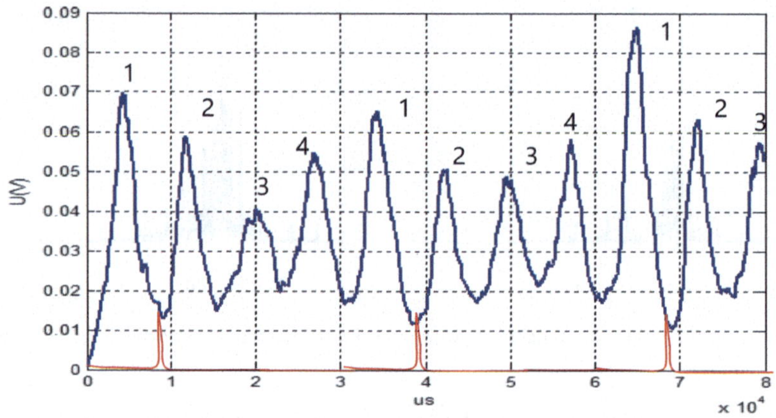

Fig. 5.38 Envelope signal of the waveform generated by the 4-tooth face mill (blue curve) and tachosignal of the rotating mill (red curve); 1, 2, 3, and 4 signify the tooth number, which generates vibrational impulse when striking the workpiece

5.1.2.5 Identification of the Cutting Tool Precision

AE measurements could be useful for the identification of the new tool precision by narrowband envelope analysis of the raw time waveform from the acoustic sensor. The magnitude modulation of the envelope signal can be used as the fault development indicator.

Figure 5.38 demonstrates that the end mill tooth No.1 generates 50% more intensive signals compared to the tooth in opposite side of the mill. The reason for this phenomenon is that the tooth No.1 is different from the other tooth. From the given vibrogram, it is possible to judge about the quality of the new milling tool and to improve it before use. For such an evaluation, it is necessary to accomplish vibrodiagnostic investigation on the stand presented in Fig. 5.39. This stand contains not only the sensor of the vibrational level, but also an additional sensor of the mill tachosignal generator. This stand enables to identify which tooth generates the higher level vibrations.

Figure 5.40 provides the general view of the set-up of the technical state evaluation sensors of the mill.

5.1.2.6 Identification of the Cutting Surface Roughness

CNC milling machine tool Optimum Optimill F150 (Fig. 5.41) was used for machining three types of the most popular industrial materials: stainless steel 1.0037—St37-2/S235JR, steel 1.4057 and aluminum alloy. A few cutting speeds n and feeds f were chosen for the experiments and surface roughness was measured three times; hence, Fig. 5.41 provides the average results.

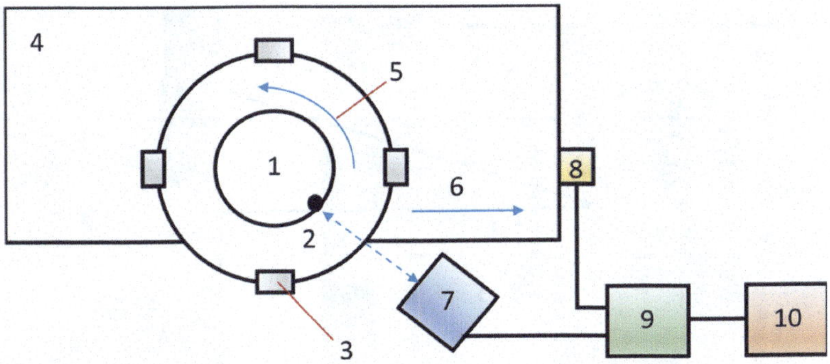

Fig. 5.39 Structural scheme of the vibrodiagnostic stand for mill technical state evaluation: 1-mill axis, 2-mark of tachosignal, 3-mill tooth, 4-workpiece, 5-direction of mill rotation, 6-feed direction of the workpiece, 7-shaper of tachosignal, 8-accelerometer, 9-oscilloscope, 10-PC with vibrodiagnostic analysis software

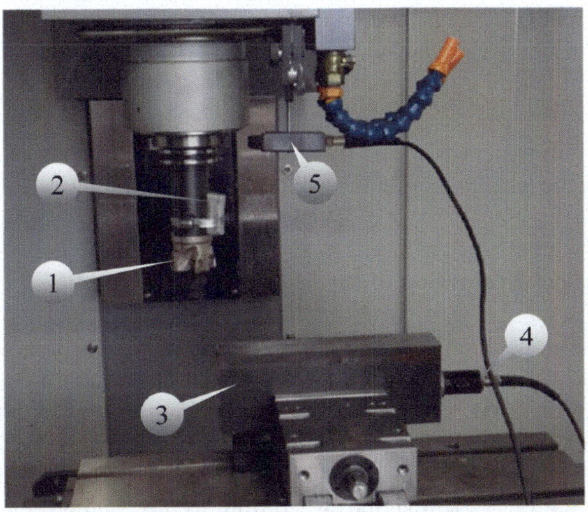

Fig. 5.40 General view of the set-up of the technical state evaluation sensors of the mill: 1-mill, 2-Wi-Fi vibrational level sensor, 3-workpiece, 4-accelerometer BK-310A, 5-transmitter of tachosignal

Parameters for milling surface with new and worn-out tool are presented in Table 5.2

Figure 5.42 presents the dependence of workpiece roughness on the duration of the harvested energy accumulation process under different cutting regimes for steel, stainless steel, and aluminum alloy milling.

The experimental results show that the accumulation process of the vibration energy harvested by the worn-out mill is 2–3 times faster compared with the accumulation process from the new mill independently from the material of the workpiece.

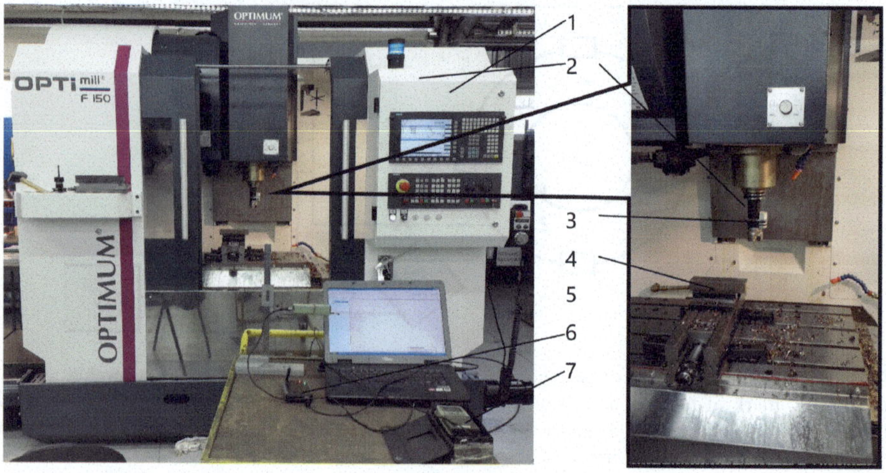

Fig. 5.41 Industrial set-up for the experiment: 1 CNC milling machine tool Optimum Optimill F150, 2—4-tooth mill $D = 50$ mm, 3 electrical energy harvester with the energy accumulation and wireless transmitter, 4 workpiece, 5 PC, 6 diagnostic signal (Wi-Fi) receiver, 7 roughness measurement device

In contrast, after treatment with the worn-out mill, the roughness of the workpiece surface exceeds the roughness obtained with the new mill by $2 \div 3$ times.

5.1.2.7 Statistical Evaluation of the Milling Process Quality

The cutting conditions in milling directly influence the production, which concerns not just time but also quality. Therefore, finding a suitable compromise among the surface quality, condition, and productivity of the cutting tool is one of the main objectives of the *CNC* machining centre. The proposed new method for the identification of milling tool condition is based on the excited mechanical vibration energy transformation to an electrical signal. According to this method, the accumulation time of harvested energy, which correlates with the cutting tool wear level, depends on such parameters as the type of workpiece material, speed (rev/min), and feed (mm/min) of the tool cutting. The proposed experiment of three factorial influence estimation was carried out to determine the relative value of the time when the identification signal, informing about tool condition, is formed and the evaluation of the interaction of the parameters. For a statistical experiment, the results of three types of workpiece materials (stainless steel 1.0037-St37-2/S235JR, steel 1.4057, and aluminum alloy), two levels of cutting speeds (low level 1000 rev/min) and high level 2000 rev/min) and three levels of the feed values (low level 100 mm/min, medium level 200 mm/min, and high level 300 mm/min) were used. The interval

Table 5.2 Milling with new and worn-out tool parameters

No	Tool condition	Cutting regimes			Surface roughness				Loading time T, s
		Spindle speed n, rpm	Feed rate f, mm/min	Cutting depth a_p, mm	1Ra, μm	2Ra, μm	3Ra, μm	ΣRa, μm	
Stell 1.00 37									
1	New	1000	100	1	0.20	0.16	0.21	0.19	22
2	Worn	1000	100	1	0.58	0.45	0.60	0.54	13
3	New	1000	200	1	0.22	0.18	0.23	0.21	29
4	Worn	1000	200	1	0.58	0.45	0.60	0.54	11
5	New	1000	300	1	0.20	0.16	0.21	0.19	23
6	Worn	1000	300	1	0.64	0.50	0.67	0.60	7
7	New	2000	100	1	0.20	0.16	0.21	0.19	25
8	Worn	2000	100	1	0.62	0.49	0.64	0.58	9
9	New	2000	200	1	0.27	0.21	0.28	0.25	25
10	Worn	2000	200	1	0.62	0.49	0.64	0.58	12
11	New	2000	300	1	0.26	0.20	0.27	0.24	22
12	Worn	2000	300	1	0.75	0.59	0.78	0.70	8
Stainless steel 1.4057									
1	New	1000	100	1	0.24	0.19	0.15	0.23	35.2
2	Worn	1000	100	1	0.69	0.54	0.72	0.65	14.3
3	New	1000	200	1	0.27	0.21	0.28	0.25	31.9
4	Worn	1000	200	1	0.69	0.54	0.72	0.65	12.1
5	New	1000	300	1	0.24	0.19	0.25	0.23	25.3

(continued)

Table 5.2 (continued)

No	Tool condition	Cutting regimes			Surface roughness				Loading time T, s
		Spindle speed n, rpm	Feed rate f, mm/min	Cutting depth a_p, mm	1Ra, µm	2Ra, µm	3Ra, µm	ΣRa, µm	
Stell 1.00 37									
6	Worn	1000	300	1	0.77	0.60	0.80	0.72	7.7
7	New	2000	100	1	0.24	0.19	0.25	0.23	27.5
8	Worn	2000	100	1	0.74	0.58	0.77	0.70	9.9
9	New	2000	200	1	0.32	0.25	0.33	0.30	27.5
10	Worn	2000	200	1	0.74	0.58	0.77	0.70	13.2
11	New	2000	300	1	0.31	0.24	0.32	0.29	24.2
12	Worn	2000	300	1	0.90	0.71	0.93	0.84	8.8
Aluminum alloy									
1	New	1000	100	1	0.61	0.48	0.63	0.57	128.0
2	Worn	1000	100	1	1.73	1.36	1.80	1.62	52.0
3	New	1000	200	1	0.67	0.53	0.70	0.63	116.0
4	Worn	1000	200	1	1.73	1.36	1.80	1.62	44.0
5	New	1000	300	1	0.61	0.48	0.63	0.57	92.0
6	Worn	1000	300	1	1.93	1.51	2.00	1.80	28.0
7	New	2000	100	1	0.61	0.48	0.63	0.57	100.0
8	Worn	2000	100	1	1.86	1.46	1.93	1.74	36.0
9	New	2000	200	1	0.80	0.63	0.83	0.75	100.0
10	Worn	2000	200	1	1.86	1.46	1.93	1.74	48.0
11	New	2000	300	1	0.77	0.60	0.80	0.72	88.0
12	Worn	2000	300	1	2.25	1.76	2.33	2.10	32.0

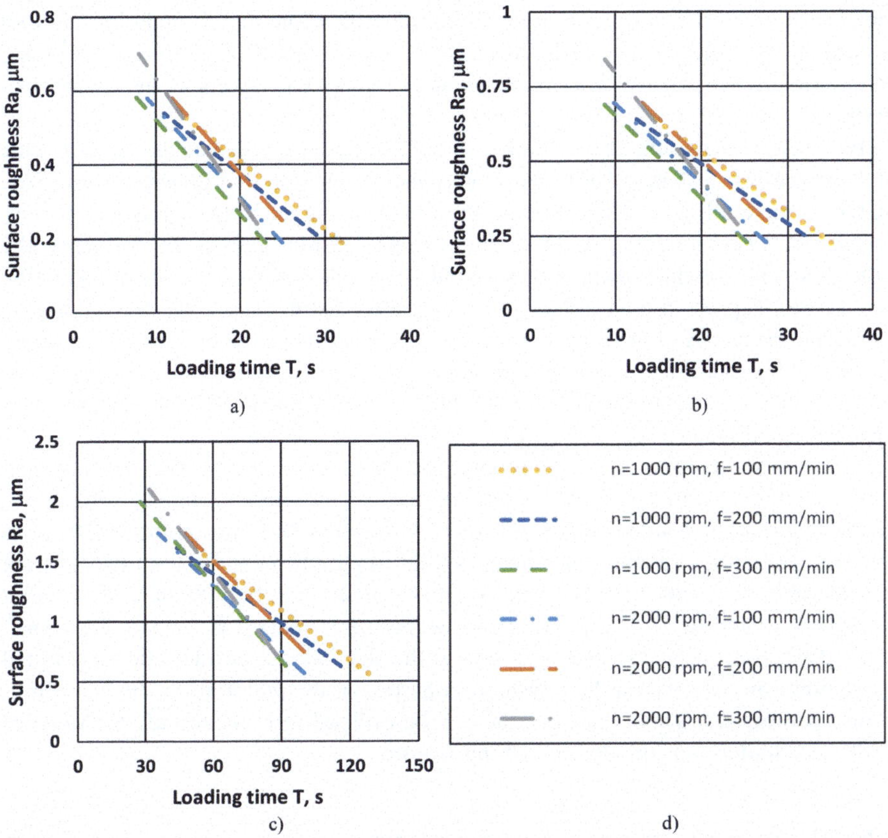

Fig. 5.42 Dependence of workpiece roughness on the duration of the harvested energy accumulation process under different cutting regimes for **a** steel, **b** stainless steel, **c** aluminum alloy milling, and **d** marking of cutting regimes

of the identification signal informing about the tool state and the formation time is appreciated in seconds. For the statistical experiment, three-factor two-level analysis of variance (Anova) statistical method was chosen to determine the experimental results of the dispersion analysis, meaning that in every population the considered properties are normally distributed and have the same dispersion. After the dispersion analysis, the following results were submitted: the sum of squares (Sum Sq), their degrees of freedom (d. f.), mean squares (Mean Sq), F criterion, and its p value. If p exceeds the chosen importance level α, then the null hypothesis is unquestioned. In other words, the experimental results provide no basis for claiming that the factor has a reliable impact on the measured variable. However, if $p \leq \alpha$, then the null hypothesis is rejected, i.e. the factor has a reliable impact on the measured variable. The purpose of the statistical experiment is to evaluate if the workpiece material, cutting speed, feed rate and their interplay have influenced a change in the time interval of the formation of the wirelessly transmitted identification signal, updating the tool condition, or if the average differences in the time interval of identification signal formation are statistically important. Table 5.3 gives the results of the analysis of milling.

It is evident that under the influence of one factor, i.e. workpiece material, *p* value of *F* criterion is less than 0.05, that is why the conclusion that different workpiece materials have influenced the time of identification signal formation is statistically significant. Other factors and their interactions did not statistically influence the formation time of the identification signal. It is relevant to have the information about both the pairs of means that are considerably distinct and those that are not, so a multiple comparison method was applied as a test that provides such information. Figure 5.43 presents a multiple comparison of marginal means of the population and suggests that the marginal mean of material 3 is significantly different from those of material 1 and material 2.

In this case, material 1 and material 2 are stainless steel and steel, the physical and mechanical properties of which are approximate similarly as the marginal mean of the identification signal formation time, while material 3, aluminum alloy, has different properties of marginal mean. While marginal mean of the materials are considerably different, from the analysis it is evident that the influence of different workpiece materials on the identification signal time interval is statistically significant. Therefore, it is essential to compose the dependence of identification signal formation time interval on the physical properties of the material and their machinability parameters. Thus, the operator will be obliged to indicate only the code of workpiece material as the distinguishing feature.

Table 5.3 Results of variance analysis of the milling process

Analysis of variance					
Source	Sum Sq.	d. f.	Mean Sq.	F	Prob > F
Material	5299.55	2	2649.78	18.82	0.0092
Cutting speed	297.68	1	297.68	2.11	0.2196
Cutting feed	177.78	2	88.89	0.63	0.5777
Material * Speed*	139.36	2	69.68	0.49	0.6426
Material * Feed	83.23	4	20.81	0.15	0.9545
Speed * Feed	1203.25	2	601.56	4.27	0.1017
Error	563.97	4	140.81		
Total	7763.97	17			

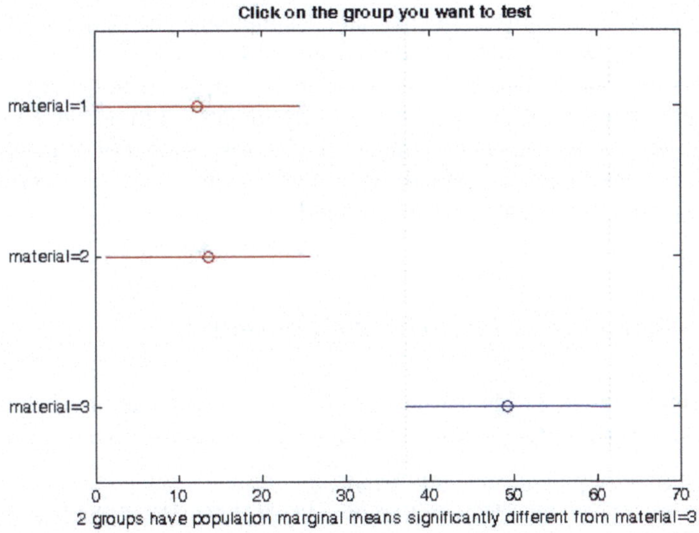

Fig. 5.43 Comparison of the marginal mean of the materials for the milling operation

5.2 Wireless IoT Vibration Sensor for Cloud Manufacturing Applications

Cloud manufacturing can manage production resources and capabilities and provide them as online services. In the cloud manufacturing and service resources are connected to the machine tool, implementing a cyber-physical system. During machining, the sensor system is used to collect signals in real time. Existing cloud manufacturing models operate centrally through a cloud manufacturing platform. A decentralized cloud manufacturing network architecture provides an alternative model for cloud manufacturing organization, which is based on the concept of an elementary work system. An elementary work system consists of a process (e.g. milling), a process implementation (e.g. a milling machine), and a human subject. The process transforms inputs (e.g. blanks) into outputs (e.g. products) using a process implementation device, and the role of the human subjects‘s is to set goals, monitor inputs and outputs, and control the device. The factory is seen as a network of elementary work systems performing various functions and is modeled as a complex adaptive production system. Cloud manufacturing is a customer-driven manufacturing model inspired by cloud computing. Its purpose is to enable ubiquitous access to distributed generation services. Instead of investing in new manufacturing capacity, direct access to manufacturing services allows product manufacturers to respond quickly to any changes in the manufacturing environment or to respond to individual product requirements. Cloud manufacturing adopts and expands the cloud computing concept to make mass manufacturing capabilities and resources more integrated and accessible to users over the Internet. Cloud manufacturing and the IoT are interlinked, as is the Industry 4.0 paradigm. The IoT, cloud computing, and artificial intelligence are related to Big Data, which is collected in real time from IoT and analyzed in cloud computing space. This process is expanding to the manufacturing process, in which the efficiency of production is closely monitored in advance by various machine learning algorithms in real time through a vast amount of data collected by various sensors. For their successful development, an universal self-powering device with wireless Bluetooth transmission is offered.

5.2.1 *Virtual Twin of Piezoelectric Transducer*

Self-powering wireless device is necessary for the implementation of cloud manufacturing. An algorithm for creating a virtual twin of such a device is proposed:

- Define necessary parameters and variables;
- Build geometry of the model with the embedded CAD tools available with Comsol multhiphysics;
- Define materials and their mechanical and physical properties for the imported or created geometry;
- Define physics interfaces for each of the model domains;
- Set model boundary conditions and coupling of defined physics;
- Set required mesh type and size, perform meshing of model geometry;
- Define study settings: steps or range, such as frequency range;
- Perform computation of the defined physical problem;
- Plot the results for evaluation in suitable format for further processing in other software packages.

Evaluation of vibrational modes and conditions for longitudinal-torsional mode coupling effect took place for the tool holder model and L&T mode frequency dependence on geometrical and material parameters [9, 10]. The surface displacement analysis of tool holders contact surface with piezoelectric transducer was carried out in Solid Mechanics (solid) module, performed in the frequency domain. For a complete tool holder with piezoelectric transducer simulation where voltage and power output from piezoelectric transducer was evaluated when end mill of the tool holder is excited was performed integrating Electrical Circuit (cir), Electrostatics (es) and Solid Mechanics (solid) modules. Fig. 5.44 shows the block diagram coupling of Comsol multiphysics modules used in FE model calculations.

The evaluation of the vibrational modes and conditions for the longitudinal-torsional mode coupling effect to take place for the tool holder model and L&T mode frequency dependence on geometrical and material parameters was performed in Solid Mechanics (solid) module of Comsol software package. For a complete tool holder with the piezoelectric transducer simulation, where the voltage and power output from the piezoelectric transducer were evaluated, when the end mill of the tool holder was excited, was performed integrating Electrical Circuit (cir), Electrostatics (es), and Solid Mechanics (solid) modules. A block diagram coupling of the

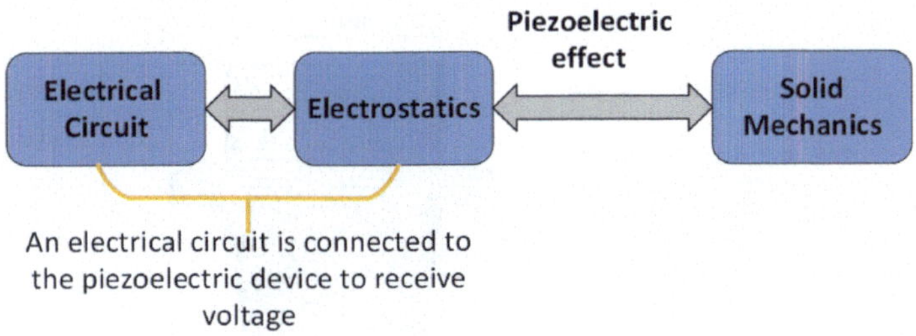

Fig. 5.44 Used physics coupling of piezoelectric transducers in COMSOL multiphysics software environment for FE calculations

modules that are used in FE model calculations is provided in Fig. 5.45 in more detail explaining the complete implemented model that is used for theoretical investigation, consisting of geometrical components, physical coupling, and inputs/outputs for/from the performed simulations.

Figure 5.46 shows the assembly of geometrical models for the research, where piezoelectric transducer voltage and power output were evaluated. It is the topmost assembly of the proposed device used in simulations, consisting of end mill tool, the tool holder (either with or without uniformly distributed helical slots on its planar surface), and ring type piezoelectric transducer prestressed by a back-mass.

The material used in the simulation for the end mill tool is molybdenum (common application in HSS cutting tools), for tool holder, carbon steel, and for back-mass, aluminum 3003. Table 5.4 shows the key properties of these materials.

PZT-5H (Lead Zirconate Titanate) has been selected for piezoelectric ceramic material, because it has the best piezoelectric material properties (exhibiting the

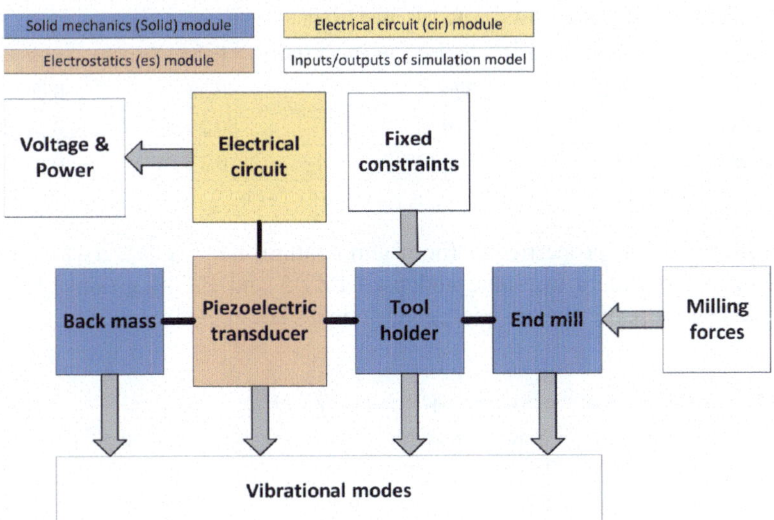

Fig. 5.45 Principal block diagram of Comsol multiphysics FE simulation model

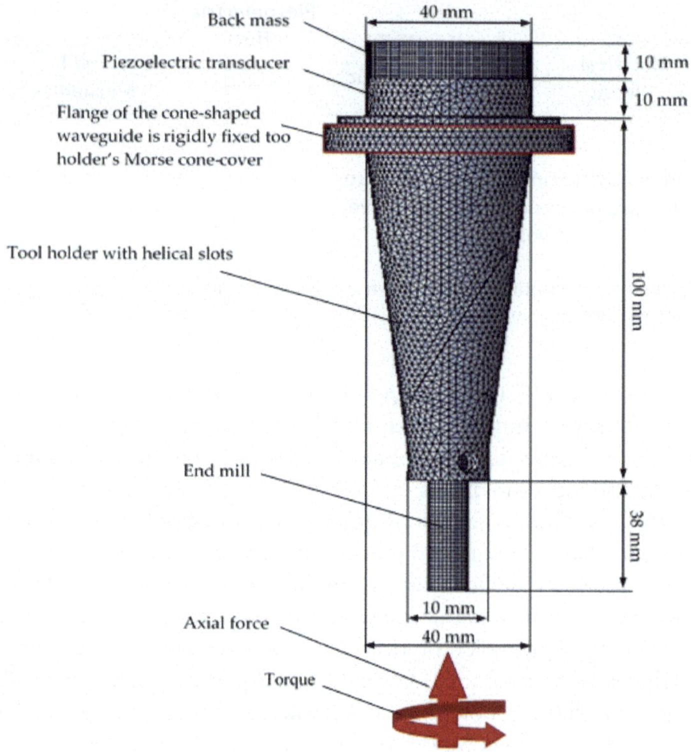

Fig. 5.46 Schematic representation of tool holder assembly node used during simulation

Table 5.4 Properties of linear elastic materials used in simulation

Materials	Density (kg/m^3)	Young's modulus (GPa)	Poisson's ratio	Loss factor
Molybdenum	10,000	320	0.38	0.001
Carbon steel	8000	210	0.28	0.01
Aluminum 3003	2700	690	0.33	0.02

highest piezoelectric properties at room temperature, i.e. k_{eff}, k_{p}, d_{31}, d_{33}) when compared to the widely available and used PZT-5 and PZT-4 groups. Table 5.5 shows the key properties of this piezoelectric material.

Table 5.5 Key properties of the piezoelectric ceramic PZT-5H

Material	Density (kg/m^3)	Young's modulus (GPa)	Poisson's ratio	d_{31}	d_{33}	k_{33}	T_c
PZT-5H	7500	200	0.3	320	650	0.75	250

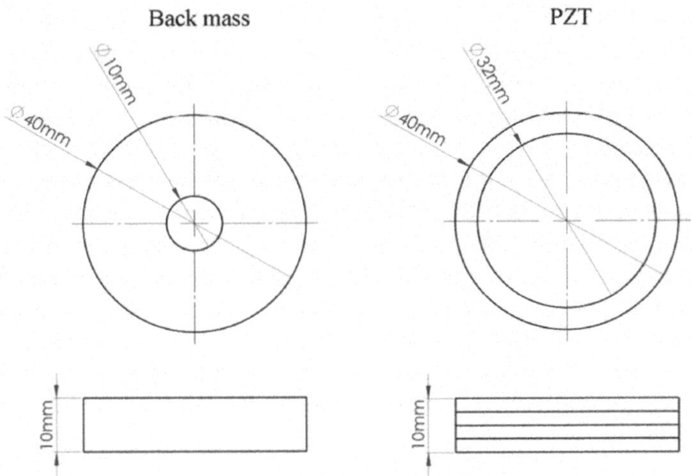

Fig. 5.47 Drawing of back-mass and piezoceramic transducer design embedded together with tool holder

The PZT-5H piezoceramic that is used in simulations is composed of four stacked rings, where each piece is 2.5 mm tall, 4 mm thick with inner radius of 32 mm and outer radius of 40 mm as provided in Fig. 5.47 when the back-mass is an aluminum disc 40 mm in diameter and 10 mm thick.

Eigenfrequency and frequency response studies are performed on the proposed FE model, to evaluate system response to the excitation in the form of surface displacement and voltage output from the piezoelectric transducer. The eigenfrequency study is performed for the first ten eigen-modes of the assemblies or the parts where the geometrical components are in the following configurations:

- End mill rigidly fixed over its cylindrical surface;
- Tool holder rigidly fixed over exterior flange surface as provided in Fig. 5.46;
- The assembly of the tool holder with end mill, where the tool holder is rigidly fixed over its exterior flange surface;
- The assembly of end mill tool, the tool holder, the piezoelectric transducer and the back-mass, where rigid constraints are implemented on relevant tool holder domains, as described in the previous step.

The performed eigenfrequency study allows to observe and evaluate the longitudinal, torsional vibrational modes, and their coupling phenomena. During the FE evaluation, the domains for fixed constraint conditions are selected for each configuration in a way to represent the actual device assembly or its positioning inside a CNC spindle as close as possible. The L&T mode coupling frequency dependence on the introduction of helical uniformly distributed slots on the planar surface of the tool holder is as well evaluated. The frequency response study over 10 ÷ 20 kHz range is performed on the tool holder with the end mill assembly and complete assembly as provided in Fig. 5.46. In each case, the models are excited by applying axial, torsional, and radial milling force components on the end mill, as it would be expected during the milling operation; depending on the model assembly, either displacement of piezoelectric transducer contact surface or its voltage and power output are measured.

As discussed previously, during frequency response analysis of the tool holder with or without uniformly distributed helical slots, the displacement in axial direction of the surface intended for the contact with piezoelectric transducer is evaluated at torsional and axial modes of vibration. During the frequency response analysis of piezoelectric transducer voltage and power output evaluation, its ground and voltage boundaries are applied to the bottom and top surfaces. The polling is in axial direction; thus, the piezoelectric transducer has d_{33} main operating mode. The piezoelectric transducer voltage output V_{out} and power output P_{out} are evaluated accordingly in the FE model [11]:

$$Vout = abs(cir.R1\_v), \tag{5.15}$$

and

$$Pout = realdot(cir.R1\_i, cir.R1\_v), \tag{5.16}$$

where $cir.R1\_v$ and $cir.R1\_i$ are the parameters of voltage and current amplitudes, as measured over the applied electrical load.

The performed eigenfrequency and frequency response studies show whether the L&T mode coupling effect can be achieved by manipulating the tool holder planar geometry and how it can be utilized to achieve the charge generation properties of a piezoelectric transducer. The obtained voltage and power output measurements from the simulation allow evaluating whether a sufficient amount of power can be generated during milling operation with the proposed design energy harvester. This energy can be used to power ultra-low power electronics, enabling data processing and wireless transmission and how the power output can be optimized by varying tool holders' geometrical parameters.

5.2.1.1 Simulation of Rotating Shank-Type Tool

FE modeling of the milling tool was performed in order to evaluate its vibrational characteristics and possibility to simplify the 3D model of the tool to a simple cylindrical shape, thus reducing requirements for computational resources and removing unnecessary complexity of the simulation at the same time avoiding any relevant negative effect on the results. For this purpose, two geometrical models, i.e. a two-flute end mill tool and its simplified cylindrical shape representation, have been created with Solidworks software. Both geometrical models are provided in Fig. 5.48.

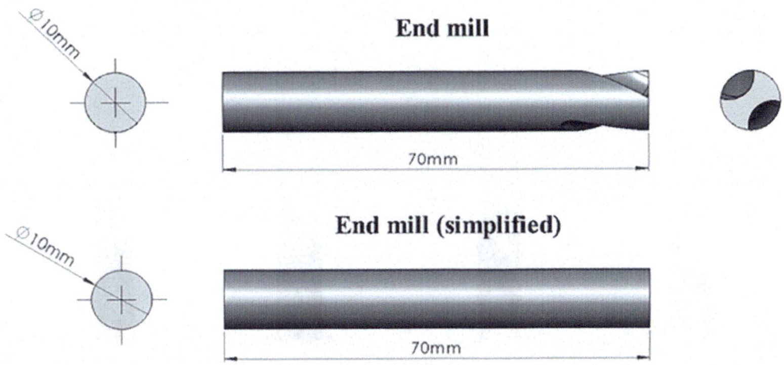

Fig. 5.48 Geometrical models of end mill tool and proposed its simplified version

For both models, the main geometrical parameters have been kept identical. Molybdenum has been selected as a material for both end mill models, as it is a common material used for high speed steel (HSS) cutting tools. The material properties are presented in Table 5.4.

The fixed boundary conditions that have been used during the simulation represent the actual expected clamping conditions of the end mill tool, when assembled together with cone-shaped tool holder. The fixed boundary conditions, eliminating displacement in x and y directions have been used for both end mill tool models and are displayed in Fig. 5.49a. Both end mill models have been meshed using free tetrahedral elements (Fig. 5.49b, c).

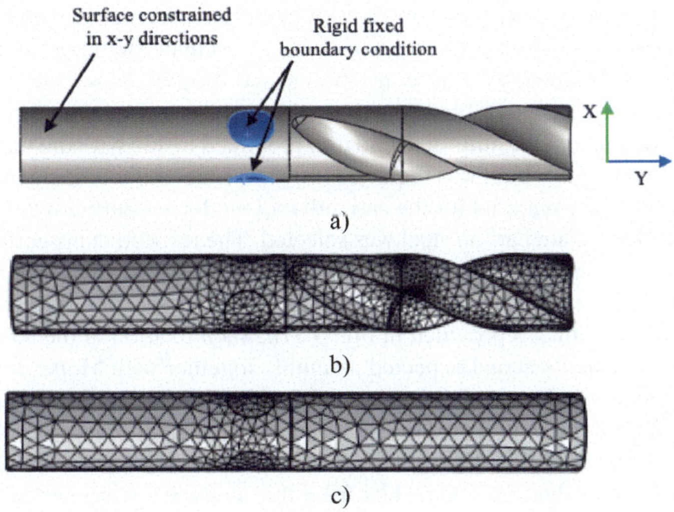

Fig. 5.49 **a** Boundary conditions set for both end mill models, **b** meshed two flute end mill and **c** meshed simplified end mill models using free tetrahedral elements

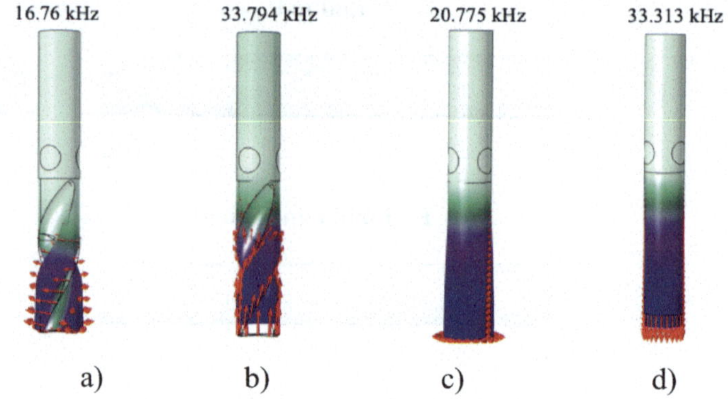

Fig. 5.50 Eigenmodes: **a**, **c** torsional and **b**, **d** axial for two flute and simplified end mill models

The eigenfrequency study has been performed for both end mill models to investigate torsional and axial mode of vibration, as the focus to achieve the L&T mode coupling. The results of the performed eigenfrequency simulation showed that torsional and axial modes of vibration were at 16.76 and 33.794 kHz for the actual end mill geometrical model, and for simplified cutter model at 20.775 and 33.313 kHz. The results of the study are presented in Fig. 5.50.

As presented in Table 5.6, the obtained eigenfrequency study results show that the difference between axial vibrational mode frequencies for the conventional and simplified end mill model is 1.4%, though the torsional mode frequency between the two end mill models is 19%. This can be expected, as the replacement of the actual two-flute geometry by a cylinder increased the model's torsional stiffness.

Taking into account that the end mill has a relatively small part of a larger assembly, a further investigation was performed in order to evaluate the axial and torsional vibrational mode frequency for the conventional and simplified end mill geometrical models when they are assembled together with the cone-shaped tool holder. The assembly CAD model of a simplified end mill tool fixed inside with the tool holder is provided in Fig. 5.51a, which was imported into the virtual model for eigenfrequency study evaluation. The material for the end mill tool model was left molybdenum, and for the tool holder, plain carbon steel was selected. The respective properties of both materials used in simulation can be found in Table 5.4.

During FE eigenfrequency analysis, the tool holder was rigidly fixed over its flange surface domains, as provided in Fig. 5.51b. Such fixation of the tool holder is intended to replicate its actual expected assembly together with Morse cone.

As in previous simulation step, the assembly models with conventional and simplified end mills have been subjected to eigenfrequency study in order to evaluate their torsional and axial vibrational mode frequencies. There, the first ten eigenfrequency modes have been evaluated. The results from this study are presented in Fig. 5.52, where axial and torsional eigenmode frequencies can be seen. For assembly model with simplified two-flute end mill geometrical model, the torsional vibrational mode

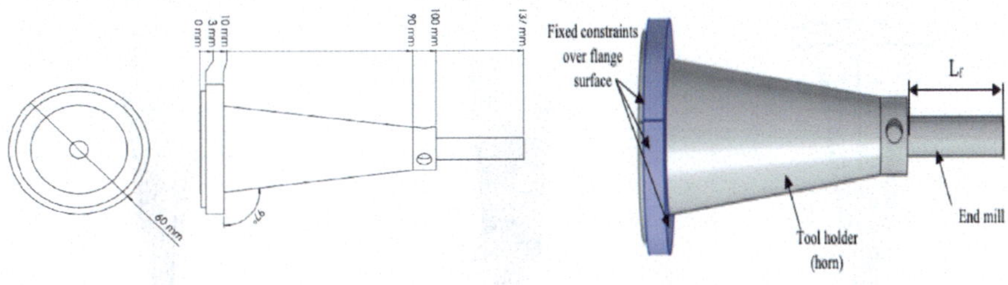

a) b)

Fig. 5.51 Geometrical assembly model of simplified end mill tool with tool holder (**a**) and tool holder with end mill tool geometrical model rigidly fixed constraint boundary conditions L_f = 37 mm (**b**)

Table 5.6 Frequency difference between torsional and axial modes for conventional and simplified end mill tool models

Eigenmode	Conventional end mill	Simplified end mill	Difference %
Torsional	16.76 kHz	20.775 kHz	19.3
Axial	33.794 kHz	33.313 kHz	−1.4

is achieved at 14.491 kHz and axial mode at 15.286 kHz, while for conventional two-flute end mill geometrical model torsional eigenmode is achieved at 14.226 kHz and axial mode at 14.823 kHz. In this case, the difference between torsional mode frequencies is 1.5%, and between axial mode frequencies it is 3.9%. These results are provided in Table 5.7 as well. From the obtained results, it is evident that using a simplified end mill model in assembly for FE studies with horn-type tool holder does not result in a significant impact on the system's vibration response. Although such tool model simplification enables to reduce the complexity of the final FE model, thus reducing the computational time and the required resources for the simulation. For this reason, in further simulation steps, only simplified end mill model shall be used.

5.2.1.2 Simulation of Waveguide with Helical Slots

Cone-shaped tool holder, sometimes, in scientific literature as well referred to as "horn", is usually used in machining processes, where it is applied to concentrate specific mode vibrations, exciting the tool, which helps to reduce the cutting forces and improve the surface quality of the machined workpiece. In such applications, the motion generated at the tool-workpiece interface is usually of longitudinal, torsional or longitudinal and torsional composite form, which can be achieved either by utilizing a transducer that is able to synchronically generate both eigenmodes at once or by means of introduction of geometrical features on the surface of the tool

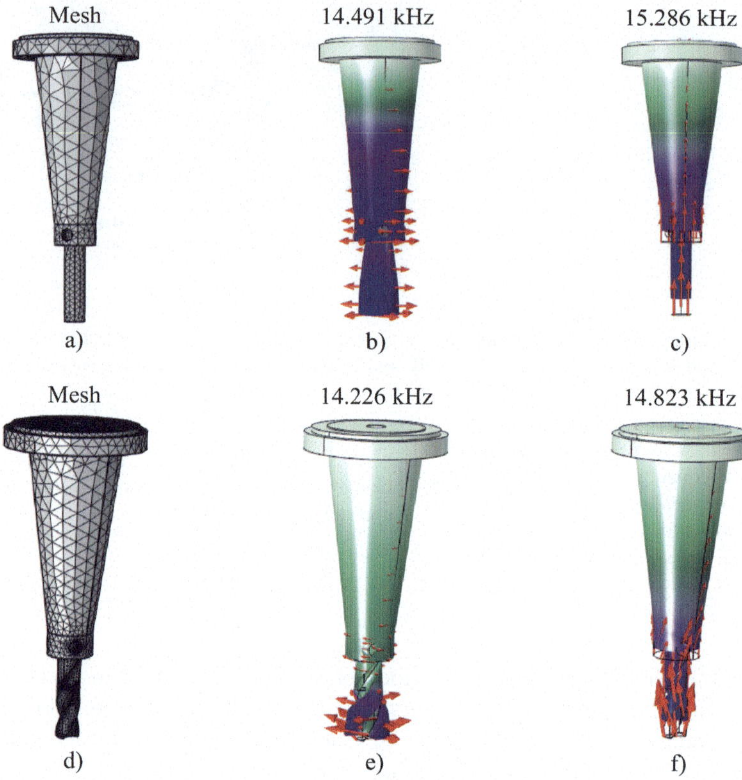

Fig. 5.52 Mesh of assembly with simplified **a** end mill model: **b** torsional and **c** axial vibration eigenmode study results. Mesh of assembly with conventional **d** model: **e** torsional and **f** axial eigenmode study results

Table 5.7 Frequency difference between torsional and axial modes for tool holder assembly models with two-flute and simplified end mills

Vibrational mode	Conventional end mill	Simplified end mill	Difference %
Torsional	14.226 kHz	14.491 kHz	1.5
Axial	14.823 kHz	15.286 kHz	3.9

holder that enable the transformation of the longitudinal motion by the transducer into the $L\&T$ form at the tool.

This part of simulation work is performed in order to investigate whether the approach where the introduction of geometrical features in the form of slots on the planar surface of the horn can be used in reverse action. In such a way, when the end mill tool is excited by torsional vibrations, once these vibrations are transferred to the tool holder, they shall be partially transformed into longitudinal vibrations. In order to perform such investigation, a horn-type tool holder design has been created

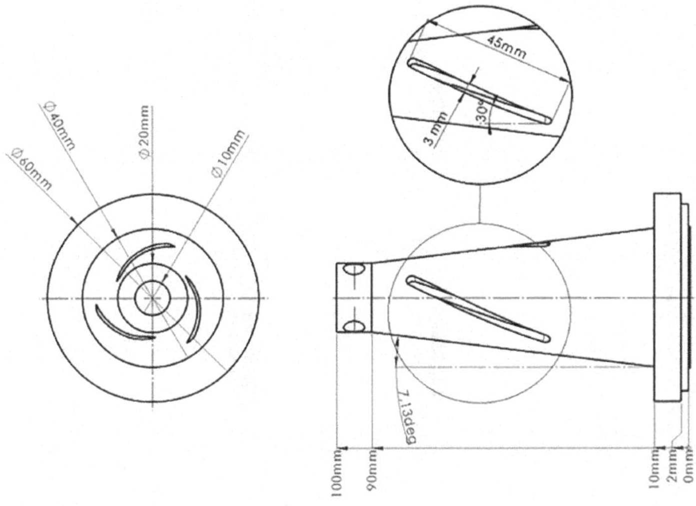

Fig. 5.53 Horn type tool holder assembled together with a simplified end mill model design with helical slots formed on its conical surface

by using Solidworks CAD software package with helical slots formed on its conical surface (Fig. 5.53). This tool holder model was imported into Comsol for FE studies to perform eigenfrequency and frequency response analysis.

The tool holder, presented in Fig. 5.53, has three uniformly distributed helical slots, formed on its conical surface, which are 45 mm in length, 3 mm in width and 3 mm in depth, with a 30° angle, formed with the tool holder's longitudinal axis. The tool holder, as in the previous studies, is assembled together with a simplified end mill tool's model. The selected material for the tool holder is plain carbon steel, and for the end mill tool, molybdenum. As a first step of the simulation, the eigenfrequency study was performed. During the eigenfrequency analysis, the tool holder's fixed boundary condition was set over its whole flange surface, as defined in the previous simulations Fig. 5.51.

The results from the performed modal analysis are presented in Fig. 5.54. There the obtained results show a frequency of 13.9 kHz for the torsional vibrational mode and 15.4 kHz for the longitudinal/axial vibrational mode.

The close frequency of the torsional and the longitudinal eigenmodes shows the introduction of the mode coupling effect where the horn can be excited to resonate in both modes simultaneously.

Further, the eigenfrequency study has included a parametric analysis evaluating how the use of different materials for the tool holder influences the formation of the torsional and axial modes of vibration. For this study, three different materials, namely, steel, aluminum, and brass, have been selected. Their mechanical properties are presented in Table 5.8.

The performed parametric study results are presented in Fig. 5.55. These results indicate that the softer material is selected for the tool holder, the higher decrease of an

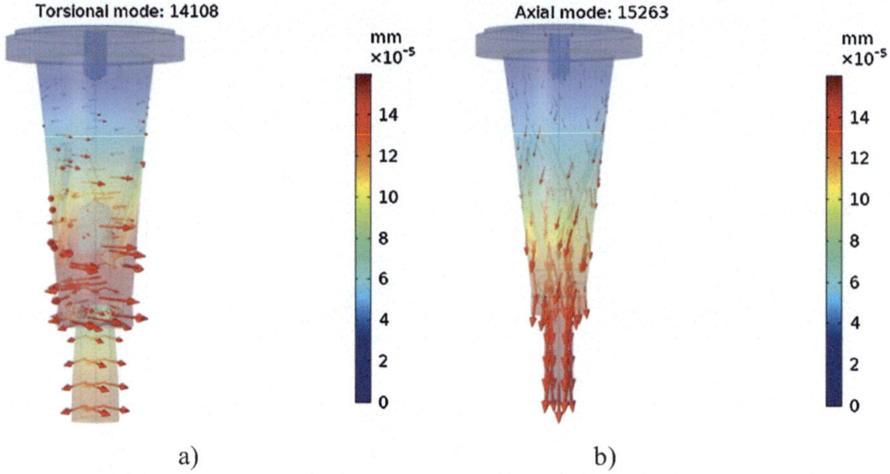

Fig. 5.54 Tool holder with helical slots **a** torsional and **b** axial modes of vibration

Table 5.8 Mechanical properties of steel, aluminum and brass materials

Materials	Density (kg/m^3)	Youngs Modulus (GPa)	Poisson's Ratio
Steel	8000	210	0.28
Aluminum	2700	69	0.33
Brass	8730	113	0.34

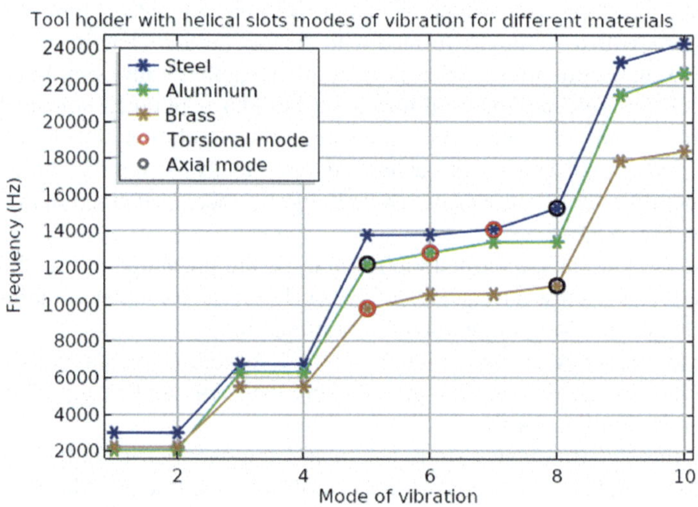

Fig. 5.55 Tool holder with helical slots torsional and axial vibrational modes frequency dependence on material type

Table 5.9 Geometrical parameters of the slots used during parametric study

Parameter	Start value	Step value	End value
Tilt angle	0°	10°	70°
Length	0 mm	10 mm	70 mm
Width	1 mm	1 mm	10 mm
Depth	1 mm	1 mm	6 mm

L&T mode frequency is observed. This is important in cases where the application required lower L&T mode coupling frequency, but the reduction of the structure stiffness is acceptable.

The efficiency of conversion from longitudinal to torsional vibrations and vice versa is dependent on the geometrical parameters of the slots, such as tilt angle, length, width, and depth. Changing these geometrical parameters affects the increase or the decrease in difference (bandwidth) between torsional and longitudinal eigenmode frequencies. Eigenfrequency study has been extended to evaluate how changes in these geometrical parameters of the slots, formed on the conical surface of horn type tool holder, affect the longitudinal and the torsional eigenmodes and their coupling. The material, which has been selected for the horn-type tool holder, is plain carbon steel.

The parametric sweep for the denoted geometrical parameters of the slots has been performed, evaluating how their variations affect the formation of the torsional and the axial mode frequency. The variations of these parameters during eigenfrequency study are presented in Table 5.9. The study results are presented in Fig. 5.56 as well as Tables 5.10 and 5.11, separating the parameters individually.

From the results, it is clear that the increase in the geometrical parameters of the slots has negligible effect on the axial mode frequency, though for torsional mode, specifically, the increase of depth leads to a significant decrease in mode frequency and widening of the frequency gap between the torsional and the axial mode, thus resulting in L&T mode degeneration. This is due to the increased depth of the slots, when the stiffness in the torsional direction of the tool holder is decreased, it leads to higher torsionality. For the application of the tool holder use during the milling operation, the depth parameter was selected 3 mm, ensuring relatively high torsionality and close L&T mode coupling frequency, while retaining sufficient structure stiffness. Other parameters of the helical slots for the tool holder have been set to 45 mm in length, 3 mm in width, and 30° tilt angle.

In the next part of the study, the impact that the tilted slots have on the generation of the longitudinal vibrations during L&T mode coupling was evaluated. This has been done by performing a frequency response study over 10 ÷ 20 kHz frequency range, when the end mill tool is subjected to the end milling excitation forces.

For this purpose, two tool holder models (Figs. 5.51a and 5.53), assembled together with the simplified end mill design, were imported into the simulation model. In this study, the denoted two tool holder models were subjected to the axial, tangential, and radial milling component forces, applied on the surface boundaries of the end mill, as provided in Fig. 5.51a. The applied axial, tangential, and radial forces

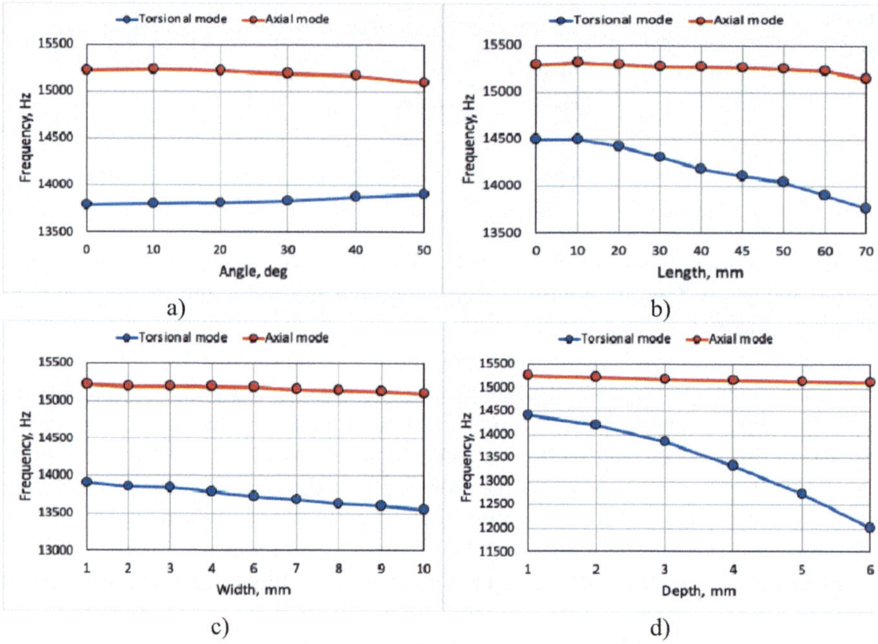

Fig. 5.56 Tool holder torsional and axial eigenmode frequency dependence on **a** tilt angle, **b** length, **c** width, and **d** depth of the slots

Table 5.10 Tool holder torsional mode frequency dependence from slot parameters

Parameter	Torsional mode frequency at start value	Torsional mode frequency at end value	Difference, %
Tilt angle	13,790 Hz	13,905 Hz	+ 0.9
Length	14,499 Hz	13,771 Hz	−5
Width	13,914 Hz	13,556 Hz	−2.5
Depth	14,422 Hz	12,030 Hz	−16.5

Table 5.11 Tool holder axial mode frequency dependence from slot parameters

Parameter	Axial mode frequency at start value	Torsional mode frequency at end value	Difference, %
Tilt angle	15,228 Hz	15,094 Hz	−0.9
Length	15,301 Hz	15,148 Hz	−1
Width	15,222 Hz	15,107 Hz	−0.8
Depth	15,268 Hz	15,109 Hz	−1

Table 5.12 Selected milling process parameters for calculation of force components acting on the milling tool

Description and symbol	Value	Unit
Milling depth, a_p	1	mm
Cutter diameter, d_0	10	mm
Feed per tooth, f_z	0.05	mm/tooth
Radial depth of cut, a_e	0.4	mm

were calculated according to Eqs. (5.17–5.19). A simplified approach to calculate tangential, radial, and axial end milling cutter forces according to empirical formulas has been proposed in [12]:

$$F_c = 9.81(96.6)a_e^{0.88} f_z^{0.75} a_p z d_0^{-0.87} \tag{5.17}$$

$$F_r \approx (0.80 \sim 0.90) F_c \tag{5.18}$$

$$F_a \approx (0.35 \sim 0.4) F_c \tag{5.19}$$

where $\boldsymbol{F_c}$ is the tangential milling force (N), $\boldsymbol{F_r}$—radial milling force (N), $\boldsymbol{F_a}$—axial milling force (N), $\boldsymbol{a_p}$—milling depth (mm), $\boldsymbol{d_o}$—cutter diameter (mm), z—number of cutter tooth, $\boldsymbol{f_z}$—feed per tooth (mm/z).

Milling process parameters (milling depth, feed per tooth, and radial depth of cut), used in the calculation of these machining forces, exerted on the end mill tool, have been selected as applicable to rough milling operation of the steel workpiece and are provided in Table 5.12.

During the frequency response study, the average displacement value of the tool holder surface in longitudinal (Y) direction was measured. The view of this tool holder surface is provided in Fig. 5.57. It is the intended contact surface with piezoelectric transducer.

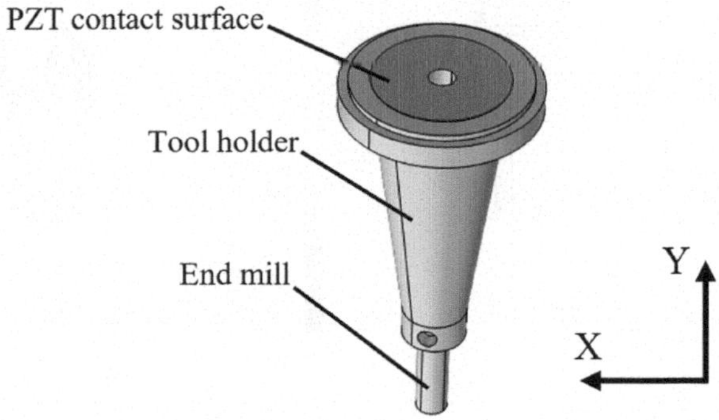

Fig. 5.57 Tool holder surface (blue color) used to measure the longitudinal displacement

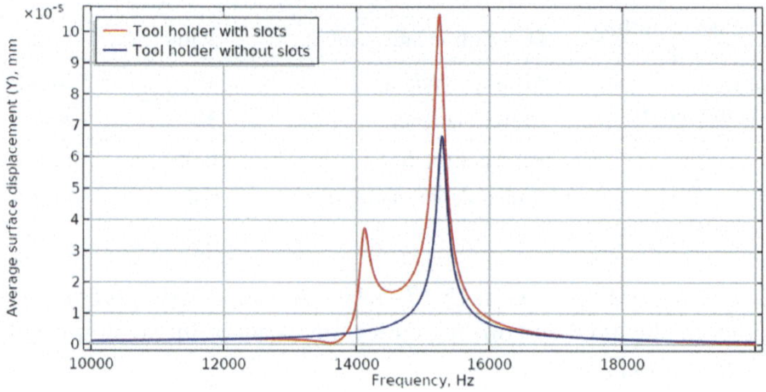

Fig. 5.58 The comparison of the average surface displacement measurement results for the tool holder with and without helical slots

Study results, presented in Fig. 5.58, indicate that the introduction of the slots not only causes the displacement in the longitudinal direction at the torsional mode (in red), but leads to the L&T mode coupling effect as well. It can be seen that the introduction of helical slots leads not only to the formation of the L&T mode coupling, but also to the increase in the longitudinal surface displacement at the axial mode of vibration of the tool holder assembly, which is almost two times higher. This can be understood as the effect of the helical slots, which increases the torsionality of the tool holder, partially allowing to transform the radial and the torsional force components, acting on it, into the contraction and the extension of the horn, expressed as additional longitudinal motion.

This is clearly visible in contour displacement visualizations that are presented in Fig. 5.59, where both tool holders are excited at the torsional mode; the tool holder

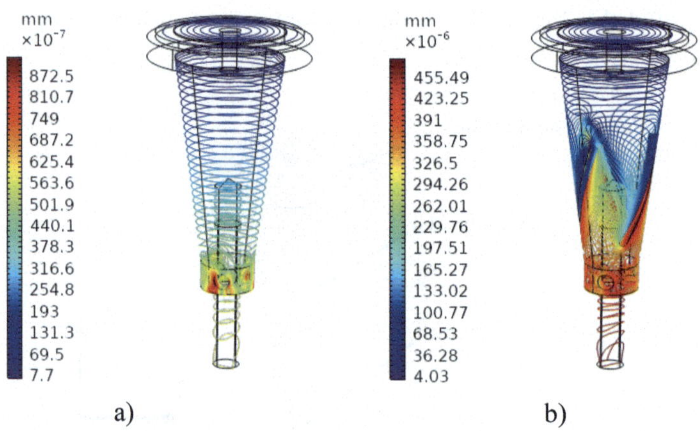

Fig. 5.59 Surface longitudinal displacement at the torsional mode for **a** the tool holder without helical slots and **b** the tool holder with helical slots

Table 5.13 Milling force exerted on the end mill component values of the slots used for the frequency response parametric study

Force	Start value	Step value	End value
Axial	0 N	5 N	20 N
Tangential	0 N	5 N	20 N
Radial	0 N	5 N	20 N

with helical slots experiences more than five times higher surface deformations in the longitudinal direction, which is transferred to the PZT contact surface.

Further on, during the frequency response study, the surface displacement in the longitudinal direction for the tool holder with helical slots was evaluated when milling forces varied. The variations of the milling force component values according to Table 5.13 are presented in Fig. 5.60.

The obtained results indicate that in case of the tool holder, excited at the axial mode of vibration, the increased amplitude of the axial milling force component has the highest impact on the formation of the longitudinal vibrations, while at the torsional mode of vibration, the increase in the radial and torsional milling force component leads to higher change in the longitudinal displacement.

5.2.1.3 Simulation of Energy Harvesting Process

Up to this point, the research findings show that the tool holder with uniformly distributed helical slots on its conical surface enables to transform dominant broadband torsional vibrations partially, which are present during the milling operation, into longitudinal motion, which is complimenting the already existing axial vibrations. These longitudinal vibrations in the design of the proposed device are to be used for deforming an axially polarized d_{33} piezoelectric transducer that is embedded together with the tool holder, thus producing electrical charge.

In order to evaluate the piezoelectric transducer energy generation properties a frequency response analysis over 10 $\div$ 20 kHz frequency range, which covers both torsional and axial vibrational mode frequencies, has been created and performed. This study has been performed on two different geometrical models of the tool holder, i.e. with and without helical slots. During the simulation, the tool holder has been assembled together with the piezoelectric transducer and pre-stressed by a back-mass. The view of the assembly model with relevant milling force components applied on the end mill is provided in Fig. 5.46.

During the FE study, the simplified end mill tool model is subjected to the axial, radial, and torsional milling force components, calculated according to formulas (5.17 $\div$ 5.19), allowing to simulate the milling operation. The fixed boundary conditions are set over the whole surface area of the tool holder's flange. The voltage response from the piezoelectric transducer is measured and normalized over the defined frequency range for both tool holder models, enabling to compare the energy generation differences that have been obtained from the piezoelectric transducer. During this study, the 40 $\times$ 32 $\times$ 10 mm size PZT-5H ceramic piezoelectric d_{33}

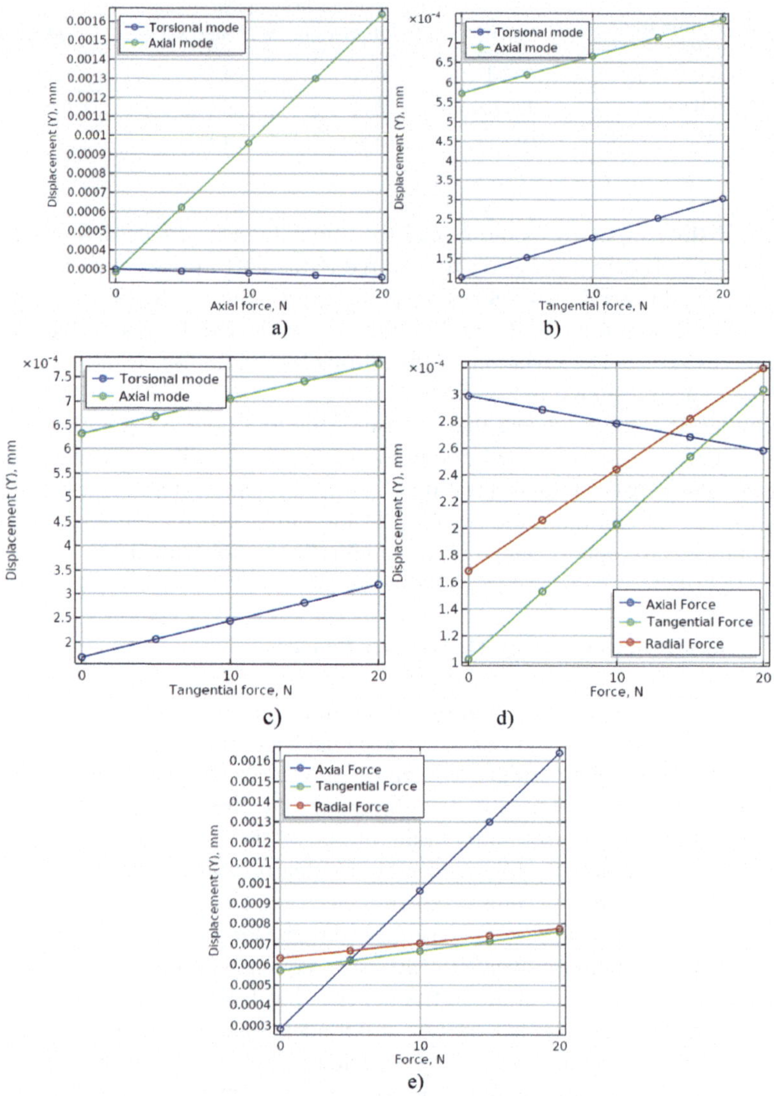

Fig. 5.60 Tool holder planar surface longitudinal displacement at torsional and axial modes of vibration depending on change in **a** axial force component, **b** tangential force component, **c** radial force component; surface displacement dependence on milling force components when the tool holder is excited at **d** torsional mode and **e** axial mode

type transducer, comprising of 4 ring-shaped elements, was selected (Fig. 5.47a) and fixed together with the tool holder on the surfaces, opposite to the end mill tool position. The PZT-5H material properties are provided in Table 5.5. The piezoelectric transducer was pre-stressed by the $40 \times 10 \times 10$ mm size aluminum back-mass. Back-mass design is provided in Fig. 5.47b.

The size and design of the piezoelectric transducer were selected according to the maximum surface axial displacement results, which were obtained during tool holder torsional resonant mode, to optimize its deformations and voltage output.

The first and the second subscripts “33”, used in piezoelectric “*d*” constants of the selected piezoelectric transducer, define the direction of the field and the direction of the resulting strain for charge constant. For piezoelectric voltage constant, the first subscript defines the direction of the generated voltage, and the second one indicates the direction of the applied force. Thus, the selected piezoelectric transducer, when deformed in its axial direction, will produce a charge.

Besides, the updated geometry with a PZT and back-mass rings, in order to simulate the voltage output from the piezoelectric transducer, two additional COMSOL Multiphysics software packages were introduced into the existing simulation model, i.e. electrostatics and electrical circuits.

For the study of the frequency response simulations, the piezoelectric transducer has been connected to a resistor load, which varied from $10\,\Omega$ to $100\,k\Omega$, where the voltage and the power, which was generated across this load, has been measured at torsional and axial vibrational mode frequencies of the tool holder. This evaluation study enables to assess and select an optimum matching load resistor, ensuring optimal power generation properties from the piezoelectric transducer. The results that have been obtained from this study (Fig. 5.61) show that the maximum power output from the piezoelectric transducer at the torsional and axial vibrational modes is obtained when a resistor load of $31.6\,\mathrm{k}\Omega$ is connected to the piezoelectric transducer. This load resistor value shall be used in all other studies when evaluating the output from the piezoelectric transducer.

In the next step of the study, the load resistor value that was used in the model has been selected according to the results from Fig. 5.61a with voltage and power output from the piezoelectric transducer that was measured over the frequency range of 10 ÷ 20 kHz.

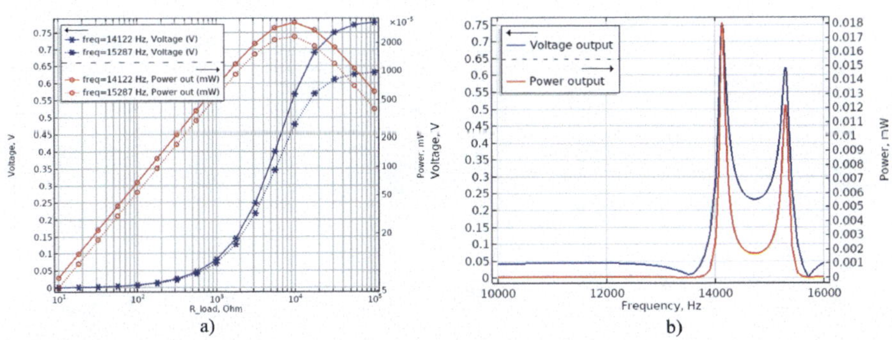

Fig. 5.61 Piezoelectric transducer generated voltage and power output dependence on the load resistor value (**a**) and over 10 ÷ 20 kHz frequency range (**b**)

Table 5.14 Voltage and power output from piezoelectric transducer when the tool holder is excited at torsional and axial vibrational modes

Measured output	Torsional mode, 14,125 Hz	Axial mode, 15,297 Hz	Difference, %
Voltage, (V)	0,75 V	0,62 V	17
Power, (mW)	0,017 mW	0,012 mW	29,4

From the study results, presented in Fig. 5.61b, it can be observed that the tool holder with uniformly distributed slots, embedded with axially polarized piezoelectric transducer, will generate the highest voltage and power output, when the tool holder is excited to resonate at its torsional mode of vibration. These result values are presented in Table 5.14. From the obtained results, it is clear that the tool holder with helical slots will experience the L&T mode coupling when excited; the L&T mode coupling results in frequency bandwidth between the torsional and axial vibrational modes, where a significant power output from the piezoelectric transducer can be expected under the excitation conditions.

During the simulation, the obtained power output measurement results from the piezoelectric transducer, when the tool holder is excited at the longitudinal and torsional mode coupling frequency, is enough to power ultra-low power application microcontrollers that are available in the market.

A frequency response study has been updated by importing the tool holder's geometrical model without helical slots and performing measurement of voltage and power output from the piezoelectric transducer over 10 ÷ 20 kHz frequency range. The obtained voltage and power results were compared to a tool holder with helical slots. The graphical study results for voltage output comparison are presented in Fig. 5.62a and power output between the tool holder with and without helical slots in Fig. 5.62b.

The summary of the results can be found in Table 5.15.

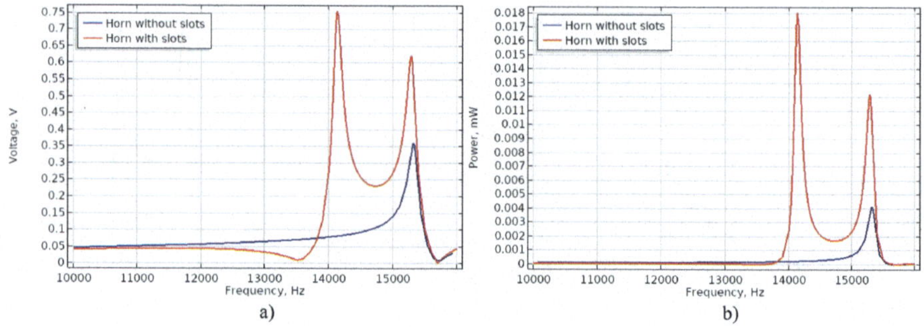

Fig. 5.62 Piezoelectric transducer generated voltage output for the tool holder with and without helical slots (**a**) and power output dependence on the load resistor value (**b**)

Table 5.15 Comparison of measured voltage and power output from piezoelectric transducer for tool holder with and without helical slots

Tool holder	Torsional mode		Axial mode	
	Voltage (V)	Power (mW)	Voltage (V)	Power (mW)
Without helical slots	0,09	0,00,095 mW	0,36 V	0,004
With helical slots	0,75 V	0,017 mW	0,62 V	0,012 mW

It is clear from the results that the transducer that is embedded together with the tool holder with helical slots on its surface generates significantly higher power output as compared to the assembly, where instead a tool holder without helical slots is used; this supports the previously obtained results presented in Fig. 5.58 where the highest surface displacement in the longitudinal direction is achieved when the tool holder with helical slots is excited to resonate at its torsional mode. This shows that these geometrical features, i.e. helical slots, introduce the longitudinal and torsional mode coupling effect, enabling to transform torsional vibrations partially at the end mill/tool interface into longitudinal vibrations that are transferred through the tool holder, deforming piezoelectric transducer and producing higher voltage output, when compared to a traditional design tool holder. The L&T mode coupling as well enables to increase the frequency range of the transducer, at which, when excited, a significant amount of power can be generated.

As for the tool holder without helical slots, the excitation at the torsional mode of vibration does not lead to a significant transformation of this motion into longitudinal deformations; the voltage output from PZT is more than eight times lower, and basically, an insignificant amount of power is generated in this mode.

In case where the tool holder with helical slots is excited at its axial mode of vibration, the piezoelectric transducer is generating 50% more voltage and three times more power as compared to the output from the piezoelectric transducer, embedded with a traditional design tool holder, because in this case, the torsional vibrations are still partially transformed into longitudinal vibrations, supplementary deforming the piezoelectric transducer.

The frequency response study, where the voltage output from the piezoelectric transducer embedded in the tool holder with helical slots is measured, has been further developed to evaluate the change of the voltage output from the piezoelectric transducer when the component values of the milling forces that are acting on the end mill tool vary. For this study, the milling force component values have been changed, as provided in Table 5.16.

Table 5.16 Range of milling forces used for parametric study

Force	Start value	Step value	End value
Axial	0 N	5 N	20 N
Tangential	0 N	5 N	20 N
Radial	0 N	5 N	20 N

Parametric study results are presented in Fig. 5.63. It can be seen from the obtained results that in case of the tool holder, excited at the axial mode of vibration, the increased amplitude of an axial milling force component has the highest impact on the formation of voltage output from the piezoelectric transducer. While at torsional mode of vibration, the increase in the radial and torsional milling force component leads to a higher rate of change in voltage output from the PZT.

The voltage rate of change dependence on the milling force component is provided in Table 5.17. At the torsional mode of vibration, the tangential force component has the highest effect on the increase in voltage that has been generated from the

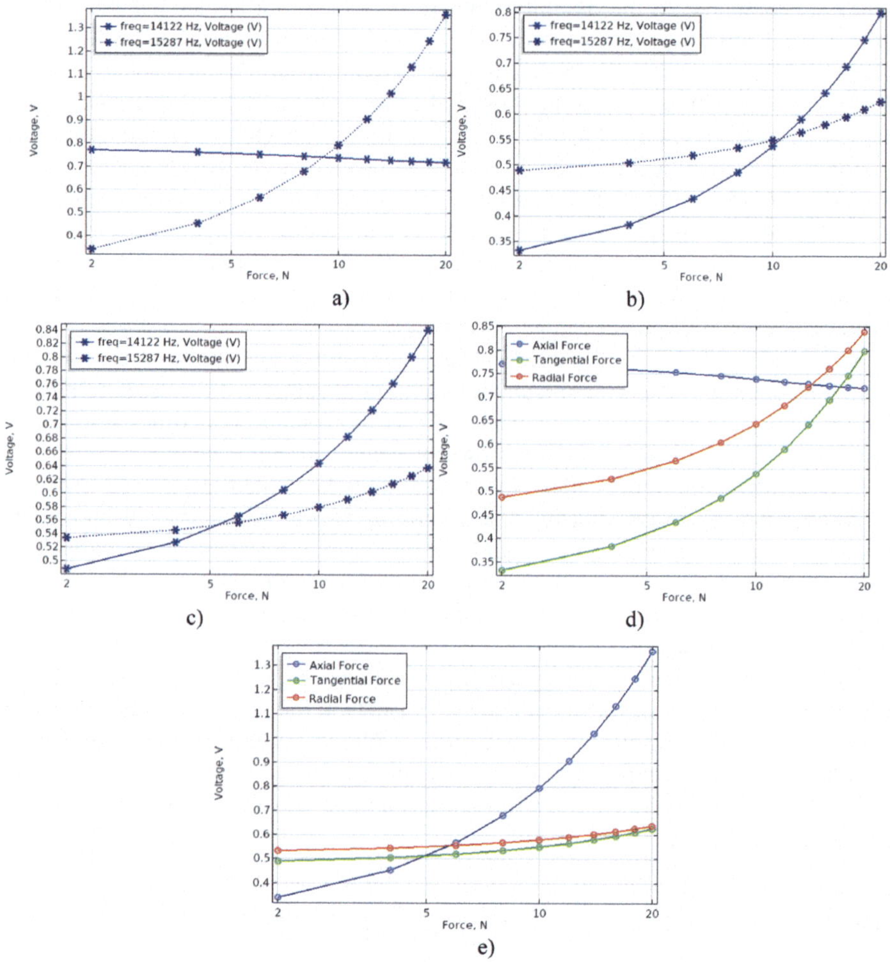

Fig. 5.63 Piezoelectric transducer voltage output when the tool holder is excited at torsional and axial modes of vibration dependence on **a** axial, **b** tangential, and **c** radial force components; voltage output dependence on milling force components when the tool holder is excited at torsional **d** and axial **e** eigenmode

Table 5.17 Voltage output from piezoelectric transducer rate of change dependence on milling force components at torsional and axial modes of vibrations

Milling force component	Voltage rate of change (V/N)	
	Torsional mode	Axial mode
Axial force	−0,002	0,05
Tangential force	0,025	0,007
Radial force	0,02	0,0057

piezoelectric transducer while at the axial mode of vibration, the force that is acting in longitudinal direction of the end mill tool influences the voltage output the most.

5.2.2 *Rotating Shank-Type Tool Condition Monitoring*

The design of a device for monitoring condition of rotating shank-type tools was proposed. The proposed device, according to the obtained FE study results, was comprised of a cone-shaped tool holder with three uniformly distributed helical slots on its conical surface, a piezoelectric transducer, and a PCBA (printed circuit board assembly). A CAD 3D model of the device that was designed with SolidWorks software is presented in Fig. 5.64, providing the cross-section and exploded views with the elements of the assembly that are defined in Table 5.18 [13].

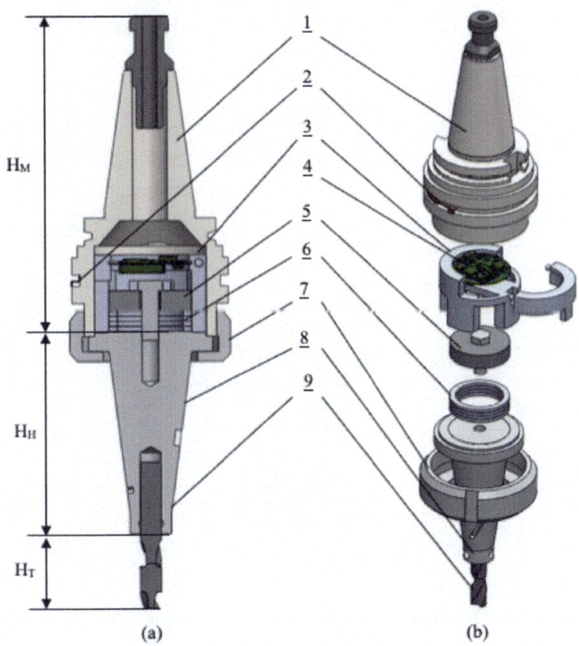

Fig. 5.64 Energy harvester device assembly section view (**a**) and energy harvester device assembly exploded view (**b**)

Table 5.18 Elements constituting rotating shank type tool condition monitoring device

Number	Component description
1	Holder's Morse cone for assembly inside CNC center
2	Antenna for wireless data transmission
3	PCBA holder inside tool holder
4	PCBA with data processing and transmission components
5	Back-mass
6	Stack type piezoelectric transducer
7	Flange for assembling tool holder with Morse cone cover
8	Cone shaped tool holder with helical slots
9	End mill tool

The general geometrical parameters of the proposed vibrational energy harvester are provided in Fig. 5.64a, where $H_M = 156$ mm, which is the height of the Morse cone, $H_H = 100$ mm is the height of the cone-shaped tool holder, and $H_T = 36$ mm is the free length of the end mill. The 3D model of horn type tool holder made from steel C45 1.0503 is presented in Fig. 5.64b.

A more detailed drawing of the energy harvester assembly model is provided in Fig. 5.65.

The size and the shape of the tool holder are identical, as used during FE modeling (Fig. 5.54). In this case, three uniformly distributed helical slots of 45 mm in length,

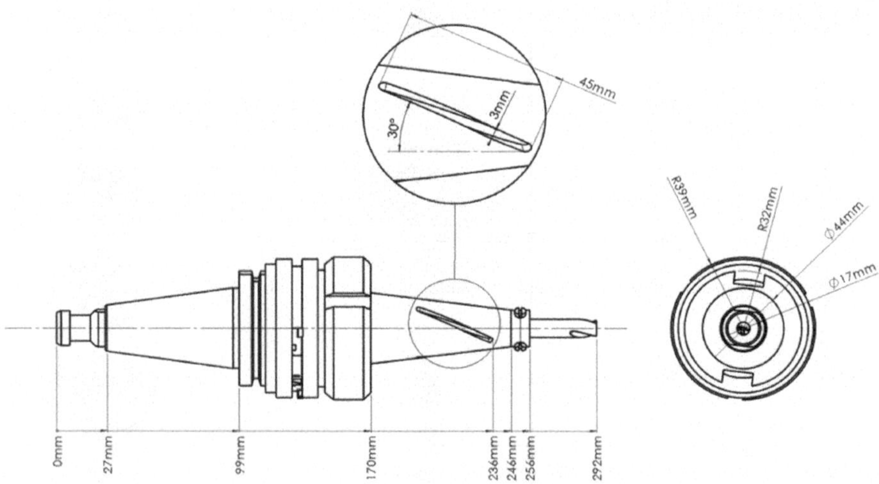

Fig. 5.65 Detailed assembly drawing of designed energy harvesting device

3 mm in width, 3 mm in depth and having a 30° angle with the longitudinal axis of the tool holder are introduced on its cone-shaped surface.

5.2.2.1 Sensor Node Printed Circuit Board Assembly Design

The sensore node consists of the PZT-5H stack-type piezoelectric transducer and aluminum 3003 back-mass where the piezoelectric transducer on one side is pressed against the surface of Morse cone-shaped tool holder and on the other side pre-stressed by aluminum back-mass. Inside the Morse cone, together with the piezo-electric transducer and a back-mass, the PCBA (printed circuit board assembly) is included. PCBA was designed to enable power management, data processing, and wireless communication of a sensor node. The 3D model of the designed PCBA is presented in Fig. 5.66.

As the proposed device is designed to operate on a tough power budget, it is important that all of the electronics are selected taking low power budget requirements into account. For this reason, an MCU ULP (ultra-low power) microcontroller MSP-430G2553 from Texas Instruments has been selected, which has great performance characteristics for the use in energy harvesting applications. The MCU-MSP430 operates at the range of $1.8 \div 3.6$ V, has a standby current of less than 0.5 μA and active current of 230 μA/MHz, its wakeup time is less than 1 μs. Other options that are available on the market for microcontrollers that are designed for low power application are Atmel ATMega 8-bit MCU series, NXP ARM Cortex-M0-based LPC111x series, ST STM32 ARM family MCUs and Nordic Semiconductor Bluetooth system-on-chip modules with ARM Cortex-M4 microcontrollers. Power performance characteristics for some of the mentioned microcontrollers are provided in Table 5.19. The MSP430 series from TI has been selected due to its comparable performance characteristics, as compared to the available alternatives, as well as because of the existing knowledge of the microcontroller architecture.

For wireless communication, a MLT BT-05 BLE 4.0 (Bluetooth low energy) serial communication module is included. This Bluetooth module comes with the operating

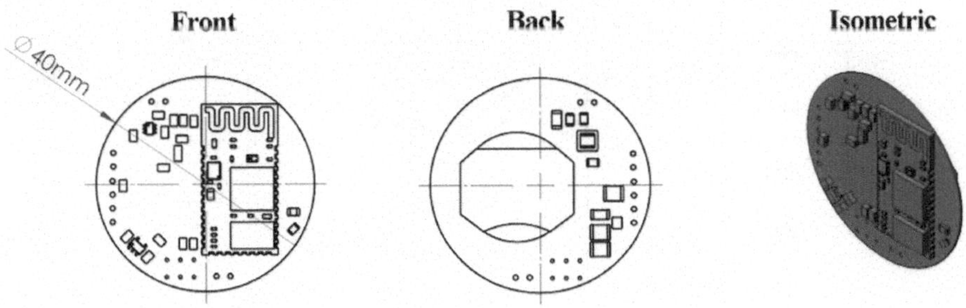

Fig. 5.66 Front, back, and isometric views of PCBA with MCU and Bluetooth module; back view of the designed PCBA show introduced placement for coin type battery

Table 5.19 Power performance characteristics for some of market available low power microcontrollers

MCU Manufacturer	MCU Type	Operating voltage	Current consumption		
			Active	Stand by	Off mode
Texas Instruments	MSP-430G2553	1.8–3.6 V	230 μA/MHz	< 0.5 μA	< 0.1 μA
Microchip (Atmel)	ATmega128L	2.7–5.5 V	5.5 mA at 4 MHz, 5 V	N/A	0.15 μA
NXP	LPC111x	1.8–3.6 V	150 μA/MHz	6 μA	220 nA
ST	STM32L4	1.71–3.6 V	120 μA/MHz	35 μA/MHz	330 nA

voltage between 3.6 $\div$ 6 V, standby power consumption of around 90 $\div$ 400 μA, has fast response speed of 400 ms, as well ensuring effective communication distance of 7 $\div$ 10 m from the transceiver with a maximum transmitter power of 6 dBm. Though the distance of more than 10 m (up to 60 m) is possible, the connection quality substantially decreases when the range increases, exceeding 10 m. On the opposite side of the PCBA, an optional external power supply in the form of a 3 V coin type battery is implemented for the use during the experimental investigation of uninterrupted data transmission.

The electrical schematics of the designed device PCBA are presented in Fig. 5.67. The schematics include a voltage multiplier, which is composed of Schottky diodes (D1A and D1B) and capacitors (C2, C3, C5, C6), to which the generated voltage from the piezoelectric transducer is being fed during the machining operation. The Voltage from the voltage multiplier is fed to charge the capacitor C4. The charge level is measured by the MCU U2 10bit ADC input and compared to a set threshold of the charge level value, which is set in the MCU. In case the measured C4 capacitor voltage level increases above this set voltage threshold level, a load, consisting of

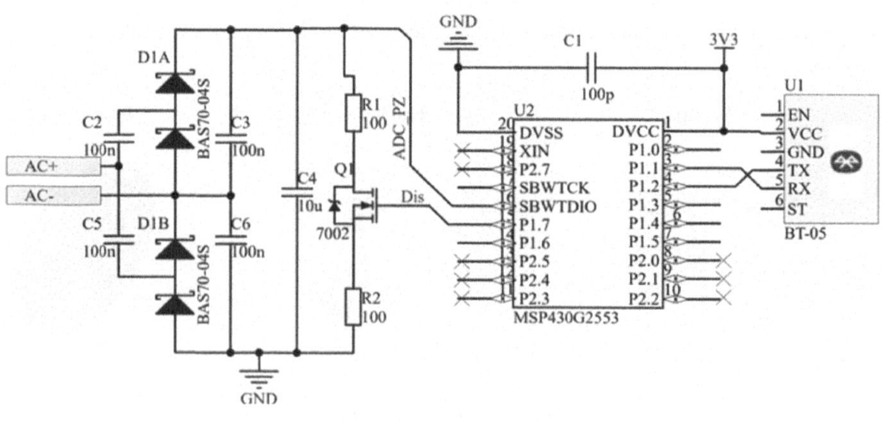

Fig. 5.67 Electrical schematics of the prototype PCBA used with the designed sensor node

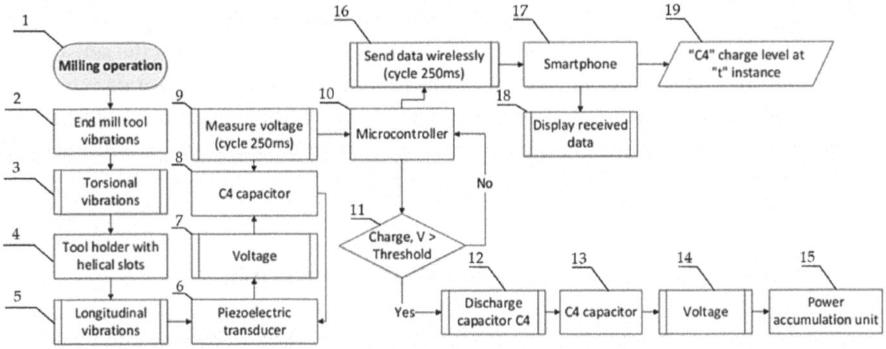

Fig. 5.68 Working process flow of the wireless energy harvesting sensor node used to detect end mill tool condition wear state

resistors R1 and R2, is switched by a Q1 N-channel field transistor, discharging the C4 capacitor. Once the C4 capacitor is discharged, another measuring cycle is initiated.

The voltage discharged from the C4 capacitor is fed to a power accumulation unit, which in the proposed design is a supercapacitor, where it is stored and can be used to power up the electronics of the sensor, providing the self-powering capability. The event of the capacitor discharge is registered by the MCU, and the instance of the event information is sent wirelessly via Bluetooth to another Bluetooth-enabled device, such as a smartphone, where time intervals between them are being logged and evaluated. The principle of operation of the developed device, to be used during milling operation, is presented graphically in the flowchart (Fig. 5.68).

As provided in Fig. 5.68, during the milling operation *1*, the predominantly random torsional vibrations exciting the cutting tool (*2.* and *3.*) are transmitted to the tool holder with helical slots *4*. At the tool holder, these torsional vibrations are partially transformed into longitudinal vibrations *5* and transferred to deform an axially polled piezoelectric transducer *6*. The voltage (*7*) from the piezoelectric transducer *6* is converted into a DC signal and continuously fed to the C4 capacitor *8*. During milling, the charge level of the capacitor C4 is measured *9* by an embedded microcontroller *10* at every 250 *ms* time interval. The microcontroller performs the following tasks: it compares the charge level of the capacitor C4 with a predetermined value *11*, in case the measured capacitor charge level exceeds the predetermined threshold, the microcontroller initiates the discharge of the capacitor *12*. Here, the capacitor *13* voltage *14* is discharged into the power accumulation unit *15*, which is used as a power source by the sensor itself and the charging cycle of the capacitor C4 *8* is restarted. In addition to controlling the discharge of the capacitor, the microcontroller *10* also initiates wireless data transmission *16* via Bluetooth to the smartphone *17*. The data transmitted contains information on the charge level of the capacitor at the time of measurement. The smartphone is used here to display the received data *18* and to store it on a local hard drive *19* for later processing and analysis.

Thus, the proposed sensor node design not only enables the energy harvested by the piezoelectric transducer to be used as an alternative power source of the sensor, but also to measure and record the change of the generated voltage over time, expressed as the change in the charge level of the capacitor. Here, the exponential increase of the capacitor charge level over time can be related to the gradual wear of the end mill tool.

5.2.2.2 Sensor Node Architecture and Working Principle

The architecture of the designed device is provided in Fig. 5.69a together with its flow diagram that is depicting its operation during milling, presented in Fig. 5.69b.

The working principle of the proposed device is based on the energy harvesting during the machining operation, implementing rotating shank-type tools. During the operation, as the tool wear condition appears and increases, the torsional forces that are exciting the tool increase as well [13]. These vibrations that are excited in the tool are transferred to the tool holder, where, due to the introduced helical slots, they are partially transformed into longitudinal vibrations, which are transferred to the

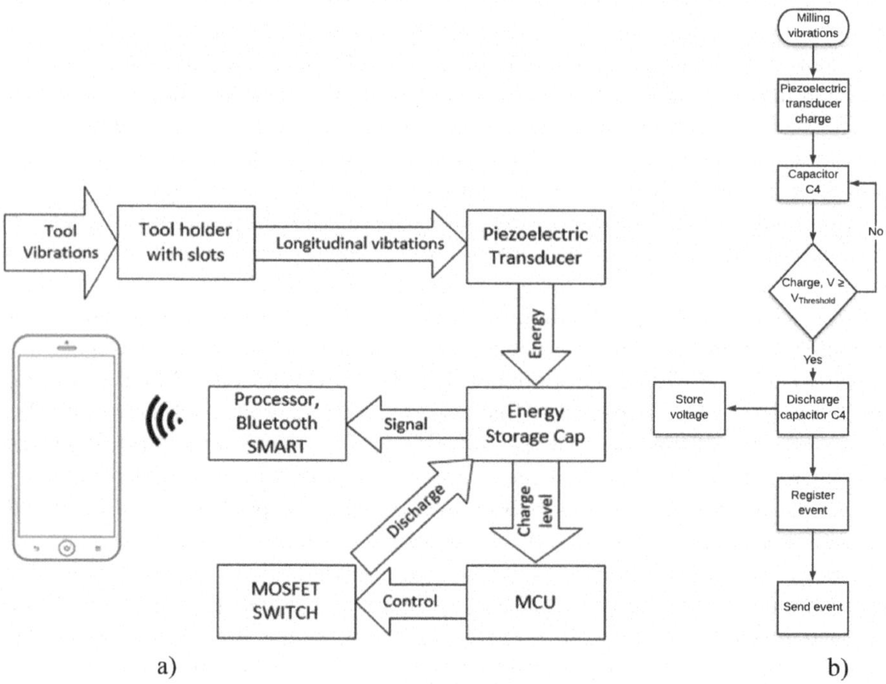

Fig. 5.69 Designed wireless sensor node architecture (**a**) and operational process flow of the designed sensor node (**b**)

interface with a piezoelectric transducer, deforming it and producing voltage that is fed to the C4 capacitor.

In this scenario, an increase in the end mill tool wear condition leads to the increased deformation of the piezoelectric transducer, thus increasing the amount of the generated voltage and reducing the C4 capacitor charging time, this way relating the increase in the end mill tool wear to the decrease in the C4 capacitor charging time. In the sensor node, the capacitor charge level is measured by the MCU and is compared to the set voltage threshold level. In case this threshold is reached, the capacitor C4 is discharged into a power storage unit, i.e. a supercapacitor or other power cells, and the capacitor discharge event is registered. Then, the information about the capacitor discharge event is sent wirelessly to a defined auxiliary smart device, such as a smartphone.

An important functionality of the sensor node is to differentiate between the RUNNING (actual operation) and the IDLE milling condition states. The sensor's process flow for differentiating between these two states is presented in Fig. 5.70.

If the capacitor is charged and its charge level exceeds the specified voltage limit, it will be discharged. In the case of capacitor discharge event, the "counter" value is increased by +1, and the timer is initiated; the "counter" value is used to compare to the "imp. count" value, which is set and represents the number of consecutive capacitor discharge events that are required for the status to be evaluated as actual

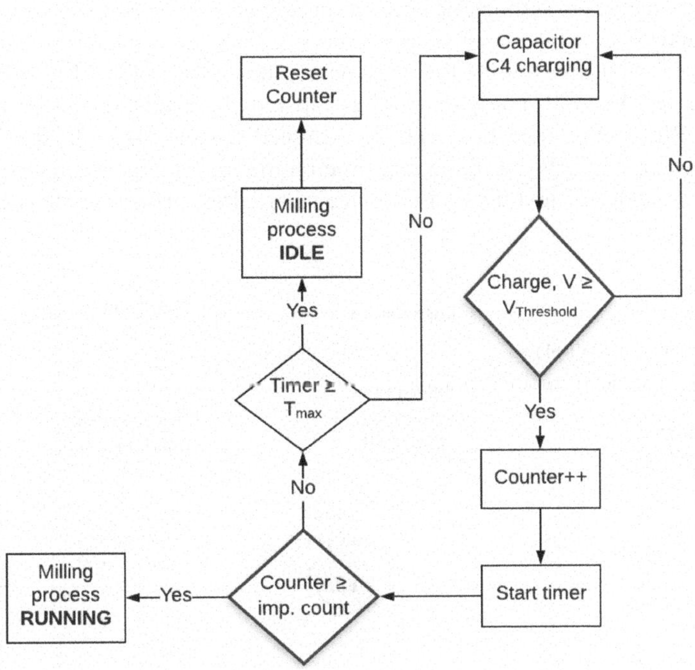

Fig. 5.70 Wireless sensor node decision on the flow of operation status process

milling operation, i.e. RUNNING. However, if the "counter" value is lower than the set threshold, the sensor checks if the Timer value has reached or exceeded Tmax. If Tmax is exceeded, it evaluates the milling operation as not running and goes to the IDLE state. Otherwise, the sensor waits for consecutive capacitor discharge events.

In this case, the Tmax value is set and represents the maximum time interval, during which, if no capacitor discharge event is present, the sensor has to go to IDLE/sleep state.

The primary communication device, to which the information about the capacitor discharge event during the milling operation is sent from the proposed wireless sensor over LAN or PAN network, is defined as an edge device. There, the communication with the edge device is performed over RF using WI-FI, Bluetooth, Bluetooth Smart, ZigBee other communication protocol, depending on the constraints that are set by the required data throughput, the distance from the transceiver, the environment as well as the available device power budget.

In general, the edge devices provide local on-site data pre-processing, data feedback, and display, at the same time acting as a gateway to the internet (Fig. 5.71).

Through the edge device, the pre-processed data from the wireless sensor node can be sent to the Cloud services for the higher-level data analytics and archiving in remote servers. Such Cloud services enable the user to access and view the collected information in the form of the generated clear data reports over dedicated application. The cloud service interface can as well be used to interact with the device itself, allowing to change its settings or perform other tasks, such as OTA (over the air) firmware updates.

In this research, the role of the edge device has been assigned to a Bluetooth enabled iOS or Android smartphone that is capable of establishing and maintaining a wireless Bluetooth connection with the designed sensor, where the data received from the sensor is displayed in the real time on the smartphone's screen and at the same time written as a text file on the device's hard drive. This text file is saved and

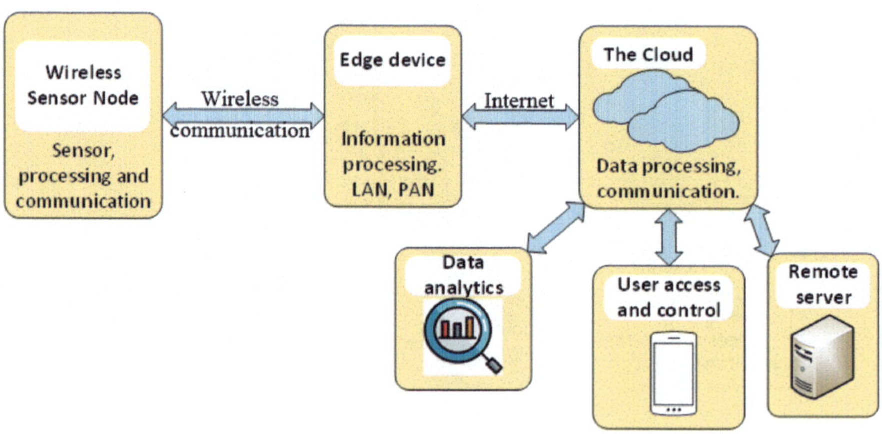

Fig. 5.71 Wireless sensor's node network architecture

can be exported for further data processing to other devices, such as PC. Thus, in the following experimental research, the sensor data is not sent via internet to any kind of Cloud service, but it is stored on a smartphone. The proposed architecture of the device supports such extensions for the future applications. The operation flowchart for the edge device is presented in Fig. 5.72.

There, the condition of the edge device is ON, meaning its transceiver is turned on, and it is ready to communicate information with the sensor. If the information about the capacitor discharge event from the sensor is received, it is stored in the reference value denoted as $E0_T$. The $E0_T$ stores the first received time value, representing the capacitor discharge event instance information, which is represented by the time value after IDLE condition. Subsequently received capacitor discharge event information, which is as well represented by time, is stored in value $E1_T$ container. These time values represent the time instance of the capacitor discharge event.

This process continues until both $E0_T$ and $E1_T$ values are registered, and once this is achieved, the time difference between $E1_T$ and $E0_T$ values is calculated by subtracting them and set to DIFF holder, which has to be checked in comparison to

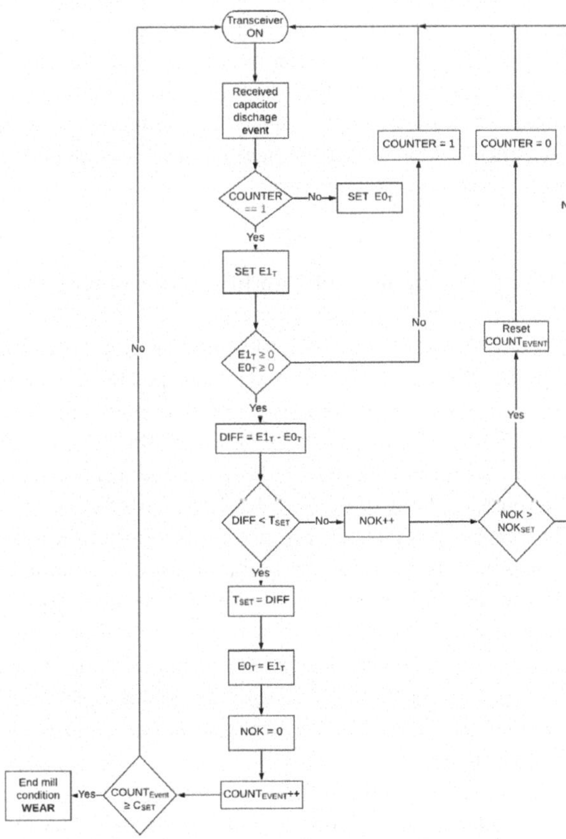

Fig. 5.72 Edge device operation process flow

T_{SET}. T_{SET} is a set threshold reference value, which is representing the minimum time interval between the two consecutive capacitor discharge events that can be regarded as a possible tool wear condition occurrence. If this condition is not fulfilled, the process is restarted.

If T_{SET} is higher than the calculated DIFF value, T_{SET} value is updated with DIFF value, and $E0_T$ is updated with $E1_T$ value. This is done so that the consecutive capacitor discharge event occurrences would be compared to the previous event instance, because if the tool condition is starting to wear out, the time interval between two capacitor discharge events will start to reduce after each consecutive capacitor discharge event.

In case all previous conditions are fulfilled, the $COUNT_{EVENT}$ value will be incremented in a continuous loop, until it exceeds set C_{SET} value. If this condition is true, the tool will be evaluated as worn out, and the edge device will send information back to the operator or the machine, informing it about the tool condition in order to prepare for a tool change. There, the C_{SET} value is a set value, representing a threshold of the consecutive decreases in the capacitor discharge events, after which the tool condition can be indicated as worn out.

In case all previous conditions are fulfilled the $COUNT_{EVENT}$ value will be incremented in a continuous loop until it exceeds set C_{SET} value. If this condition is true, the tool will be evaluated as worn out and the edge device will send information back to the operator or the machine informing it about the tool condition in order to prepare for a tool change. Here C_{SET} value is a set value representing a threshold of consecutive decreases in capacitor discharge events after which the tool condition can be indicated as worn out.

5.2.2.3 Design of Tool Condition Monitoring Device

In order to investigate how the introduction of helical slots affects the vibrational response of the tool holder, two prototypes have been manufactured, according to the prepared 3D CAD model (Fig. 5.64b). Both tool holders were manufactured from the C45 1.0503 steel, as defined during the FE simulations in the research.

The difference between the prepared prototypes was that one had uniformly distributed helical slots (Fig. 5.65), machined on its cone-shaped surface, while the other was kept without any geometrical alterations. These two manufactured tool holders have been used during the vibrational response study to evaluate whether the introduction of helical slots on the surface of the tool holder results in the increase in longitudinal displacement of the surface, which is intended to be in contact with the piezoelectric transducer. The experimental setup block diagram used for this study is presented in Fig. 5.73a, while the actual view from the performed experiment is presented in Fig. 5.73b, c. The experimental set-up that was used for both tool holders was identical.

As presented in the block diagram of the experimental set-up, the end of the tool holder, where the end mill tool was going to be fixed, was used as a contact surface with the piezoelectric actuator, exciting the tool holder. The piezoelectric generator

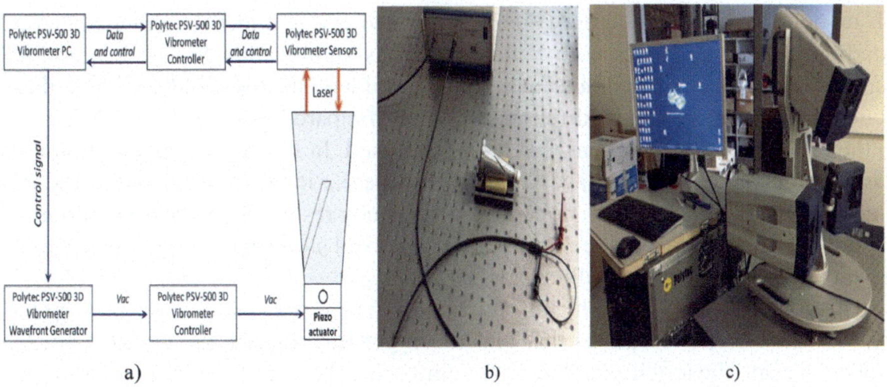

Fig. 5.73 Tool holder vibrational response test set-up block diagram (**a**), view with fixed piezoelectric actuator (**b**) and view of Polytec PSV-500 3D laser Doppler vibrometer (**c**)

was connected to a waveform generator, exciting it by a chirp type signal in the 0 ÷ 50 kHz frequency range.

A Polytec PSV-500 3D laser Doppler vibrometer was used to perform non-contact surface displacement measurements. These measurements were made on the tool holders' surface, which is dedicated to contact with the piezoelectric transducer, see Fig. 5.73a. The results from the performed vibrational response experiment are presented in Fig. 5.74.

It can be seen from the findings that in case of the tool holder with helical slots, if it is excited to resonate at its axial mode, the measured surface displacement amplitude is twice as high, when compared to the obtained results for the tool holder without helical slots, when it is as well excited to resonate at its axial mode. The vibrational response study results agree with the results that have been obtained during the

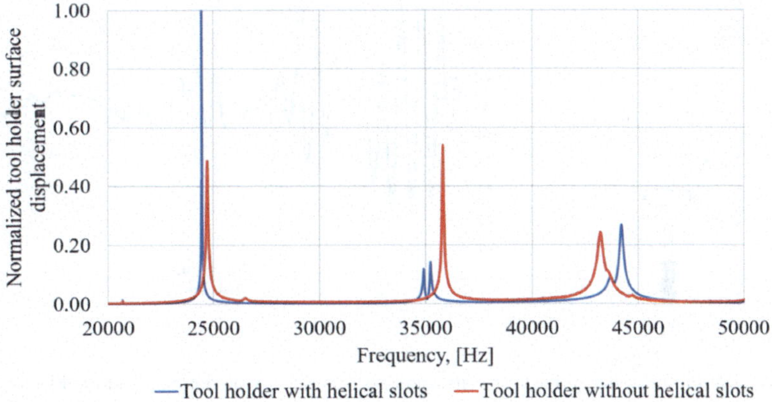

Fig. 5.74 The measured surface displacement amplitudes for the tool holder with and without helical slots excited at axial eigenmode

FE modeling of the tool holder (Fig. 5.59), showing that under the same excitation condition, the tool holder with helical slots has significantly higher longitudinal surface displacement amplitudes, when compared with the tool holder without these helical slots. The frequency differences, when compared to the FE model, are due to the different fixing positions that have been used. In FE model, the tool holder is fixed over its outer flange surface, (Fig. 5.51b) whereas in the experimental study, it is placed on its own free weight; nonetheless, the study results show the trend observed during the FE studies that the introduction of helical slots on the tool holder lead to the increase of longitudinal vibrations.

This is achieved because the introduction of helical slots enables partial transformation of torsional vibrations that have been generated at the input surface of the tool holder into the longitudinal motion, reinforcing the already existing longitudinal vibrations. These combined longitudinal vibrations are transferred through the tool holder, deforming a piezoelectric transducer.

The second stage of the experimental research was performed to investigate the vibrational response of the developed device during the impact test. For this, the cone-shaped tool holder with three uniformly distributed helical slots on its cone-shaped surface was assembled together with three flute end mill, axially poled PZT-5H piezoelectric transducer, aluminum 6082 back-mass and a wireless sensor with an embedded system, all of which were installed inside the Morse cone mounted on the spindle. The whole device was assembled inside the Leadwell V-20 CNC milling centre's spindle, as provided in the experimental set-up block diagram in Fig. 5.75a.

In order to determine the vibrational response and the resonant modes of the assembled device, an impact test has been performed, where the teeth edge of the end mill tool has been excited mechanically. The response of the excitation was

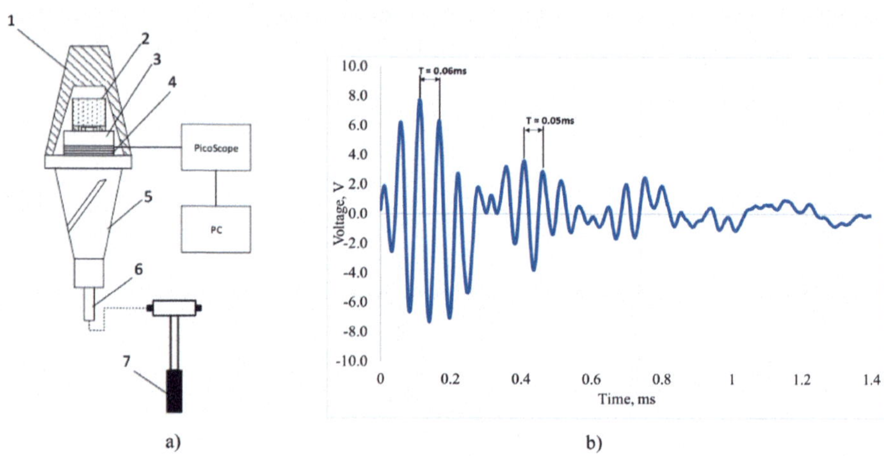

Fig. 5.75 **a** Vibration energy harvester impact test set-up: 1-Morse cone, 2-embedded wireless sensor node, 3-aluminum back-mass, 4-axially poled piezoelectric transducer, 5-horn type tool holder with helical slots, 6-three flute end mill tools, 7-device excitation tool and **b** voltage output from piezoelectric transducer

recorded in the form of voltage output from the piezoelectric transducer. For this reason, the piezoelectric transducer output to the wireless sensor electronics was disconnected and instead connected directly to a PicoScope oscilloscope, which was used to capture the change in the voltage, generated by the piezoelectric transducer during the impact test. This piezoelectric transducer output signal is recorded by a dedicated PicoScope software package and can be analysed later.

The results of the voltage output over the time from the piezoelectric transducer during the impact test, which were recorded by a dedicated PicoScope software package, are presented in Fig. 5.75b.

From the presented results, a very interesting phenomenon has been observed, i.e. the amplitude modulation of the exponentially decreasing transient vibrations, as well known as the beating phenomenon. The beating phenomenon is an amplitude modulated response, i.e. a combination of two vibration signals with slightly offset frequencies from one another. From these results (Fig. 5.75b), it is visible that two close longitudinal and torsional vibrational modes interfere constructively and destructively, causing signal amplitude modulation in a regular pattern.

This observed beating frequency can be expressed by formula:

$$f_{\mathrm{b}} = |f_{\mathrm{T}} - f_{\mathrm{L}}| \tag{5.20}$$

where $\boldsymbol{f_T}$ and $\boldsymbol{f_L}$ are torsional and longitudinal eigenmode frequencies of the tool holder.

Such high-frequency spontaneous excitation of the milling tool's axial vibration is similar to the vibrational milling, where the tool is excited by additional high-frequency vibrations. In such case, a versatile method of transforming the milling vibrations into voltage is proposed, where the energy of the cutting tool's torsional vibrations is captured by a piezoelectric transducer, while at the same time being a low-frequency oscillating milling method, which does not require the need for an additional piezoelectric actuator.

This widely applicable cutting technology, which is able to handle a broad range of machining shapes and materials, is ideal for cutting difficult-to-cut materials. It is as well the state-of-the-art technology, which reduces risks associated with the machining these materials, such as the nesting of chips and built-up edges.

In the proposed prototype, the torsional vibration generated during the milling operation lead to the simultaneous axial deflection and elongation of the slotted cone-shaped tool holder, i.e. the elongation and contraction of the structure, which is consistent with Bayly's model [14]. Then, the excited longitudinal vibrations are transferred through the body of the tool holder, deforming the embedded piezoelectric transducer and generating charge.

5.2.2.4 Web-Based Systems Development for Data Transmission

The third part of the experimental research was focused on evaluating the energy harvesting properties of the developed device under actual milling conditions and its dependence on the milling process parameters. The milling process is described as an interrupted cutting process, because during this process, the rotating teeth of the end mill tool are subjected to an impact force cycle, when they enter and exit the workpiece, exciting the tool and machine tool vibrations. The intensity of these vibrations induced in the end mill tool depends on the manufacturing regimes and the condition of the tool itself, such as its tooth cutting edge wear. This means that it is not possible to avoid the milling tool vibrations, though such vibrations could be harvested and converted to electrical voltage for other uses.

Exploiting machine vibrations is one of the most effective forms for harvesting the environmental ambient energy, which can further be used to power small sensor, eliminating the need for wired connections and frequent maintenance, when compared to the use of other types of power supply.

The experimental test set-up block diagram, which was used to evaluate the energy harvesting properties of the developed device during the milling operation, is presented in Fig. 5.76a with an actual view inside CNC milling centre, as presented in Fig. 5.76b.

There, the full energy harvester assembly has been prepared with a spindle of the Leadwell V-20 CNC milling centre. The device configuration matched, as provided in Fig. 5.64, consisting of three flute HSS end mill fixed inside the tool holder, and on the other side, the axially polled piezoelectric transducer, the back-mass and an

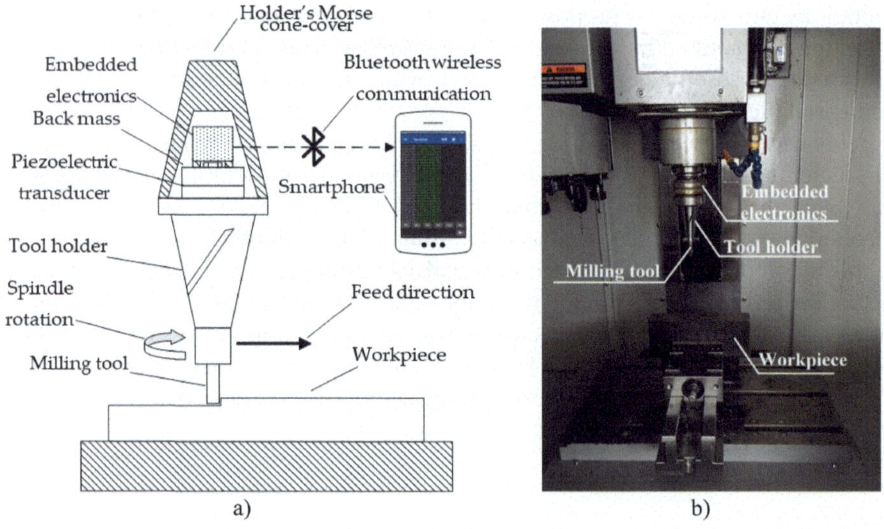

Fig. 5.76 Vibration energy harvester device for energy harvesting during the milling operation test set-up scheme (**a**) and vue inside CNC milling center (**b**)

Table 5.20 The selected milling process parameters used during the experiment

Parameter	Spindle speed	Feed speed	Feed per tooth	Depth of cut
Value	1210 RPM	148 mm/min	0.041 mm/tooth	1 mm

embedded system with a wireless communication capability, embedded inside the Morse cone.

During the actual milling operation, the two tool holders, one with helical slots and one without, were used. Two different configurations of the device allowed to assess how strongly the helical slots on the planar surface of the tool holder affect the amount of energy collected by the piezoelectric transducer during the milling process.

During the milling operation, a block L × H × B 250 mm × 50 mm × 50 mm of 1.0037 steel has been machined as a workpiece, for which the milling process parameters were selected according to the used end mill tool and workpiece material, as provided in Table 5.20.

During the milling operation, the wireless sensor was configured to send the information representing the actual C4 capacitor (Fig. 5.66) charge voltage level via Bluetooth every 500 ms to a Bluetooth-enabled smartphone. The received information was then stored inside the hard drive of the smartphone in a form of a text file. The text file included the received message with a timestamp and C4 capacitor charge level at the instance, when it was sent from the sensor. This information, stored inside the smartphone, was extracted for further data processing and evaluation.

A wireless sensor node was configured to discharge capacitor C4 if its voltage level reached or exceeded set threshold of 0.7 V, this would reset and repeat the capacitor charging process. The results recorded on the smartphone from the experiment during the milling operation for capacitor charging levels over the time are presented in Fig. 5.77.

Figure 5.77a represents extract from these results showing capacitor charging period during milling operation when tool holder without helical slots is implemented and Fig. 5.77b shows capacitor charging rate where tool holder with three uniformly

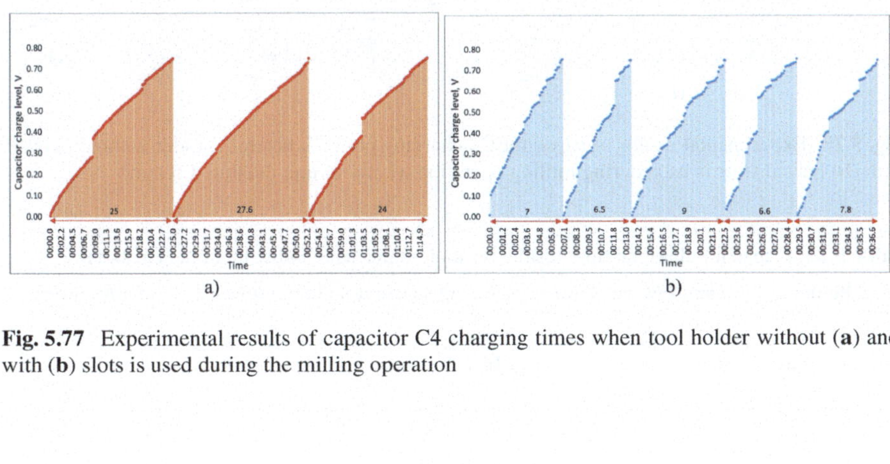

Fig. 5.77 Experimental results of capacitor C4 charging times when tool holder without (**a**) and with (**b**) slots is used during the milling operation

Table 5.21 Capacitor charging time results for tool holder with and without helical slots

Tool holder	Depth of cut	Average capacitor charge time	Difference
With slots	1 mm	7.25 s	3.44 times
Without slots	1 mm	25 s	

distributed slots is assembled with our device. From the obtained results it can be concluded that the average capacitor charging period to the set 0.7 V threshold is 7.25 s for tool holder with helical slots and in case tool holder without slots is used the average capacitor charging time is 25 s. The results show that the charging of capacitor is faster by more than 3.4 times for tool holder with helical slots, this means that up to 3.4 times more vibrational energy is harvested during milling operation over the same time interval if device is implementing the tool holder with helical slots. Results of the experimental study are summarized in Table 5.21.

In the next step, the experiment was repeated but this time the milling process parameter—milling depth has been increased. The milling depth was increased from 1 mm (used in previous experimental step) to 1.5 mm leaving all other process parameters and test set-up itself identical as defined in Fig. 5.73 and Table 5.20. The experiment was repeated for both assemblies, with tool holder that has helical slots and for tool holder without helical slots. Results for this experiment iteration with increased milling depth are presented in Fig. 5.78. Results of the experimental study are provided in Table 5.22.

The obtained results show that when the milling depth is increased to 1.5 mm the average capacitor C4 charging time to the set 0.7 V threshold is 4.7 s for tool holder with helical slots and 18.7 s for tool holder without helical slots. The results show

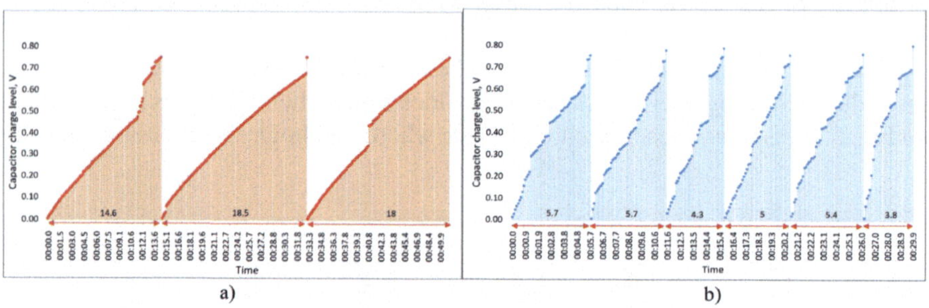

Fig. 5.78 Experimental results of capacitor C4 charging period when tool holder without (**a**) and with (**b**) helical slots is used during milling operation where milling depth is 1.5 mm

Table 5.22 Capacitor charging time results for tool holder with and without helical slots

Tool holder	Depth of cut (mm)	Average capacitor charge time (s)	Difference
With slots	1.5	4.7	3.97 times
Without slots	1.5	18.7	

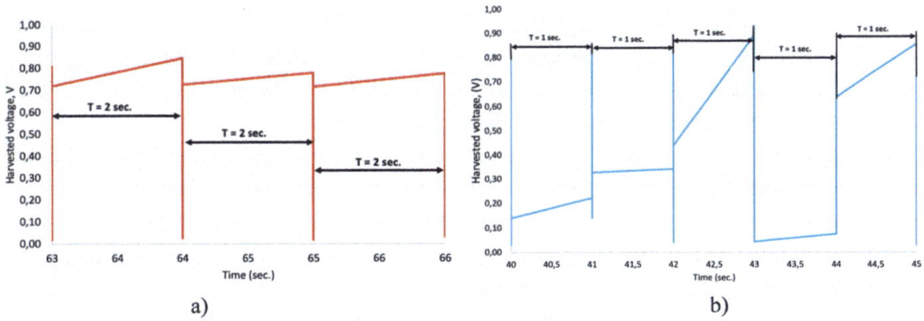

Fig. 5.79 Experimental results of capacitor C4 charging times when tool holder without (**a**) and with (**b**) helical slots is used during milling operation when spindle speed is increased from 1210 to 1500 rpm

that for device implementing helical slots the capacitor charging time is 4 times faster than device using tool holder without helical slots, that is four times more vibrational energy is harvested during milling operation over the same time period.

In both cases of performed experiments the capacitor charging time decreases as compared to results from Fig. 5.77, meaning more vibrational energy is harvested when milling depth parameter is increased from 1 mm to 1.5 mm. For tool holder without helical slots it decreases by 30% and for tool holder with helical slots by 35%. This means that increase in milling depth results in higher torsional vibrations excited in the end mill tool during operation.

The last iteration of the experiment was performed with another milling process parameter, spindle speed, increased from 1210 ÷ 1500 rpm, while other process parameters were maintained with milling depth at 1.5 mm. The experiment was performed to evaluate vibration energy harvesting properties for both tool holders with and without helical slots, and results are presented in Fig. 5.79.

Results show that with increase in spindle speed from 1210 to 1500 rpm there is a significant decrease in capacitor charging time until set threshold for both tool holders, with and without helical slots. Recorded average C4 capacitor charging period for device assembled with tool holder without helical slots was 1.8 s and for tool holder with helical slots it was 0.44 s. Results show that as in latter experiment step the difference between implementing tool holder with helical slots results in 4 times faster capacitor charging as compared when the device implements tool holder without helical slots. Results are summarized in Table 5.23.

Table 5.23 Capacitor charging time results for tool holder with and without helical slots

Tool holder	Spindle speed (RPM)	Depth of cut (mm)	Average capacitor charge time (s)	Difference
With slots	1500	1.5	0.44	4.09 times
Without slots	1500	1.5	1.8	

Important observation is that capacitor C4 charging time decreased significantly for tool holder with or without helical slots with increase in milling depth and spindle speed, which means that process parameter has a significant effect on amplitude of vibrations excited in the end mill tool during operation.

The increase in amount of harvested energy can be anticipated with increase in spindle speed, because it leads to increased frequency of tool teeth interaction instances with the workpiece, thus leading to increased frequency of the milling tool excitation events. And increase in milling depth leads to higher forces exerted on the milling tool cutting edge during its impact cycle.

In general, the increase in harvested energy where tool holder with uniformly distributed helical slots is used during milling operation can be attributed to the appearance of L&T mode as observed during frequency response evaluation of the developed device (Fig. 5.73). When the device is excited at frequency of L&T mode coupling the radial and tangential milling force components acting on the end mill tool are transferred to tool holder where they are transformed into longitudinal deformations exciting piezoelectric transducer, where this effect is not present for tool holder without helical slots.

5.2.2.5 Experimentation with Physical Twin of Sensor Node

The last part of the experimental study was focused on evaluating whether the proposed device can be used to detect end mill tool wear condition during actual milling operation. The tool wear condition is to be expressed as change—increase in C4 capacitor charging rate. The working principle of the proposed wireless sensor node is that with increased wear of the tools cutting edges capacitor C4 charging time decreases, thus forming relationship between capacitor charging rate and end mill tool condition. As provided in Fig. 5.71 the auxiliary smart device—“edge device” which is receiving capacitor discharge event information from the sensor node can evaluate whether the capacitor charging time is gradually decreasing, reaching a value which can be regarded as a tool wear condition requiring for the operator or the machine to plan a tool change event before the tool condition negatively affects the quality of the parts being machined or reaches critical tool failure.

The set-up for this experimental study step has been left the same as defined in Fig. 5.76a where the sensor node has been assembled together with cone-shaped tool holder implementing three uniformly distributed helical slots, Morse cone, and an end mill tool (Fig. 5.65), all assembled inside Leadwell V-20 CNC center. The capacitor C4 discharge threshold level has been left the same as well, equal to 0.7 V. If capacitor charge exceeds this threshold, it is discharged and information about this event is sent wirelessly over Bluetooth. As in previous steps the receiving device used was a Bluetooth-enabled smartphone.

The workpiece was changed to an aluminum 6082 block $L \times H \times B$ 300 × 25 × 150 mm to be machined with two flute carbide end mill tools having a 25° helix corned radius. The selected end mill tool is designed for machining specifically of

Table 5.24 Milling process parameters used during experiment set up to evaluate end mill tool wear condition

Parameter	Spindle speed	Feed speed	Feed per tooth	Depth of cut
Value	4000 rpm	300 mm/min	0.037 mm/tooth	1 mm

aluminum alloys. According to selected workpiece material and end mill tool milling process parameters were set as provided in Table 5.24.

During the experiment, two identical design end mill tools were used. The difference between the used cutters is that one of the end mills condition was new, while the other tool had its wear condition artificially and gradually formed, this tool wear condition was achieved by increasing the damage to the cutting edges of the tool's teeth. For the one end mill damage for its cutting edges was formed at three locations, where crater wear was formed ≤ 0.05 mm, flank wear ≤ 0.5 mm and corner wear $\leq$ 0.05 mm, magnified view of these end mill tool tooth wear conditions is presented in Fig. 5.80.

The results of experiment evaluating capacitor charging time during milling operation for new and worn-out end mill tool conditions are presented in Fig. 5.81.

When completely new end mill tool is used during milling operation the average capacitor charging time until the set 0.7 V threshold is 40 s (Fig. 5.81a) while for milling tool with worn-out cutting edges as provided in Fig. 5.79, the average capacitor charging time decreases to around 7 s (Fig. 5.81b), these results show that for worn-out end mill tool during milling operation capacitor charging time is around 6 times faster when compared with use of new end mill tool. Experimental study results are presented in Table 5.25.

As already discussed this is achieved because the increase in milling tool teeth cutting edge wear leads to higher torsional forces exerted on the tool during its impact force cycles with the workpiece during milling operation, these torsional

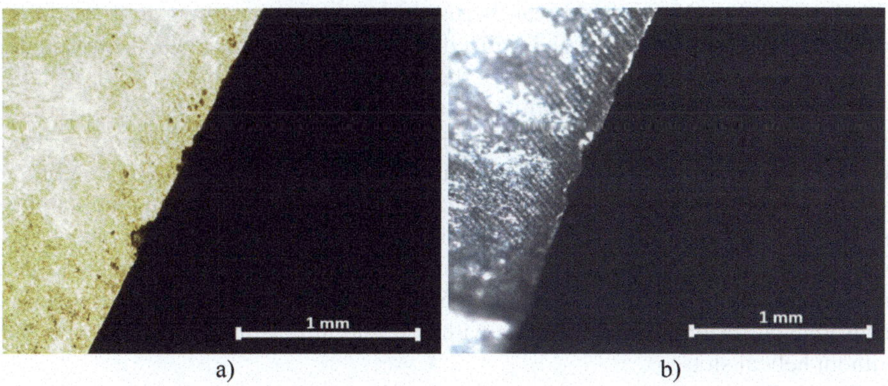

Fig. 5.80 **a**, **b** View of edge wear artificially formed on the end mill used during experiment evaluating worn-out tool condition

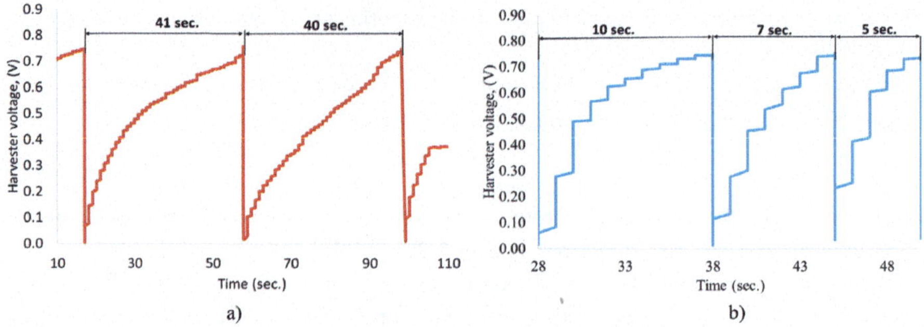

Fig. 5.81 Average capacitor charging times where **a** new and **b** worn-out end mill tool is used during milling operation

Table 5.25 Capacitor charging time comparison for worn out and new end mill tools

End mill tool condition	Average capacitor charge time (s)	Difference
New	7	5.7 times
Worn out	40	

forces increases with increase of tool wear condition, also leading to increase in longitudinal vibrations converted from torsional vibrations by the cone-shaped tool holder that deform and excite the piezoelectric transducer charging C4 capacitor.

From the obtained results, it is possible to conclude that the proposed wireless sensor node design with cone-shaped tool holder implementing uniformly distributed helical slots can be used in milling operation to detect degradation in milling tool condition—wear, which can be expressed as property of increase in capacitor charging rate. During milling operation with increase in wear of the milling tool a significant increase in charge generated by piezoelectric transducer is also observed. This harvested ambient energy can be used not only as a milling tool condition indicator, but it can be stored and utilized later to power the wireless sensor node, enabling self-powering functionality of the device.

After the milling operation with worn out and new end mill tools was performed (worn-out condition for milling tool is defined in Fig. 5.80), the workpiece surface roughness has been evaluated in order to understand the impact introduction of the helical slots on the tool holder have on dynamic characteristics and dynamic stability of the machining process. Surface roughness measurement of the workpieces was performed with Mitutoyo SJ-210 portable surface roughness tester performing R_a test point measurements. Table 5.26 and Fig. 5.82 present workpiece surface roughness measurement results from machining with new end mill using tool holder with and without helical slots.

The surface roughness results presented for workpiece machined with new end mill tool show that surface roughness mean value where tool holder with helical slots is used is 18.6% lower than for tool holder without helical slots, but the range of

Table 5.26 Workpiece surface roughness parameters for tool holder with and without slots employing new end mill tool during machining operation

Tool holder type	Min, Ra (μm)	Max, Ra (μm)	Mean, Ra (μm)	Range, Ra (μm)
Without slots	0.114	0.486	0.183	0.372
With slots	0.15	0.391	0.225	0.241

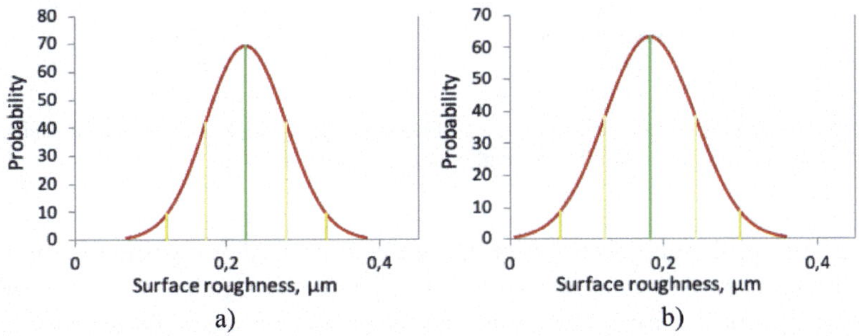

Fig. 5.82 Normal distribution of workpiece surface roughness measurement results after milling with new end mill for tool holder without (**a**) and with (**b**) slots

measured surface roughness values (spread) is 35% lower. Thus, it seems that due to introduced torsionality—stiffness reduction of the tool holder when machining with new end mill tool, lower surface quality for the workpieces can be expected, but the surface quality variance shall be lower.

Table 5.27 and Fig. 5.83 present surface roughness of the workpiece measurement results after it was machined with a worn-out tool using tool holder with and without helical slots.

As the worn-out end mill tool is used for machining aluminum workpiece, surface roughness results for tool holder with helical slots show, that the mean measured values are almost the same when compared to results obtained for tool holder without helical slots and the spread of measured values (variance) is reduced even more, by 45.6% when tool holder with helical slots is used. Now these results show that as the wear of end mill tool increases the beating phenomena as seen in Fig. 5.75b during impact test becomes more dominant in tool holder with helical slots, creating combination of longitudinal and torsional mode frequency response in the tool, somewhat replicating effects used in ultrasonic machining, leading to workpiece surface quality

Table 5.27 Workpiece surface roughness parameters for tool holder with and without slots employing worn-out end mill tool during machining operation

Tool holder type	Min, Ra	Max, Ra	Mean, Ra	Range, Ra
Without slots	0.16 μm	0.485 μm	0.275 μm	0.325 μm
With slots	0.188 μm	0.365 μm	0.265 μm	0.177 μm

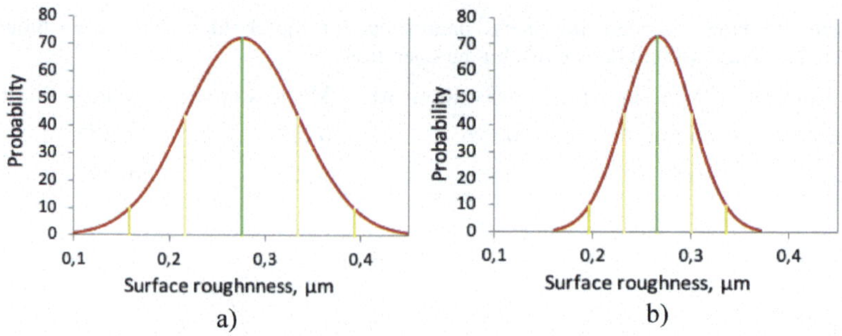

Fig. 5.83 Normal distribution of workpiece surface roughness measurement results after milling with worn-out end mill using tool holder without (**a**) and with (**b**) slots

improvement, prolonging useful lifetime of the tool. Further investigation is necessary to be performed evaluating, helical slot geometrical parameters influence on tool holder stiffness during machining operation and the dynamic response of the tool holder. After positive results were obtained during experimental research, patent proposal has been submitted [13].

5.2.2.6 Self Powering Wireless Sensor Node Applicability for Cloud Manufacturing

Manufacturers are harnessing ever-increasing amounts of data from equipment and suppliers to improve decision-making. The ability to analyze data from every link of the manufacturing value chain helps companies to better control manufacturing processes, keep production running smoothly, and reduce costs. Monitoring is an important part of manufacturing process control and management. It plays a crucial role in ensuring agility in a manufacturing system, process robustness, responsiveness to client demands, and achievement of a sustainable production environment. Recent developments in information systems and computer technology allow for the implementation of new philosophies that integrate various monitoring applications into one complex system connected through company-wide IT systems and with systems operating throughout the whole supply chain. Accessing manufacturing and other industrial data in the cloud via wireless devices such as self-powering sensors provides many benefits.

The most efficient way for monitoring the technological processes is related to the cloud manufacturing possibilities, where not only appropriate software, but the attention of high-level specialists could be found. The idea behind cloud manufacturing processes, including the involved components and the useful cloud-based services, is presented in Fig. 5.84. The upper layer, denoted in the picture, represents the factory floor setting, the deployed machinery, equipment, and smart tools, as well as the final consumers of the cloud services, i.e. the factory employees or other industrial partners. Wireless sensor nodes mounted on various parts of machine tools transmit

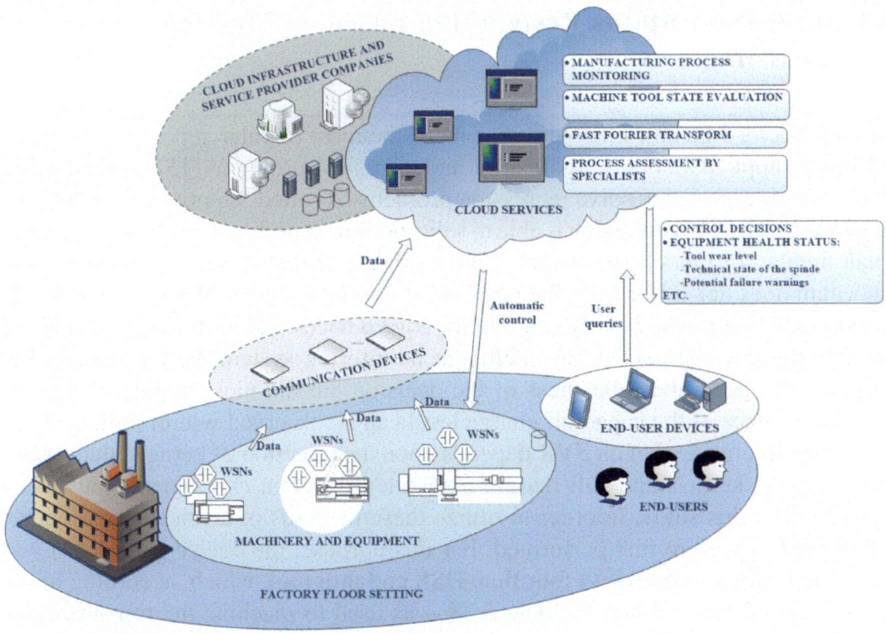

Fig. 5.84 Cloud-based services application for manufacturing

the sensed data to the communication devices, through which, the collected data is further transmitted to cloud resources for further processing. The cloud infrastructure and services are provided by the 3rd party companies, assuring the customer enterprises unlimited amount of the required storage and processing resources as well as the applications and most acceptable pricing and usage options. The beneficial cloud services could be mainly focused on the monitoring and control of the manufacturing process: the predictive maintenance and condition-based monitoring of the technological processes, the FFT for vibration and acoustic signal analysis or even real-time automatic control of the devices. The information about the cutting regimes and machine tool characteristics could be easily presented for the end-users, assuring not only the preferred visual representation of the results, but the variety of end-user devices as well. For the complete evaluation of technological process by every industrial partner other special software and qualified specialists would be available too.

5.3 Tool Wear Status Recognition Based on Machine Learning

Support Vector Machines (SVMs) are one of the most popular supervised learning algorithms applied for both classification and regression problems [15]. The approach where SVMs are used to solve regression problems is called Support Vector Regression (SVR). The SVR approach is able to solve nonlinear problems with a comparably small number of model parameters. Unlike other machine learning algorithms, the algorithm does not suffer from the problem of overfitting [16]. Moreover, the SVR-based prediction model is very suitable for edge devices due to its decision-making time. In the development of an intelligent monitoring system for the cutter wear process, the speed and robustness of the decision are the most important factors, because changes of the capacitor charge level can be observed within milliseconds. Since the effectiveness of an SVR depends upon the selection of kernels, the parameters of those kernels and soft margin parameter, different experiments have been carried out in this study. In order to utilize these methods of artificial intelligence, milling test cuts were first performed. For this purpose, the device (Fig. 5.64) was assembled with a sharp (new) four flute HSS end mill tool, which, according to the process parameters defined in Table 5.28, was used to machine the top surface of a 1.0037 type steel. The experimental study was carried out by machining the top surface of the workpiece 61.

61 times continuously, starting with a sharp (new) end mill tool, gradually (over milling operation) achieving its wear. During the milling of the top surface of the workpiece, once the machining was started, data from the sensor node with the capacitor charge level were sent every 250 ms. A smartphone with Bluetooth connectivity was used for the receiver to visually display the data on the screen in real time and store it for later processing. Each time milling operation of the workpiece top face was completed, its surface roughness was measured and logged at 15 different points using Mitutoyo SJ-210 surface roughness tester (Mitutoyo America Corp., USA). A flowchart of the experimental process, showing the steps involved in each milling iteration carried out during the experiment, is given in Fig. 5.85. Two parameters were recorded during the experiment: the capacitor charge level during continuous milling and the workpiece surface roughness measurements after each milling iteration. Both parameters recorded at the sensor node were fed as input data to an SVM-based prediction model to assess whether they can be used to detect the gradual tool wear in real time during milling operation, which is expressed by the relationship between the change of the capacitor charge level and the increase in workpiece surface roughness.

Table 5.28 Milling process parameters used during experiment

Parameter	Spindle speed, n	Feed speed, v_f	Feed per tooth, f_z	Axial depth of cut, a_p	Radial depth of cut, a_e
Value	1210 RPM	148 mm/min	0.031 mm/tooth	1 mm	9,8 mm

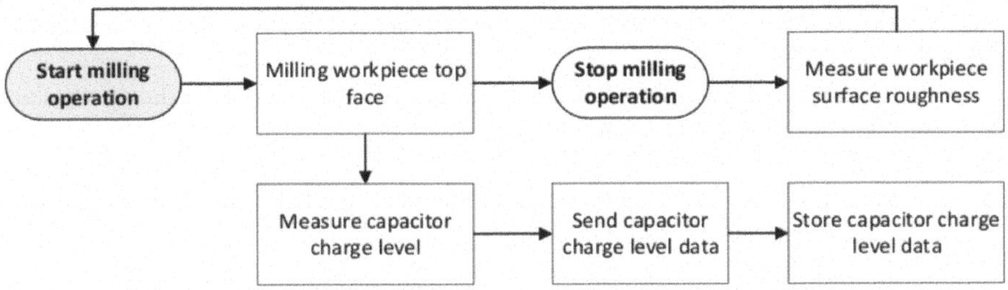

Fig. 5.85 The flowchart of process steps used during experiment execution

Each milling iteration of the top surface of the workpiece lasted on average 10 min, during which 2400 data points were recorded to determine the charge level of the capacitor and 15 different surface points were taken to measure the average surface roughness after the milling operation. The average surface roughness values are considered as the output of the SVR model. However, the raw data representing the charge level of the capacitor, measured every 250 ms, are not suitable as input data for the model. Therefore, seven common statistical measures [17] have been calculated from the distribution of the capacitor charge level.

Feature *Avg* is the simple average value of all 2400 data points, denoting capacitor charge level. Variation *Var* and standard deviation *Sd* are calculated accordingly (Table 5.29).

The autocorrelation function is a useful characteristic for finding recurring patterns. This characteristic indicates the degree of similarity between values of the same variables over two time intervals. This concept has been used for defining the attribute *ACorr*, which refers to the average autocorrelation value calculated between two measures of the capacitor charge level at times x_t and x_{t-k} $\mathrm{x}_{\mathrm{t-k}}$[17]:

$$ACorr = \frac{1}{n-1}\sum_{i=1}^{n} ACF(x_i, x_{i-k}), \quad k = 1, 2, 3 \ldots. \tag{5.21}$$

Table 5.29 Calculated statistical features used as SVM model input data

Feature name	Explanation
Avg	Average value of the capacitor charge level values
Var	Variability value of the capacitor charge level values
Sd	Standard deviation of the capacitor charge level values
ACorr	Autocorrelation value of the capacitor charge level values
M_4Avg	4 data point simple moving averages of the capacitor charge level values
InterQ	Interquartile value of the capacitor charge level values
Energy	Absolute energy of the capacitor charge level values

where value k-is the time interval (the lag), which represents autocorrelation between values that are one time interval apart.

The feature M_4Avg calculates moving averages. In this case, four data points are taken and their average is calculated [18]:

$$MAF = \frac{\left(x_i + x_{i+1} + \cdots + x_{i+(M-1)}\right)}{M} \tag{5.22}$$

and

$$M_4Avg = \frac{1}{n-(M-1)} \sum_{i=0}^{n-(M-1)} MAF_i \tag{5.23}$$

where n-data points, where M is the size of the sliding window, and in our case $M = 4$.

Another quite informative characteristic is interquartile *InterQ*, which calculates the difference between the third quartile and the first quartile for a data:

$$InterQ = Q_2 - Q_1 \tag{5.24}$$

where Q_1 is the first quartile, and Q_3 is the third quartile.

Feature *Energy* is the sum of the squared data values [19]:

$$Energy = \sum_{i=0}^{n-1} (x_i)^2. \tag{5.25}$$

Three specific measures have been derived using expert's knowledge (Table 5.30):

The end of the cycle is determined if the difference between data points is relatively large $\Delta x_i > h\Delta \mathrm{x_i} > \mathrm{h}$. The most appropriate threshold value for h = 150 has been determined experimentally. The average cycle length *Avg_cycle* is calculated by taking into account all recorded lengths at the capacitor charge level. It has been observed that higher values of workpiece surface roughness (*R*a) have lower values of average capacitor charge cycle. For example, for a roughness of 1.959, the average capacitor cycle length is 59 time intervals (1 time interval = 250 ms), which is 59 × 250 ms = 14,750 ms = 14.759 s, meanwhile for a roughness of more than 4, the cycle is very small averaging about 1.750 s. The relation between the decrease in the average capacitor charging cycle time and the increase in surface roughness is provided in Fig. 5.86.

The obtained results show that there is a negative correlation (r = −0.743) between the length of the capacitor charging cycle and the surface roughness of the workpiece, which is due to the wear of the cutting edge of the milling tool. In this case, the charge level of the capacitor at the time of the measurement was expressed in integers, where

Table 5.30 Pearson correlation coefficient values

	Avg	Var	Sd	ACorr	M_4Avg	InterQ	Energy	BigV	$\text{Signal}_{\text{jump}}$	Avg_cycle
Roughness	−0.739	0.641	0.574	−0.817	−0.767	0.825	0.812	0.811	0.889	−0.743

where *BigV* provides the percentage amount of very high values of the capacitor charge level, $x_i > 360, i = \overline{1, n}$. It has been noticed that the amount of such values has a positive relationship with surface roughness and correlation coefficient is equal to 0.811.

$Signal_{jump}$ provides the sum of squared differences $(\Delta x_i)^2$, including the condition: the value of $(\Delta x_i)^2(\Delta x_i)^2$has to be greater than 0.9 of the quantile of differences between data points, $Q_{\Delta x_i}(p)$, $p = 0.9$:

$$Signal_{jump} = \sum_{i=0}^{n-1} (\Delta x_i)^2 > Q_{\Delta x_i}(0.9), \quad where\ \Delta x_i = (x_{i+1} - x_i) \quad (5.26)$$

where Q_Q—quantile function, p—probability value m $0 < p < 1$. This feature highly correlates with the output (Table 5.29).
Avg_cycle is the average length of one capacitor charge cycle, until the set threshold level.

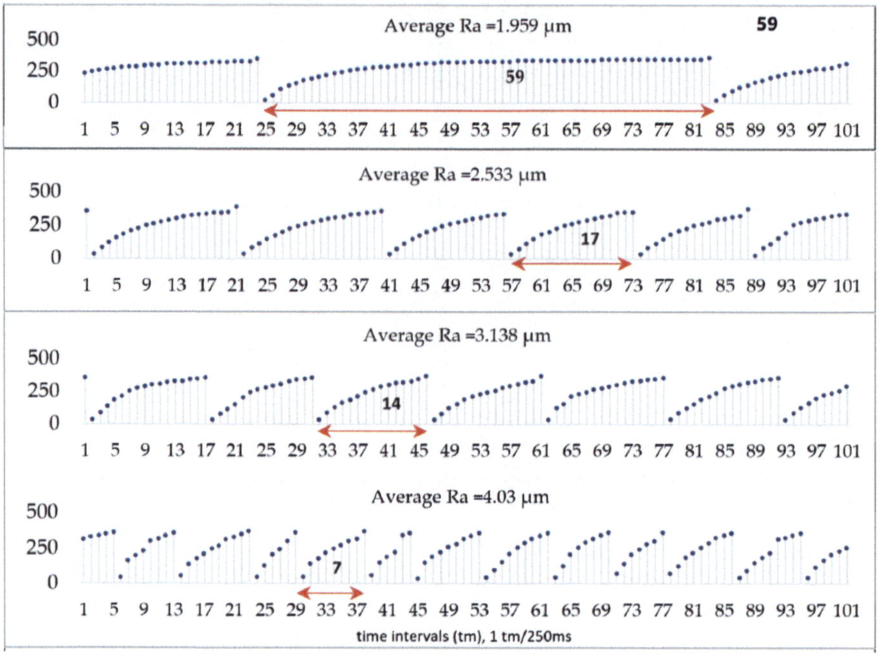

Fig. 5.86 Capacitor charge cycle duration dependence on the surface roughness of the workpiece: 14.759 s versus $Ra = 1.959$ µm, 4.25 s versus $Ra = 2.533$ µm, 3.25 s versus $Ra = 3.138$ µm, 1.75 s versus $Ra = 4.03$ µm

one unit equals 0.0015 V, and the MCU was set to discharge the capacitor when it reaches an integer value of 350, that is when its charge level equals 0.5 V. During the milling operation, when the charge on the piezoelectric transducer capacitor voltage reaches or exceeds the set threshold value, the capacitor is discharged and the cycle repeats itself.

Ten features have been included for the prediction task and the correlation coefficients show that the most informative features are *ACorr, InterQ, Energy, BigV,* and $Signal_{jump}$. The most irrelevant feature ($r = 0.574$) is the standard deviation of the capacitor charge level.

5.3.1 *Support Vector Machines Algorithm Adaptation for Milling Force Prediction*

All modeling experiments were carried out using the Python programming language in Jupyter notebook in the Google Colab environment. The fit of the SVR (Support Vector Regression) model was evaluated by calculating the coefficient of determination and prediction error.

R^2 (coefficient of determination) is commonly used to evaluate model performance. R^2 is the regression score, which is a statistical measure of how close the data are to the fitted regression line. In regression, it is a measure of how well the regression predictions approximate the real data. When R^2 equals 1, it indicates that the regression predictions perfectly fit the data.

$$R^2 = \frac{SSR}{SST} = \frac{\sum_{i=1}^{m} \left(y_i - \hat{y}_i\right)^2}{\sum_{i=1}^{m} (y_i - \overline{y})^2} \tag{5.27}$$

where *SSR* is the sum of squares of residuals, *SST* is the total sum of squares, y_i is the actual value, $\hat{y}_i$ is the predicted value, and $\overline{y}$ the mean value.

The provided results (Fig. 5.87) indicate that R^2 value for RBF-SVR model varies from 0.930 to 0.975, depending on the number of kernels, varying from $1 \div 4$. These results denote that the RBF-SVR model explains all the variability of the response data. More R^2 scoring variations can be observed with the Polynomial SVR model, ranging from 0.838 to 0.911, respectively. The regression score of the Linear SVR model is more or less stable at around 0.77.

Three error measures for time-series prediction are usually calculated: the root mean square error (RMSE); the Mean Absolute Deviation (MAD), and the Mean Absolute Percentage Error (MAPE). In our experiments, MAPE is calculated to evaluate the prediction accuracy of SVR models. MAPE is a relative error measure that uses relative errors to compare the predicted accuracy between time-series models. The formula for calculating the MAPE is provided below [18]:

$$E_M = \frac{1}{n} \sum_{i=1}^{n} \left| \frac{y_i - \hat{y}_i}{y_i} \right| \cdot 100 \tag{5.28}$$

where n is the number of time point, y_i is the actual value at a given time period i, and $\hat{y}_i$-the predicted value.

The data used to test the model (capacitor charge level values over time) are obtained from 31 different milling operations. The average MAPE value of the SVM model with a radial basis function kernel and $C = 4$, predictions are equal to 2.420%. The SVM with a polynomial kernel and $C = 4$ resulted in an average MAPE value of 5.431%, while the highest error was observed with the linear kernel of 8.608%. The predicted and real (actual) values of the surface roughness during the testing are presented in Fig. 5.88.

As the data can be considered as a time series, various additional features such as entropy, "peak to peak" distance, seasonality and trend can be calculated for prediction. The Seasonal-Trend Decomposition by Loess (*STDL*) method [19] can be applied to time series, because it can decompose a time series into seasonal, trend, and remainder components [20]:

$$Y_t = T_t + S_t + R_t \tag{5.29}$$

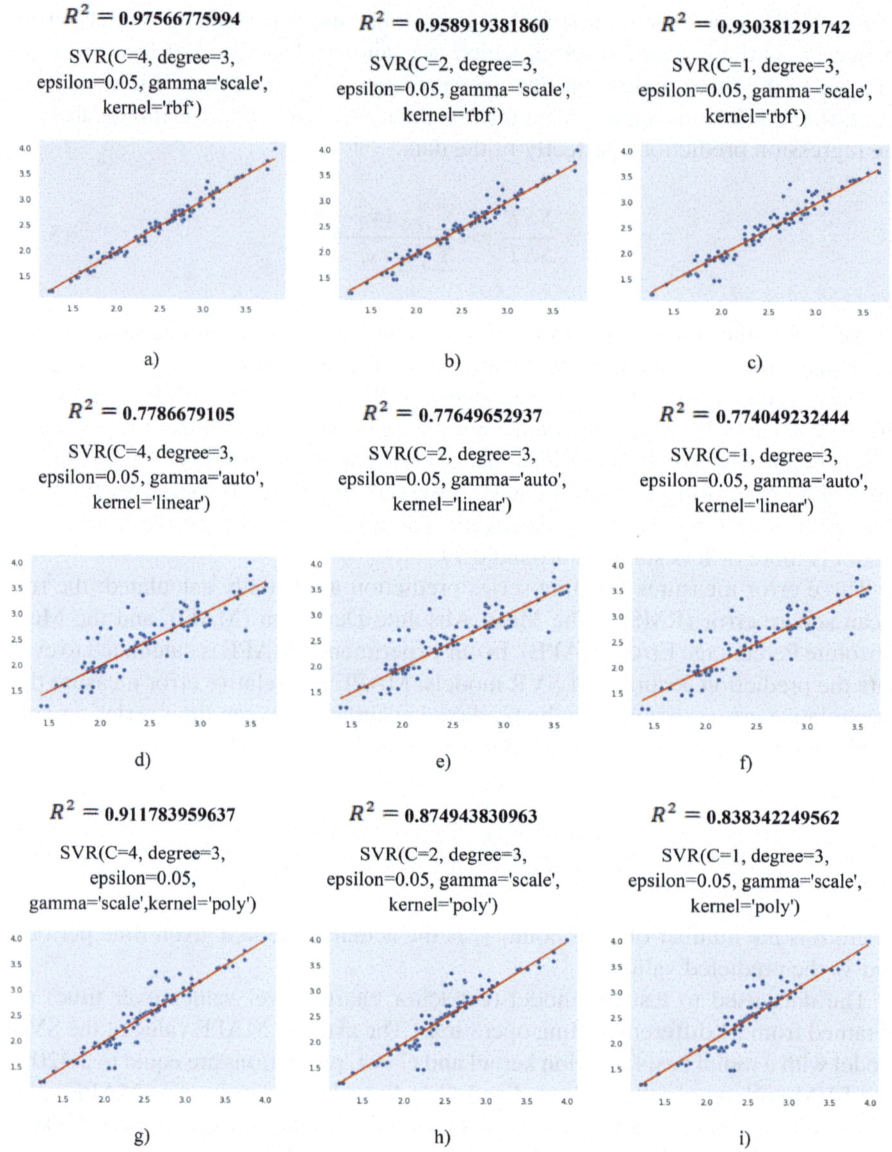

Fig. 5.87 Coefficient of determination R^2 value for RBF-SVM model depending on the number and the type of kernels. **a** 4 rbf kernels, **b** 2 rbf kernels, **c** 1 rbf kernel, **d** 4 linear kernels, **e** 2 linear kernels, **f** 1 linear kernel, **g** 4 polynomial kernels, **h** 2 polynomial kernels, **i** 1 polynomial kernel

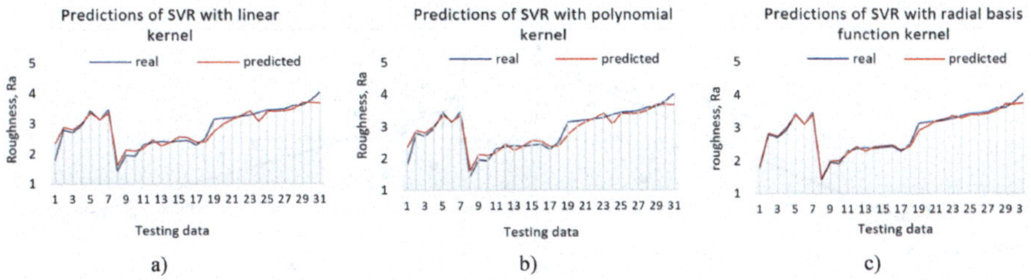

Fig. 5.88 Testing results of the SVR model with different kernels: **a** linear, **b** polynomial and **c** radial basis functions

where T_t-is the trend component, S_t-is the seasonal component representing, for example, the annual cycles, and R_t is an irregular (remainder).

STDL model diagram for the capacitor charging level seasonal trend when workpiece surface roughness is $Ra = 4.03$ μm and $Ra = 3.21$ μm and $dp = 200$ (number of presented data points) is provided in Figs. 5.89 and 5.90, respectively. *STDL* parameters: seasonal period = 12, seasonal window = periodic, seasonal degree = 0, trend degree = 1, low pass degree = 1, robust loess fitting = False. The experimental results with different model parameters exhibit almost no seasonality, therefore we can conclude that the *STDL* model is not useful for data analysis.

The feature *SE* (spectral Shannon Entropy), often is applied to time series [20]:

$$SE = -\int_{-\pi}^{\pi} \hat{f}(\lambda) \log \hat{f}(\lambda) d\lambda \tag{5.30}$$

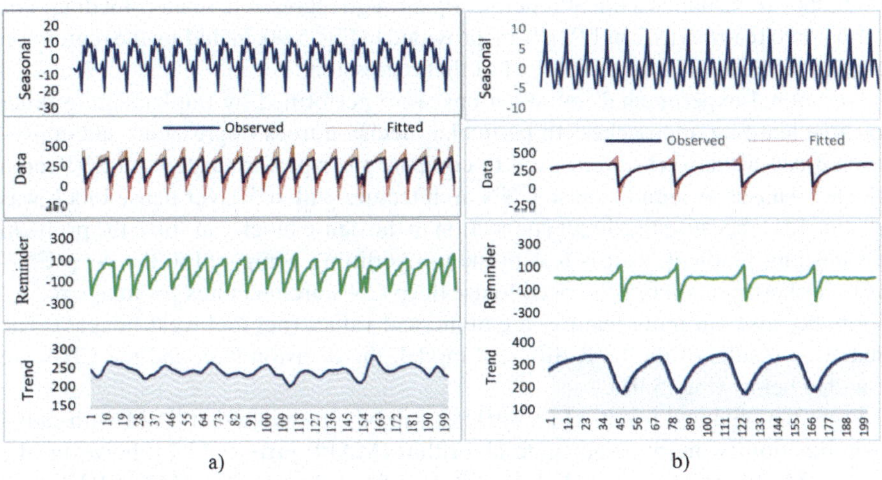

Fig. 5.89 STDL of the capacitor charge level data, when workpiece surface roughness $Ra = 4.03$ (**a**) and $Ra = 3.21$ (**b**)

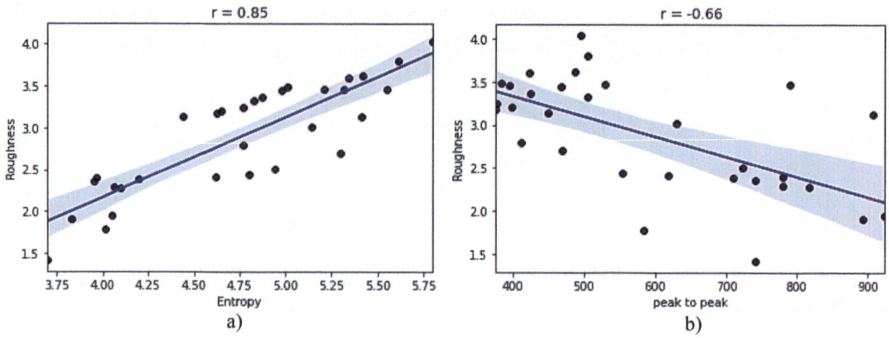

Fig. 5.90 Relationship between data entropy value and roughness (**a**) and peak-to-peak value and roughness (**b**)

here $\hat{f}(\lambda)$ is an estimate of the spectral density of the data. It measures the predictability of the time series.

Large *SE* values are calculated when the time series is difficult to forecast, while small values indicate a high signal-to-noise ratio.

Another popular time series feature is "peak-to-peak" which calculates the distance between two peaks: lowest and highest [17]:

$$PtoP = |\max(X) - \min(X)| \tag{5.31}$$

The entropy feature has provided promising results for our data, resulting in a significant value of correlation coefficient r = 0.858. The peak-to-peak (*PtoP*) calculation is less informative and has an inverse correlation with the output value, r = −0.660. To visualize a linear relationship through regression, scatterplot diagrams of those two features (*SE* and *PtoP*) are provided in Fig. 5.90, including the regression line and the 95% confidence interval of that regression.

Additional experimental investigations were performed by implementing other machine learning approaches. In particular, decision trees (regression) and convolutional neural networks were used to compare their performance with SVR on a selected dataset. A simple dense CNN architecture with a 5-layer dense block was selected [21], because the direct connection in the dense block can solve the problem of vanishing gradient, as it is less prone to overfitting compared to the deep CNN [16]. Furthermore, there is no need to use deep CNN architectures for image recognition, because our input features are numerical values (not tool wear images). The prediction results of the SVR different model, the decision tree, and the CNN are provided below (Fig. 5.91).

From the obtained results (Fig. 5.91), it can be concluded that SVM with radial basis function is the most accurate algorithm (MAPE error 2.42%); however, the average MAPE error is only slightly different from the results of DT (3.02%) and

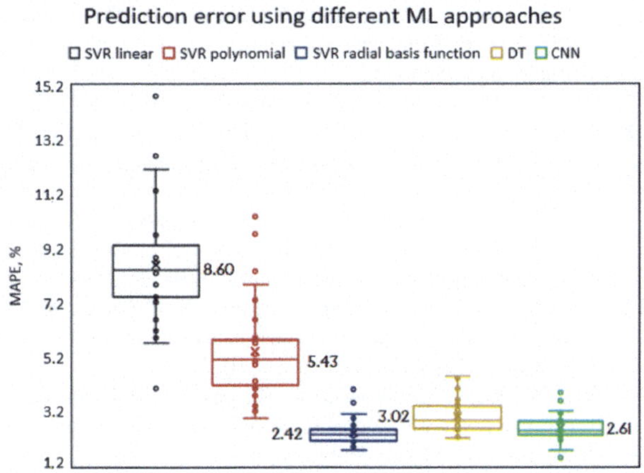

Fig. 5.91 Comparison results for different ML algorithms MAPE (SVR with different kernels, DT, and CNN) represented using boxplot

CNN (2.61%), but the final decision should be made considering two factors: accuracy and performance speed. Convolutional neural networks have shown their superiority in terms of accuracy; however, the larger number of parameters and the complex architecture make this an extremely time-consuming approach. Besides, CNNs are more efficient in solving problems with a huge amount of instances and attributes. For these reasons, the SVM model is preferable for this problem, noting that the prediction error is 7.28% lower than that of CNNs.

By exploring the computed empirical features, it was observed that time series features such as autocorrelation, interquartile, absolute energy, entropy are the most relevant for solving the problem. However, according to the correlation coefficient, the most informative feature is the specially created feature $Signal_{jump}$ ($r = 0.889$) used for determining signals' jumps due to the difference in data points at the 90% confidence level.

References

1. Ostasevicius V, Markevicius V, Jurenas V, Zilys M, Cepenas M, Kizauskiene L, Gyliene L (2015) Cutting tool vibration energy harvesting for wireless sensors applications. Sens Act A-Phys 223:310–318
2. Grazeviciute J, Skiedraite I, Ostasevicius V, Jurenas V, Bubulis A (2006) Ultrasound application in turning process. In: Proceedings of 6th international conference on vibroengineering, pp 152–154
3. Ostasevicius V, Jurenas V, Zukauskas m (2014) Investigation of energy harvesting from high frequency cutting tool vibrations. Mechanika 5:500–505
4. Ostasevicius V, Jurenas V, Karpavicius P, Baskutiene J (2017) Self-powered wireless sensor system application for cutting process control. Mechanika 23(3):456–461
5. Ostasevicius V, Jurenas V, Markevicius V, Gaidys R, Zilys M, Cepenas M, Kizauskiene L (2016) Self-powering wireless devices for cloud manufacturing applications. Int J Adv Manuf Tech 83:1937–1950

6. Ostasevicius V, Milasauskaite I, Dauksevicius R, Baltrusaitis V, Grigaliunas V, Prosycevas I (2010) Experimental characterization of material structure of piezoelectric PVDF polymer. Mechanika 6(86):78–82
7. Rao SS (2007) Vibration of continuous systems. Willey, New York, p 744
8. Mystkowski A, Ostasevicius V (2020) Experimental study of macro fiber composite-magnet energy harvester for self-powered active magnetic bearing rotor vibration sensor. Energies 13(18):22
9. Ostasevicius V, Jurenas V, Augutis V, Gaidys R, Cesnavicius R, Kizauskiene L, Dundulis R (2017) Monitoring the condition of the cutting tool using self-powering wireless sensor technologies. Int J Adv Man Technol 88(9–12):2803–2817
10. Ostasevicius V, Karpavicius P, Jurenas V, Cepenas M, Cesnavicius R, Eidukynas D (2020) Development of universal wireless sensor node for tool condition monitoring in milling. Int J Adv Man Technol 110(3–4):1015–1025
11. Ostasevicius V, Mystkowski A, Karpavicius P, Jurenas (2020) Investigation of piezoelectric-transducer application for vibrational energy harvesting in milling operation. IEEE Conference Proceedings of MSM. Bialystok, p 4
12. Li H, Shin CY (2006) A comprehensive dynamic end milling simulation model. J Man Sc Eng Trans ASME 128(1):86–95
13. Ostasevicius V, Karpavicius P, Paulauskaite-Taraseviciene A, Jurenas V, Mystkowski A, Cesnavicius R, Kizauskiene L (2021) Machine learning approach for wear monitoring of end mill by self-powering wireless sensor node. Sensors 21(9):3137; 26
14. Ostasevicius V, Jurenas V, Karpavicius P, Bubulis A, Eidukynas D, Cesnavicius R, Cepenas M (2019) Wireless sensor to assess the quality of rotating tools. Patent Appl LT2019 535: 6
15. Steinwart I, Christmann A (2008) Support vector machines. Springer Science+Business Media, p 601
16. Zheng Q, Yang M, Yang J, Zhang Q, Zhang X (2008) Improvement of generalization ability of deep CNN via implicit regularization in two-stage training process. IEEE Access 6:15844–15869
17. Barandas M, Folgado D, Fernandes L, Santos S, Abreu M, Bota P, Liu H, Schultz T, Gamboa H (2020) TSFEL: Time series feature extraction library. Elsevier, Soft 11:1–7
18. Hyndman RJ, Athanasopoulos G (2018). Forecasting: principles and practice, 2nd edn. OTexts, Melbourne, Australia, p 380
19. Cleveland RB, Cleveland WS, McRae JE, Terpenning IJ (1990) STL: a seasonal-trend decomposition procedure based on loess. J Off Stat 6(1):3–33
20. Kang Y, Hyndman RJ, Li F (2020) GRATIS: generating time series with diverse and controllable characteristics. Stat Anal Data Min 13(4):354–376
21. Huang G, Liu Z, Maaten L, Weinberge QW (2017) Densely connected convolutional networks. In: 2017 IEEE Conference on Computer Vision and Pattern Recognition (CVPR), pp 1–10

The manufacturer's authorised representative in the EU is Springer
Nature Customer Service Centre GmbH, Europaplatz 3, 69115 Heidelberg,
Germany. If you have any concerns regarding our products, please
contact ProductSafety@springernature.com

Printed and bound by CPI Group (UK) Ltd, Croydon, CR0 4YY
22/09/2026
02220961-0011